Springer Series on Epidemiology and Public Health

Series editors
Wolfgang Ahrens
Iris Pigeot

More information about this series at http://www.springer.com/series/7251

Roy J. Shephard • Catrine Tudor-Locke
Editors

The Objective Monitoring of Physical Activity: Contributions of Accelerometry to Epidemiology, Exercise Science and Rehabilitation

Editors
Roy J. Shephard
Faculty of Kinesiology & Physical Education
University of Toronto
Toronto, ON
Canada

Catrine Tudor-Locke
Department of Kinesiology
University of Massachusetts Amherst
Amherst, MA
USA

ISSN 1869-7933 ISSN 1869-7941 (electronic)
Springer Series on Epidemiology and Public Health
ISBN 978-3-319-80604-4 ISBN 978-3-319-29577-0 (eBook)
DOI 10.1007/978-3-319-29577-0

Softcover reprint of the hardcover 1st edition 2016

Printed on acid-free paper

This Springer imprint is published by Springer Nature
The registered company is Springer International Publishing AG Switzerland

Introduction: A New Perspective on the Epidemiology of Physical Activity

There is now little dispute that regular physical activity has a beneficial effect in reducing the risk of many chronic conditions [1, 2], but it remains difficult to change population behaviour by encouraging the necessary weekly volume of physical activity [3]. One important roadblock in this task has been uncertainty about the message, and much of the general public has become cynical about public health recommendations due to frequent changes in statements about the minimum amount of physical activity needed for benefit [4].

Issues to Be Discussed

In this text, we will begin by reviewing the various approaches to the measurement of habitual physical activity adopted by epidemiologists over the past 70 years, looking critically at their reliability and validity. We will consider the urging of Janz some nine years ago that epidemiologists turn from questionnaires to objective data [5], and we will trace the evolution of the pedometer from its humble beginnings as a somewhat imprecise variant of the pocket watch to an inexpensive but reliable instrument with a capacity for the storage and analysis of data collected over many weeks. A review of its remaining limitations will prompt us to examine the possibilities of newer multi-modal approaches to activity measurement. We will then highlight issues of sampling, noting that short and seasonal periods of monitoring can give a misleading impression of activity patterns, particularly when applied to individual subjects. A comparison of subjective and objective data will reveal the extent of the misinformation gathered on the adequacy of physical activity in the current generation of city dwellers. Given the continuing limitations of many personal activity monitors, we will pose the question whether more useful information could be obtained by focusing upon the duration of inactivity rather than activity; are data on sitting times simply the inverse of activity durations, or do they provide additional information? Turning to various major causes of chronic ill health, we will then consider how far questionnaire-based conclusions need modifying in terms of the new information yielded by objective activity monitoring. Do new data answer the age-long puzzle of activity vs. appetite in the causation of obesity? Does the new instrumentation bring us closer to making an evidence-based recommendation on minimum levels of physical activity needed to maintain good health? Given the likely two- to threefold exaggeration of habitual physical activity, as reported in questionnaires [6], should the recommended minimum level of physical activity be revised downward, or is it better to leave recommendations in terms of the potential exerciser's exaggerated perceptions? And are the postulated economic benefits of enhanced physical activity magnified or diminished when viewed through the lens of an objective monitor? If we examine current instrumentation critically, what are its limitations and weaknesses? And what new approaches might overcome these problems?

Finally, are there other practical applications of simple objective physical activity monitors, such as motivators in rehabilitation programmes and as a method of examining the pattern and quality of sleep?

These are some of the questions that are reviewed in this text. We have learned much from their in-depth consideration. We trust that our readers will find equal reward from studying these issues and that the outcome will be a much greater understanding of the actions required to enhance population health and physical activity.

Toronto, ON Roy J. Shephard
Amherst, MA Catrine Tudor-Locke

References

1. Bouchard C, Shephard RJ, Stephens T. Physical activity, fitness and health. Champaign, IL: Human Kinetics; 1994.
2. Kesaniemi YK, Danforth E, Jensen PJ et al. Dose-response issues concerning physical activity and health: an evidence-based symposium. Med Sci Sports Exerc. 2001;33:S351–8.
3. Dishman RK. Exercise adherence: its impact on public health. Champaign, IL: Human Kinetics; 1988.
4. Shephard RJ. Whistler 2001: A Health Canada/CDC conference on "Communicating physical activity and health messages; science into practice." Am J Prev Med. 2002;23:221–5.
5. Janz KF. Physical activity in epidemiology: moving from questionnaire to objective measurement. Br J Sports Med. 2006;40(3):91–192.
6. Tucker JM, Welk GJ, Beyler NK. Physical activity in U.S.: adults compliance with the physical activity guidelines for Americans. Am J Prev Med. 2011;40:454–61.

Contents

Meet the Authors

Adrian Bauman, PhD Prevention Research Collaboration, School of Public Health, Sydney University, Sydney, NSW, Australia

Kevin Bragg, BSc(hons) Prevention Research Collaboration, School of Public Health, Sydney University, Sydney, NSW, Australia

Valerie Carson, PhD Faculty of Physical Education and Recreation, University of Alberta, Edmonton, AB, Canada

Jean-Philippe Chaput, PhD Faculty of Medicine, University of Ottawa, Ottawa, ON, Canada

Healthy Active Living and Obesity Research Group, CHEO Research Institute, Ottawa, ON, Canada

Sarah Connor Gorber, PhD Research, Knowledge Translation and Ethics Portfolio, Canadian Institutes of Health Research, Ottawa, ON, Canada

Richard Larouche, PhD Healthy Active Living and Obesity Research Group, CHEO Research Institute, Ottawa, ON, Canada

Željko Pedišić, PhD Prevention Research Collaboration, School of Public Health, Sydney University, Sydney, NSW, Australia

Institute of Sport, Exercise and Active Living, Victoria University, Melbourne, VIC, Australia

Faculty of Kinesiology, University of Zagreb, Zagreb, Croatia

Travis Saunders, PhD, CSEP-CEP Department of Applied Human Sciences, Faculty of Science, University of Prince Edward Island, Charlottetown, PE, Canada

Roy J. Shephard, CM, MD, PhD, DPE, LLD, DSc, FACSM Faculty of Kinesiology & Physical Education, University of Toronto, Toronto, ON, Canada

Mark S. Tremblay, PhD, DLitt, FACSM Healthy Active Living and Obesity Research Group, CHEO Research Institute, Ottawa, ON, Canada

Department of Pediatrics, University of Ottawa, Ottawa, ON, Canada

Catrine Tudor-Locke, PhD Department of Kinesiology, University of Massachusetts Amherst, Amherst, MA, USA

Chapter 1
Physical Activity and Optimal Health: The Challenge to Epidemiology

Roy J. Shephard

Abstract Epidemiologists seek associations between environmental factors, life-style influences and human health; they use current modifications of a series of guidelines enunciated by Bradford Hill to assess the hypothesis that observed associations are causal in nature. We now have a long list of medical conditions where physical activity has been suggested as having a beneficial influence in prevention and/or treatment. Questionnaire evaluations of such claims have been hampered by the limited reliability and validity of self-reports. The introduction of pedometer/accelerometers and other objective monitors has facilitated the determination of causality, allowing investigators to study the effects of clearly specified types, intensities, frequencies and durations of physical activity. Nevertheless, further improvement of monitoring devices is needed in order that epidemiologists can capture the full range of activities typical of children and younger adults. Objective monitoring does not support the hypothesis that a minimum intensity of physical effort is needed for health benefit; indeed, in sedentary individuals the largest improvements in health are often seen with quite small increases of habitual activity. There is no obvious threshold of response, but for many medical conditions available data suggests a ceiling of benefit, with no apparent gains of health once habitual activity attains a specified upper limit. Causality can never be totally proven, but objective data allows the inference that multiple health benefits will stem from moderate daily physical activity; the evidence is sufficiently strong that people of all ages should be urged to adopt such behaviour.

1.1 Introduction

The primary tasks of the epidemiologist are to examine the population prevalence of a given condition, to unearth external factors that seem to be associated with a high prevalence of this condition in particular groups of people, and to assess the

Roy J. Shephard (✉)
Faculty of Kinesiology & Physical Education, University of Toronto, Toronto, ON, Canada
e-mail: royjshep@shaw.ca

R.J. Shephard, C. Tudor-Locke (eds.), *The Objective Monitoring of Physical Activity: Contributions of Accelerometry to Epidemiology, Exercise Science and Rehabilitation*, Springer Series on Epidemiology and Public Health,
DOI 10.1007/978-3-319-29577-0_1

likelihood that such associations are causal in nature. Such information is vital in planning tactics to reduce the risk of contracting a given condition, and in managing it when it is already present.

In this chapter, we consider how this mandate of the epidemiologist is currently pursued in the context of the complex relationships between habitual physical activity and optimal health. We begin by examining definitions of physical activity and exercise. We note the limitations of questionnaires previously used to define the intensity frequency and duration of habitual physical activity. We underline that despite the new opportunities offered by objective monitors of physical activity, it remains important to allow for both reactive responses to activity measurement and seasonal variations in activity patterns. We then consider how data from objective monitors can be related to public health recommendations concerning a minimum daily dose of habitual physical activity, and emphasize that even objective monitors have limitations of reliability and validity when applied to children and young adults under free-living conditions. Medical disorders where physical activity has been thought of benefit in prevention or treatment are tabulated, and readers are pointed to new insights derived from objective monitoring; concepts of threshold and ceiling doses of physical activity are explored, and the shape of the dose/response curve is defined. Finally, the causality of observed associations is reviewed in the context of modern formulations of Bradford Hill's criteria for causal relationships.

1.2 Definitions of Physical Activity and Exercise

Epidemiologists began a close examination of relationships between exercise, physical activity, physical fitness and cardiovascular health during the late 1940s (Chap. 2), but it was not until 1985 that clarity was brought to the related literature through a formal definition of these several terms [1].

1.2.1 Physical Activity

Physical activity is positively related to physical fitness, and is characterized as "***any*** *bodily movement produced by skeletal muscles that results in energy expenditure*" [1]. The authors of this seminal article [1] recognized that the amount of energy expended in any given bout of exercise depended on the amount of muscle involved, and the intensity, frequency and duration of muscle contractions; they proposed expressing energy expenditures in units of kJ/day or kJ/week, although they recognized that measurement might need to integrated over periods as long as a year in order to obtain representative data. They further noted that total activity comprised an occupational component and various leisure activities (including sports, conditioning programmes and household chores); since 1985, both occupational and domestic components of the total have declined for most of the population in developed countries.

Notice that the original definition of Caspersen and his associates comprised ***any bodily movement***—no specific minimum was specified, although it was recognized that activities could be allocated between unspecified light, moderate and heavy categories.

1.2.2 Exercise

Although many previous authors had used the terms physical activity and exercise interchangeably, Caspersen and his associates [1] emphasized that exercise was a subset of physical activity, referring to activity that was "*planned, structured, repetitive and purposive in the sense that improvement or maintenance of one or more components of physical fitness is an objective*."

We may add that in the context of physical activity epidemiology, the exercise component is commonly supervised and has known parameters of frequency, intensity, and duration. The focus of both subjective and objective monitoring is thus upon assessing other, less structured and poorly standardized components of the week's physical activities and other behaviours.

1.2.3 International Consensus Conference Definitions

The first International Consensus Conference on Exercise, Fitness and Health was held in Toronto in 1990, with Claude Bouchard (Fig. 1.1) chairing this gathering. It adopted essentially the same definition as Caspersen and colleagues. It further defined leisure activity as "*physical activity that a person or a group chooses to*

Fig. 1.1 Claude Bouchard chaired major International Consensus Conferences on physical activity, fitness and health in Toronto (1988, 1992) and Hockley Valley, ON (2001)

undertake during their discretionary time," exercise as "*leisure time physical activity*," and training as "*repetitive bouts of exercise, conducted over periods of weeks or months, with the intention of developing physical and/or physiological fitness*" [2].

The Toronto conference made the point that whereas physical activity patterns could be used to estimate energy expenditures, the reverse was not necessarily true. This is an important issue, as some epidemiologists such as Ralph Paffenbarger and his colleagues (Chap. 2) have attempted to evaluate health in relation to the gross weekly energy expenditure of study participants. To avoid issues associated with inter-individual differences in body mass, the Toronto meeting recommended expressing the intensity of physical activity in METs (multiples of resting metabolic rate). It further noted that if activity was categorized in METs, the relative intensity of effort was age dependent (Table 1.1), although it did not address the issue that relative effort was also sex dependent in younger adults. The bounds of the four age categories in Table 1.1 were not defined specifically, but from the peak MET values that were chosen (45.5, 35, 24.5 and 14 ml/[kg min]), average ages of 25, 45, 65 and 85 years might be inferred.

Claude Bouchard chaired a second International Consensus Conference on Physical Activity, Fitness and Health, also held in Toronto, in 1992 [3]. In one of the opening sessions, opportunity was taken to elaborate further the definitions of physical activity and exercise. It suggested that in questionnaire reports, some impression of the intensity of exercise might be inferred from its frequency; for example, a report of swimming would likely show a gradation of effort from occasional involvement to regular sessions to preparation for competition.

Attention was drawn to a significant difference in the semantic descriptions of intensity between leisure activities, which usually lasted 1 hour or less (Table 1.1), and a common classification of occupational activity (Table 1.2). The latter made no reference to the age or the sex of workers, but was probably thinking in terms of middle-aged men. A given MET intensity of occupational activity was consistently rated as heavier than a leisure pursuit, because it was usually sustained for several hours at a stretch, with only short rest breaks; moreover, worksite activity might

Table 1.1 The relative intensity of physical activity in relation to age (based on recommendations of the International Consensus Conferences of 1990 and 1994 [2, 3])

Semantic description of effort	% Maximal aerobic effort	Intensity of activity, expressed in METs			
		Young adult	Middle-aged	Old	Very old
Rest		1	1	1	1
Light	<35	<4.5	<3.5	<2.5	<1.5
Fairly light	>35	<6.5	<5.0	<3.5	<2.0
Moderate	>50	<9.0	<7.0	<5.0	<2.8
Heavy	>70	>9.0	>7.0	>5.0	>2.8
Maximal	100	13	10	7	4

Table 1.2 A comparison of semantic descriptions of the intensity of effort between leisure pursuits for a middle-aged man (see Table 1.1) and one commonly used classification of occupational activity (Brown and Crowden [4])

Semantic description of effort	Leisure (middle-aged man); energy expenditure in METS	Occupational activity; energy expenditure in kJ/minute	Occupational activity; energy expenditure in METs
Resting	1		
Sedentary		<8.4	1.6
Light	3.5	8.4–14.7	2.8
Fairly light	5.0		
Moderate	<7.0	14.7–20.9	4.0
Heavy	>7.0	20.9–31.4	6.0
Very heavy		>31.4	>6.0
Maximal	10		

involve the adoption of awkward postures and the use of small muscles under adverse environmental conditions.

The second Consensus Conference underlined regional differences in interpretation of the word "sport." In North America, sport generally implied a form of exercise that involved competition, but in some parts of Europe many forms of exercise and recreation were considered as "sport," as exemplified by UNESCO's organization of a "Sport for all movement" [5].

The average person probably has 3–4 hours per day available for the pursuit of leisure activities [6, 7], although there are wide variations in this discretionary time, depending upon the duration of paid work, commuting times (long for many in major urban centres), the division of domestic work between male and female partners, and the need for self-sufficiency activities (greater in those with low incomes). Recent estimates of the daily time spent watching television suggest that 3–4 hours is a conservative estimate of the leisure time currently available to many North Americans (Table 1.3).

The 1992 Consensus Conference emphasized that the use of labour-saving devices had reduced the daily energy cost of most domestic tasks substantially below the figures listed in the classical "*Compendia of common physical activities*" [9], the one exception being the care for dependents, which sometimes still involves periods of heavy physical activity.

1.2.4 World Health Organisation Definition of Physical Activity

The World Health Organisation published its definition in 2010 [10]. This essentially reiterated earlier concepts of physical activity, describing it as: "*any bodily movement produced by skeletal muscles that requires energy expenditure*." The WHO further commented that physical inactivity was the fourth leading risk factor

Table 1.3 Average daily time use of U.S. adults in 2013, showing averages for the population as a whole and the time allocations of those who engaged in the specified form of physical activity [8]

Activity category	Population average (min/day)	Participant average (min/day)	Comments
Occupational activity	252 (M) 166 (F)	507 (M) 448 (F)	Many women worked part time
Domestic chores	80 (M) 131 (F)	124 (M) 158 (F)	Men did less indoor work—cleaning and laundry 19 vs. 49 %, food preparation and clean up 42 vs. 68 %
Leisure (socializing and communicating)	39 (M) 47 (F)	115 (M) 118 (F)	Hosting events, visiting friends
Leisure (TV watching)	179 (M) 154 (F)	223 (M) 196 (F)	
Leisure (telephone and e-mail)	6 (M) 11 (F)	38 (M) 46 (F)	
Leisure (sport and recreation)	24 (M) 12 (F)	114 (M) 75 (F)	
Care for relatives	22 (M) 32 (F)	105 (M) 137 (F)	
Care for others	10 (M) 15 (F)	101 (M) 101 (F)	

for global mortality (accounting for 6 % of deaths), and it was the main cause underlying 21–25 % of breast and colon cancers, 27 % of cases of diabetes mellitus and 30 % of cases of ischaemic heart disease.

1.3 Questionnaire Assessments of Intensity, Frequency and Duration of Activity

The assessment of the intensity, frequency and duration of physical activity is important to the epidemiologist, but the indications yielded by questionnaires have at best been crude.

In terms of intensity, as noted above inferences were sometimes drawn from the frequency and the nature of participation (occasional, regular, or training for competition). A second possibility was to anchor the intensity of effort to some symptom. For example, in the simple questionnaire devised by Godin (Fig. 1.2) and Shephard [11], subjects were asked to indicate "*How often did you participate in sports or vigorous physical activities long enough to get sweaty during leisure time within the past four months*." However, even with anchoring to such a response, investigators were unable to achieve much more than distinguish those who were periodically active from those who were not.

Fig. 1.2 Gaston Godin developed a simple physical activity questionnaire where the intensity of effort was anchored upon the perceived production of sweat

When attempting to specify the frequency of an activity, not only was there difficulty in recalling the average number of times the activity had been performed in the past month, but because many pursuits were also seasonal in nature, representative data were not obtained unless observations had been dispersed over an entire calendar year. Respondents also tended to over-state the duration of bouts of activity, because they included time allocated to changing, showering, socializing and even travel to and from an exercise venue [12]. The end-result was commonly a substantial over-estimate of the time spent on physical activity relative to objective measurements (Chap. 6); sometimes, those conducting population surveys were left with subjects who had reported activities for a total of more than 30 hours during a given day.

1.4 Precautions Needed During Objective Monitoring of Physical Activity

Objective monitors such as the pedometer/accelerometer provide much more accurate estimates of the total time that is committed to significant physical activity than do most questionnaires; in many instances, objective monitors also provide an instantaneous measure of the intensity of the activity that is being undertaken. Nevertheless, the objective information is not free of pitfalls; indeed, the observer must deal with some of the same issues that are encountered during subjective monitoring. In particular, there is often a reactive response to activity measurement, and short periods of recording are biased by weekly and seasonal variations in activity patterns. Fortunately, these issues can be countered by some simple precautions.

1.4.1 Reactive Response to Activity Measurement

If a person knows that his or her habitual activity is being monitored, there may be a temporary increase in the intensity and the total amount of activity performed, in a conscious or sub-conscious desire to impress the observer.

With questionnaire responses, the intensity, frequency and duration of effort may all be exaggerated. As Stacy Clemes (Fig. 1.3) has emphasized, the readings obtained from personal monitors also tend to be higher during the first week, particularly if the subject is able to see the counter readings [13, 14]. However, there is disagreement as to the extent of the problem; it is more important in some subjects than in others, and in any event it can easily be circumvented by preventing study participants from viewing the monitor, and by discarding readings obtained during the first week of observation.

1.4.2 Minimum Sampling Period

Errors in the assessment of physical activity inevitably weaken associations with population health. It is thus important to eliminate problems from intra-individual variations in habitual physical activity before examining inter-individual differences of activity patterns and their possible relation to health [15]. Intra-individual differences are related to day of the week, season, and weather conditions, and such influences must be countered by a careful definition of the minimum sampling period.

In the 1960s, the arbitrary recommendation of the International Biological Programme was that observers should record habitual physical activity on at least two weekdays and two weekend days [16]. Many more recent observers have chosen to record subjective or objective data over 7 consecutive days or less (for example, Blair et al. [17] and Cain and Germia [18]). Often, there has been no

Fig. 1.3 Stacy Clemes raised the issue of a reactive response to the wearing of a pedometer

preliminary discounted period to allow for a reactive response to measurement. One report suggested that the day of the week accounted for less than 5 % of the total variance; in terms of sampling, "*any three days provided a sufficient estimate*" [19]. Although the investigators found statistically significant differences in the activity of middle-aged adults on Sundays, these were not of great practical importance. Another study of middle-aged Japanese suggested that 3 days of recording were sufficient to establish the average level of physical activity for a given week with an 80 % reliability [20]. Trost et al. [21] obtained 7 days of consecutive objective monitoring; they concluded concluding that in adults an ICC of 0.80 could be obtained with 4–5 days of monitoring by a uniaxial accelerometer, and (by extrapolation of their data) that in adolescents 8–9 days was required to reach a comparable level of accuracy.

However, physical activity patterns are modified not only by the day of the week [15, 19, 22] but also the season [15, 22–25]. The simplistic approach of measuring behaviour over a single week fails to acknowledge that many of the leisure pursuits contributing to the relationship between physical activity and health are necessarily seasonal in nature. Studies collecting only 7 days of data are plainly unable to assess the magnitude of errors arising from the neglect of seasonal differences. Nevertheless, one report acknowledged that in order to capture an accurate picture of an individual's total intake of food energy, it was necessary to obtain 27 days of data in men, and 35 days in women [26]. Another report [27], based on questionnaire data, found that five 24-hour recall assessments over a 12 month period accounted for only a small fraction of the variance in physical activity of 60–70 year old adults (14 % in men and 22 % in women). This second investigation concluded that seasonal factors accounted for 11 % of the total variance in men, and 9 % in women; the remaining variance (49 % in men, 61 % in women) was attributed to "white noise." Subsequently, more rigorous mathematical analysis has discredited the idea of "white noise," and has called into question interpretations of minimum sampling times based upon this hypothesis [28].

Pedometer/accelerometer records for free-living Japanese seniors in the community of Nakanojo have demonstrated that even such simple activities as walking are influenced by seasonal changes in environmental conditions (Figs. 1.4 and 1.5). Rainfall is the most important factor in an elderly population, with the daily step count dropping exponentially from around 7000 steps/day in dry weather, to around 4000 steps/day when the rainfall is 150 mm. Other significant environmental influences include day length, mean ambient temperature, minutes of sunshine and relative humidity (Table 1.4) [24].

Plainly, there remains scope to extend such objective monitoring of seasonal and environmental effects to other age groups, and to those living in other parts of the world.

The collection of pedometer/accelerometer data continuously over an entire year, and the calculation of power functions for temporal variations in physical activity patterns has now allowed us to define precisely the number of days of sampling needed to specify a person's average annual step count with a known level of confidence (Table 1.5). Even longer periods may be needed for more detailed

Fig. 1.4 Yukitoshi Aoyagi directs the longitudinal study of physical activity of seniors living in Nakanojo, Japan

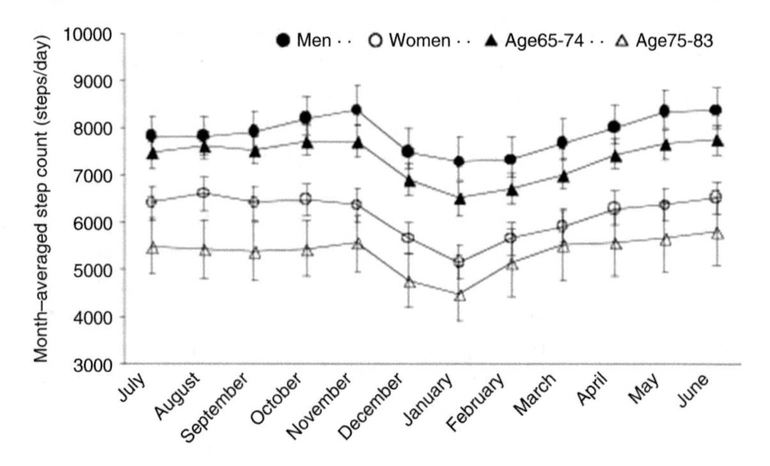

Fig. 1.5 Step counts, averaged by month, for men and women aged 65–75 and 75–84 years living in Nakanojo, Japan. Based on data of Yasanuga et al. [23]

Table 1.4 Influence of environmental factors upon pedometer/accelerometer step counts of seniors in Nakanojo, Japan, on days when rainfall <1 mm

Environmental factor	Regression equation	r^2	Statistical significance
Day length	$11.2\ (x) - 0.006\ (x^2) + 1813$	0.13	<0.01
Mean ambient temperature	$124\ (x) - 3.65\ (x^2) + 5943$	0.32	<0.01
Minutes of sunshine	$96\ (x) - 13.9\ (x^2) + 6656$	0.029	<0.05
Relative humidity	$8.1\ (x) - 0.01\ (x^2) + 6137$	0.030	<0.05

The regression equations are of the type Step count = a (Factor) + b, and are based on the studies of Togo et al. [24]

interpretation, such as the average minutes of moderately vigorous physical activity taken per day. The physical activity patterns of Japanese seniors are plainly more consistent for women than for men. If observations are made on consecutive days

Table 1.5 Number of days of observation (n) required to predict an individual's average annual step count with 80 and 90 % reliability in relation to sampling pattern

	n for 80 % reliability		n for 90 % reliability	
Sampling pattern	Men	Women	Men	Women
Consecutive days	25	8	105	37
Random days	4	4	11	9
Structured by season and day of the week	8	4	16	12

Data for seniors living in Nakanojo, Japan, based on the data of Togo et al. [28]

(as in many recent pedometer/accelerometer studies), an extended monitoring period is needed to attain a 90 % reliability of assessments. The most economical pattern in terms of the number of samples would be to make observations on a randomly selected basis throughout the year, although reliability can also be enhanced by a simpler and more practical structured approach, picking an equal number of observations from each season of the year and each day of the week.

Weekly activity patterns may be less variable for those who are employed than those who are retired, but there remains a need to repeat the same type of power function analysis that we have used in Nakanojo on different age groups. In particular, the weekly activity patterns of those with corporate employment should be compared with those working at home, or caring for children and relatives. Levin and associates [29] used a Spearman-Brown prophesy formula to evaluate three sources of activity assessment in a small group of volunteers. Fourteen visits were made to the laboratory at approximately 26-day intervals, and Caltrac accelerometer records were obtained for 48 hours prior to each visit. For 80 % confidence (an intra-class correlation coefficient (ICC) of 0.80 with averaged annual levels of physical activity), six 48-hour Caltrac accelerometer records (i.e. a total of 12 days of recording) were needed. Alternatives were to study nine of the twelve 48-hour activity records, or three of twelve 4-week activity recall records. The analysis of Matthews et al. [27] was based upon self-reports, but it concluded that for 80 % reliability, 7–10 days of assessment was required in middle-aged men, and 14–21 days in women.

1.5 Interpretation of Measurements Obtained from Objective Monitors

Modern objective monitors yield data on both the volume and the intensity of physical activity. The traditional step count provides an indication of the volume of activity undertaken during the day, and from the instantaneous impulse rate an impression is gained of the intensity of physical activity. We will consider now how this information should be interpreted.

1.5.1 Step Counts

The original target for those wearing a pedometer was to take 10,000 steps/day. We will equate this target with recent public health recommendations for a minimum daily dose of physical activity, and will make an arbitrary classification of activity in terms of daily step counts.

1.5.1.1 The 10,000 Step/Day Target

When the pedometer was first introduced (Chap. 2), its developer (Yoshiro Hatano, Fig. 1.6) suggested that for adults a count of 10,000 steps/day was an appropriate target for those seeking good health; 10,000 steps equated to an energy expenditure of 1.2–1.6 MJ/day, depending on the wearer's body size and walking speed [30].

One investigation found that 73 % of individuals who recalled a day during the previous week when they had been active for at least 30 minutes achieved a count >10,000 steps on that day [31], but a second report found that even after they had been prescribed a daily 30 minute walk, only 38–50 % of the women concerned reached a pedometer count of 10,000 steps/day [32]. Longitudinal studies have supported the health value of the 10,000 steps/day target by demonstrating such benefits as reductions of blood pressure and body mass, and enhanced glucose tolerance when initially sedentary groups had achieved this target [33, 34].

Unfortunately, many sedentary middle-aged and elderly people find 10,000 steps/day too ambitious a goal to attain and/or sustain. One study of Japanese workers who were set this target found that only 83 of an initial 730 volunteers remained active after the study had continued for 12 weeks [35]. On the other hand, 10,000 steps/day is likely an inadequate target for children and adolescents [36];

Fig. 1.6 Yoshiro Hatano developed the first mass-produced electronic pedometer during the 1960s

one report from the U.K. found counts of 12,000–16,000 steps/day in 8–10 year old children prior to any specific intervention [37].

1.5.1.2 Equating Step Counts with Public Health Activity Recommendations

A common public health recommendation for adults is to spend at least 150 minutes per week in moderate to vigorous aerobic exercise [38, 39]. If the chosen exercise is walking, a 30 minute session might equate to covering a distance of 2.5 km at a pace of 5 km/hour, and with a stride length of 0.7 m, the individual concerned would take 3570 steps; empirical data conform to this expectation, with average values of 3100 [32], 3410 [40], and 3800–4000 steps [31] during 30 minutes of deliberate activity. However, to this total must be added the count of around 4000 steps/day incurred from incidental movements when a person remains at home taking little deliberate exercise. We have set this baseline at 4000 steps/day in the elderly [41], but others have argued that in younger adults the base count should be as high as 6000–7000 steps/day (below). Plainly, this is an important number to clarify. On the basis of current information, the public health recommendation for the elderly would correspond to a pedometer/accelerometer step count of at least 7500 steps/day, and in young adults it should possibly exceed 10,000 steps/day. Most people can meet the public health recommendation, and many cannot attain a count of 10,000 steps/day, so that 10,000 steps/day may exceed the currently recommended volume of physical activity.

For children and adolescents, the public health recommendation is to take at least an hour of moderate to vigorous enjoyable exercise per day [42, 43]. In children, this might correspond to walking at a pace of 6 km/hour, 8570 steps; given a baseline of only 4000 steps/day, the total count would still exceed 12,500 steps/day.

1.5.1.3 Arbitrary Classification of Activity Patterns

Pedometer counts on days when no deliberate exercise was taken have ranged from 6000 in middle-aged [44] and elderly [40] subjects to 7400 [31] in a young and active sample. Thus, counts in the range 6000–7000 steps/day appear to reflect the minimum physical demands of daily life. Those taking less than 5000 steps/day are more likely to be classed as obese than those taking >9000 steps/day [40], and a count of <5000 steps/day has been accepted as a measure of sedentarism [36].

Based on these considerations, Tudor-Locke and Bassett [36] proposed classifying pedometer counts for adults into five categories: “sedentary” (<5000 steps/day), “low active” (5000–7000 steps/day; participating in normal daily activities but taking no deliberate exercise or sport), “somewhat active” (7500–10,000 steps/day, taking some volitional activity or facing heavy occupational demands), “active” (10,000–12,500 steps/day), and “highly active” (>12,500 steps/day).

Somewhat lower standards are probably required for the elderly. Our extensive observations on senior citizens in Nakanojo showed that counts of less than 2000 steps/day generally implied that a person was dependent, and counts of 2000–4000 counts/day were seen in those who were housebound, or almost so [41]. We proposed dividing counts for those who were not dependent into four equally sized quartiles: sedentary (2000–5000, mean 4000 steps/day), undertaking normal daily activities (5000–7000, mean 6000 steps/day), somewhat active (7000–9000, mean 8000 steps/day) and active (>8000, mean 10,000 steps/day).

1.5.2 Estimates of Exercise Intensity

Many of the current generation of pedometer/accelerometers are able to estimate the intensity of physical activity from the instantaneous rate of counting [45–47].

The Actigraph is an accelerometer that estimates the total energy expenditure (EE) using equations developed by Patty Freedson (Fig. 1.7) and her associates [48]

$$\text{EE (kJ/min)} = 0.00391\ (\text{Counts}) + 0.560\ (\text{Body mass, kg}) - 3.07$$
$$\text{EE (METs)} = 0.000795\ (\text{Counts}) + 1.44$$

These equations were established by having volunteers walk or run on the treadmill at three speeds (4.8, 6.4 and 9.7 km/hour). Freedson et al. [48] claimed a standard deviation relative to indirect calorimetry of 5.8 kJ/minute for the first equation and 1.12 METs for the second, both with an r^2 of 0.82. Sedentary behaviour has been associated with vertical accelerations yielding counts of 100/min [49] or 150/minute [50]. Activity measured with the Actigraph has been arbitrarily classed as sedentary, light (<3 METs), moderate (3–6 METs), and hard (6–9 METs), although these categories make no reference to the age or sex of the individual.

Fig. 1.7 Patty Freedson developed one widely used equation for converting instantaneous accelerometer counts into estimates of energy expenditures

Various authors [48, 51–54] have also specified ranges of instantaneous counts for moderately vigorous activity (1267–2743, n weighted average 2020 counts/minute) and vigorous physical activity (5725–6403, n weighted average 5999 counts/minute).

We have used the Kenz Lifecorder in our studies of exercise intensity. The Kenz pedometer/accelerometer incorporates an undisclosed proprietary equation that (after entering the individual's body mass) converts the step counts measured over 4 second intervals into an 11-level gradation of active energy expenditures. At the end of the recording period, which can be as long as 6 months, it is possible to estimate the number of minutes spent at each of these 11 intensities of activity. In our sample of seniors, almost no activity was recorded at an intensity >6 METs (in fact, this would have been close to 100 % of aerobic power for many of those studied). We thus tabulated minutes of exercise per day spent at an intensity >3 METs and minutes at an intensity <3 METs. The times spent at an intensity >3 METs were, for dependent individuals <2.5 minutes/day, for those who were housebound <5 minutes/day, and for those falling into the four quartiles of daily activity <7.5 (mean 5), 7.5–15 (mean 10), 15–25 (mean 20), and >25 (mean 30) minutes/day.

Future studies could profitably subdivide the times spent at intensities between 3 and 6 METs. In order to compare the Kenz Lifecorder data with indirect calorimetric estimates, the total energy expenditure can be calculated as [55]:

$$\text{Total energy expenditure} = \text{EE}_{\text{p/a}} + \text{EEminor}_{\text{p/a}} + \text{BMR} + 0.1\ \text{TEE}_{\text{p/a}}$$

where $\text{EE}_{\text{p/a}}$ is the active energy expenditure, $\text{EEminor}_{\text{p/a}}$ is the energy expenditure associated with incidental movements, the BMR is calculated from height, body mass, sex and age, and the final term (0.1 $\text{TEE}_{\text{p/a}}$) is an allowance for the thermic effect of food, based on the total energy expenditure recorded by the instrument.

Our studies of seniors [41] have demonstrated a close correlation between the overall daily step-count and the time spent in performing activities of at least moderate intensity (>3 METs) ($r = 0.964$). Correlations with ill-health are generally a little greater for the step count in women than, for the duration of activities >3 METs, and the converse is true in men. Possibly, the elderly women have more consistent but less intense patterns of physical activity than the men.

1.6 Reliability and Validity of Objective Monitoring

The main attraction of objective monitoring relative to the use of questionnaires is that the pedometer/accelerometer data offer the promise of greater reliability and validity. How far are these advantages realized in practice? We will comment on the behaviour of monitors during steady walking, the effects of variations in speed and pattern of movement, and the potential to collect data under free-living conditions.

Fig. 1.8 David Bassett has carried out extensive studies on the reliability and validity of different types of pedometer and accelerometer

1.6.1 Steady Walking

Pedometers and accelerometers generally respond reliably when recording a consistent movement pattern such as level walking. Tests on a panel of volunteers showed that the 24-hour step count determined with the Kenz pedometer/accelerometer had an intramodal reliability (Cronbach's alpha) of 0.998 between two instruments that were worn on the left and right hips, and the counting error relative to 500 actual paces taken on a 400 m track at a normal walking speed was only -0.2 ± 1.5 steps [56].

Schneider et al. [56] found that most of 10 pedometer/accelerometers were able to estimate the treadmill walking distance to within ±10 % and the gross energy expenditure to within ±30 % of the actual value when walking at a speed of 4.8 km/hour. A study by David Bassett (Fig. 1.8) and his associates [57] showed that at both moderate and slow walking speeds, the Yamax device gave the best estimate of a 4.8 km distance among five inexpensive monitoring instruments (Yamax, Freestyle Pacer, Eddie Bauer Compustep, Bean pedometer, and Accusplit fitness walker) that were tested on a walking course, with an average systematic error of about 2 %. The distance error for some of the other devices exceeded 10 %.

1.6.2 Variations in the Speed and Pattern of Walking

The validity of data with most pedometers and accelerometers depends on the speed and pattern of movement, with the best results being obtained at a normal walking pace. Treadmill step counts recorded by the Kenz Lifecorder were under-estimated at low walking speeds (an error of 8.4 % at 3.2 km/hour, and of 1.7 % at 4 km/hour

[58]; total energy expenditures also tended to be under-estimated. A second treadmill trial [55] confirmed that at a walking speed of 3.2 km/hour, the Kenz Lifecorder and the Actigraph yielded step counts that were, respectively, $92 \pm 6\,\%$ and $64 \pm 15\,\%$ of the actual values as counted by an observer, although at higher speeds (80–188 m/minute), both devices yielded readings that were within 3 % of the true figure.

Comparisons of the Yamax Digiwalker with a heel-mounted resistance pad showed an error of 460 ± 1080 steps/day during "purposeful walking" [59]. The Suzuken Calorie counter select 2 was tested against the directly measured oxygen consumption; overall, the relationship was fairly close (a Pearson correlation coefficient of 0.97, with a mean error of −3.2 to +0.1 kJ/minute). Nevertheless, this accelerometer over-estimated energy expenditures by an average of 3.8 kJ/minute during short-step walking, and under-estimated expenditures by an average of 5.1 kJ/minute during long-step walking, even with a fixed and comfortable treadmill speed of 4 km/hour [60]. Kumahara et al. [61] tested the Kenz Lifecorder in a metabolic chamber; subjects engaged in nine speeds of treadmill walking and running (ranging from 2.4 to 9.6 km/hour); the error averaged a substantial 9 % for total energy expenditure and 8 % for physical activity energy expenditure.

1.6.3 Free-Living Conditions

Recording errors are increased further if subjects choose their own activity patterns rather than walking on a fixed course and/or at a fixed pace. Inaccuracies are also introduced by a short stride length and abnormalities of gait [62]. Both pedometers and accelerometers respond poorly to cycling, skating, load-carrying, household chores, and other non-standard activities [63]. Moreover, such instruments take no account of the additional energy expenditures that arise when climbing hills or making movements against external resistance, and artifacts may occur because counts are recorded when driving in a car over bumpy ground [64].

McClain and associates [65] recognized that both the reliability and the validity of the pedometer/accelerometer were reduced on moving from the assessment of standardized laboratory exercise to free-living conditions, where they found differences averaging 1516 steps/day between the Kenz Lifecorder and the Actigraph. Dondzila et al. [66] compared the step-counts recorded during 24 hours of free living relative to a criterion measurement provided by counts on a New Lifestyles NL-100 pedometer (Table 1.6).

A comparison of Yamax Digiwalker output with direct measurements of oxygen consumption during various forms of play in children with an average age of 9.2 years yielded a correlation coefficient of 0.81; substantially better concordance was obtained with a triaxial accelerometer ($r = 0.91$) [67].

Several authors have compared pedometer/accelerometer data with what is commonly accepted as the gold standard in measurements of energy expenditure—the metabolism of doubly-labelled water (Table 1.7). Some authors have

Table 1.6 Discrepancy between counts recorded by specified pedometer/accelerometers (steps/day) and counts recorded by New Lifestyles NL-100 pedometer, during 24 hours of free living (based on data of Dondzila et al. [66])

Pedometer type	Subjects aged 20–49 years		Subjects aged 50–80 years	
	Mean error	Range	Mean error	Range
Kenz Lifecorder	−305	−710 to 100	−38	−937 to 861
Omron	949	598 to 1300	613	24 to 1191

Table 1.7 Comparisons of pedometer/accelerometer estimates of total energy expenditure with the "gold standard" (doubly-labelled water [DLW] measures of energy expenditure)

Author	Sample	Apparatus	Findings
Choquette et al. [68]	10 M, 7 F 60–78 year Sherbrooke, QC	Caltrac accelerometer 7 days of recording	No significant correlations with DLW data Caltrac under-estimates DLW by 0.8 MJ/day (M), 2.2 MJ/day (F)
Colbert et al. [69]	12 M, 44 F >65 year Wisconsin residents	Pedometer, accelerometer, and Sense-wear armband; 7–10 days of recording	Low correlations with DLW data (0.5–0.6) for all 3 devices. Systematic error on expenditure of 2.1 – 1.7 MJ/day (Crouter equation), +1.4 MJ/day (Freedson equation); 95 % limits of regression prediction ± 60 %. No comment on sex differences.
Fogelholm et al. [70]	20 overweight middle-aged Finnish women	HR monitor, pedometer, Caltrac accelerometer	Active energy expenditure Error lowest with Caltrac 0.08 ± 1.61 MJ/day relative to DLW on expenditure of 4.1 MJ/day
Gardner and Poehlman [71]	Elderly U.S. claudicant men 68.7 ± 7.3 year	Pedometer, accelerometer regression equations	Error of pedometer vs. DLW ± 516 kJ/day (95 % limits ± 66 %) Error of accelerometer ± 320 kJ/day (95 % limits ± 41 %)
Rafamantanantsoa et al. [72]	25 M; 48 ± 10 year Japanese	Accelerometer, 3 day and 14 day records	Correlations of accelerometer with DLW 0.78, 0.83; under-estimates of accelerometer −2.3, −2.4 MJ/day
Starling et al. [73]	32 M, 35 F 45–84 year Vermont residents	Caltrac accelerometer 7 days of recording	Caltrac under-estimates DLW by 2.1 MJ/day (M), 1.6 MJ/day (F)

found a good correspondence of the two data sets [70, 71]. However, the doubly-labelled water technique determines the average energy expenditure over a 2-week period, and particularly in elderly individuals who engage in relatively little physical activity, the overall score is heavily influenced by the individual's resting energy expenditure. Perhaps for this reason, other investigators have found substantial under-estimates of total weekly energy expenditures when using objective monitors (Table 1.7).

More information is needed concerning the performance of the various available objective monitors under free-living conditions. For the present, we may infer that pedometer/accelerometers work reasonably well when assessing old people, provided that they do not have a severe anomaly of gait. In such populations the accuracy of data may be sufficient for epidemiological evaluations of the amount of physical activity associated with health benefits [46]. However, the information that is obtained on the absolute energy expenditures of younger adults under free-living conditions remains much more questionable. Values yielded by a CSA monitor, a Tritrac accelerometer and a Yamax Digiwalker under-estimated doubly-labelled water figures for the total energy expenditure of 13 healthy young women by 59, 35 and 59 % respectively [74]. Plainly, there remains substantial scope to enhance the accuracy of estimates of physical activity, possibly by combining simple pedometer/accelerometers with other data on body posture, heart rate, breathing rate or the rate of sweating.

1.7 Medical Conditions Potentially Modified by Intensity and/or Volume of Habitual Physical Activity

1.7.1 Conditions of Interest

The International Consensus Conferences (above) provide a relatively comprehensive list of medical conditions where an increase of habitual physical activity might have preventive or therapeutic value (Table 1.8). Information is drawn from the topics discussed at the International Consensus Conferences on Physical Activity, Fitness and Health of 1988 [2] and 1992 [3].

1.7.2 Need for Enhanced Objective Monitors

In almost all of the investigations reviewed at these conferences, epidemiologists relied upon the evidence of questionnaires rather than objective monitors, although as early as 1988 the first of these conferences expressed the hope that there would soon be low cost objective devices that could record movement patterns or physiological responses at 1 minute intervals, storing this information for 12–24 hours [2]. The 1992 conference [3] underlined the need for additional research to develop "*objective monitors of physical activity (such as improved motion sensors) better suited to epidemiological investigations.*"

Table 1.8 Medical conditions where an increase of habitual physical activity (PA) may have preventive or therapeutic value, and conclusions reached at the International Consensus Conferences (ICC) of 1988 [2] and 1992 [3]

Medical condition	ICC 1988	ICC 1992
Atherosclerosis	Influence of PA unclear	No human studies of PA and progression or regression of lesions
Coronary heart disease	Regular PA reduces risk factors and cardiac events	Regular PA reduces risk, independently of other risk factors
Coronary rehabilitation		Regular PA enhances functional recovery; structured programmes may reduce recurrences and mortality
Hypertension	PA beneficial in borderline and moderate essential hypertension	PA gives small reduction in resting blood pressure
Stroke		Little direct evidence of prevention from PA
Peripheral vascular disease		Functional improvement with PA, but most studies uncontrolled
Type I diabetes		PA confers physiological and psychological benefits
Type II diabetes	PA improves insulin sensitivity	Diabetes related to sedentary lifestyle, PA beneficial in treatment
Obesity	PA reduces body fat, improves insulin sensitivity and blood lipids	Moderately obese who are successful in losing weight usually exercisers. PA conserves lean tissue
Osteoarthritis	Moderate PA may be helpful	Little evidence on relationship of arthritis to exercise (except if injured)
Osteoporosis	PA important in prevention and rehabilitation	Weight-bearing or resisted activity essential for bone health
Low back pain	Preventive value of PA not yet established	Value of PA in preventing back pain unclear
Chronic airway obstruction	Unclear if POA reduces frequency of asthma; can enhance function in COLD	PA may alter pattern of breathing, reduce perceptions of dyspnea
Renal disease		PA may enhance functional capacity in end-stage renal disease
Bladder control		No benefit from general PA, but pelvic floor exercises may help
Neuromuscular disorders		Lack of valid research
Cancer	PA reduces risk of colon cancer, possibly breast and reproductive tract in women	PA reduces colon cancer, evidence for breast, male and female reproductive tracts equivocal
Surgery	Pre-operative increase of PA beneficial	Active individuals show fastest recovery from surgery

(continued)

Table 1.8 (continued)

Medical condition	ICC 1988	ICC 1992
Mental health	PA can reduce anxiety, depression, reduce tension, enhance sleep	PA increases self-esteem and psychological well-being. PA may reduce depression and anxiety
Substance abuse		PA related gains in mental health may reduce substance abuse
Reproductive health	Reversible suppression of menstrual function with heavy PA	Reversible suppression of reproductive hormones with intense PA in both sexes, but findings confounded by negative energy balance
Pregnancy	Moderate PA not harmful	Moderate PA well-tolerated by mother and foetus, with decreased risk of gestational diabetes
Optimal growth	Active children may have favourable lipid profile	PA increases bone mineralization, controls body fat and other cardiac risk factors
Aging	PA counters loss of aerobic power and strength	PA has small but important effects on cognitive function
Quality of life (QOL) and independence		Important but neglected. Methods of measurement of QOL need refining

At a third conference, intended to specify dose-response relationships between physical activity and these various conditions, the need for accurate objective monitoring of physical activity was yet more evident [75, 76]. However, it was underlined by Lamonte and Ainsworth (Fig. 1.9) that current electronic motion sensors were limited in their ability to discriminate specific types of physical activity, often involved inconvenient measurement procedures [76], failed to reflect energy expended in uphill walking [77] and often gave erroneous information under free-living conditions [78, 79]. There was thus a pressing need to develop enhanced motion sensors that incorporated information on ventilation, heart rate and increases in body temperature.

1.7.3 New Insights from Objective Monitoring

The use of objective monitors has given new insights into the relative value of activity and fitness-based indices, and concepts of thresholds and ceilings of activity for benefit. For some health issues, the main benefit was seen with a modest level of physical activity, but for other benefits, gains increased progressively as more activity was undertaken.

Fig. 1.9 Barbara Ainsworth has played a leading role in the evaluation of various objective motion sensors

1.7.3.1 Relative Value of Activity and Fitness Indices

One issue discussed at the 2001 Conference was whether habitual physical activity or the attained level of physical fitness was more important as an index of the health benefits of exercise; possibly, they may act upon differing components of health. A comparison based on questionnaire data that reported three or more levels of physical activity led to the conclusion that there was a closer association of benefit with aerobic fitness (as measured by treadmill endurance time) than with the reported physical activity [80]. This is counter-intuitive, since the chosen measure of physical fitness depends in part on body build and genetic factors rather than habitual activity, and it could be argued that closer correlation with aerobic fitness is simply a reflection that this parameter is being measured more accurately. In support of this criticism, we recently compared the correlation of one measure of atherosclerosis (a deterioration of pulse-wave velocity) with aerobic power (as measured by a test of walking speed) and habitual physical activity (as monitored objectively by a Kenz pedometer/accelerometer). In our comparison, the association was greater for the motion sensor than for the measure of peak aerobic power [81]; in a multiple regression analysis, step count, duration of activity >3 METS and maximal walking speed accounted for 11, 7 and 4 % of the total variance in pulse wave velocities.

1.7.3.2 Thresholds of Benefit

Intuitively, one might presume that any physical activity would yield better health than total inactivity, even if the individual undertook only a relatively low intensity of activity. However, some epidemiologists, notably Jeremy Morris (Chap. 2), have suggested that there is a threshold intensity of questionnaire-reported habitual physical activity, at least in terms of cardiovascular disease, below which no benefit is realized [82]. Objective monitoring provides the detailed gradation of activity needed to examine this question more precisely.

In terms of cardiovascular disease, objective information has been obtained by using measurements of pulse wave velocity as a surrogate of cardiovascular disease in elderly individuals [83]. The cardio-femoral pulse wave velocity showed a negative correlation both with daily step count ($r = -0.23$) and with the total daily duration of moderate physical activity ($r = -0.18$). Moreover, in terms of a possible threshold, in fact the largest change of vascular distensibility in this elderly population was seen on moving from the least active quartile (averaging 3570 steps/day, and 4.8 minutes/day of moderate activity), to the next most active quartile (averaging 5838 steps/day, and 12.2 minutes/day of moderate activity). In confirmation of a low threshold of benefit, Sugawara et al. [84] examined 103 post-menopausal women, finding that carotid arterial stiffness was inversely related to the duration not only of vigorous physical activity (>5–6 METs, depending on age), but also to the duration of moderate (>3–4 METs) and light (<3–4 METs) activity. Likewise, Gando et al. [85] demonstrated that the carotid/femoral pulse wave velocity was correlated with triaxial accelerometer determinations of the time that the older members of a group of 538 unfit but otherwise healthy subjects allocated to moderate (>3 METs, $r = -0.31$), light (<3 METs, $r = -0.39$) and sedentary ($r = 0.44$) activities, but was not correlated with the time spent in vigorous physical activity. Again, a longitudinal trial in 274 sedentary, obese young adults found that an increase of moderate physical activity over a year of observation was associated with a decrease of pulse wave velocity [86]. Finally, Andrea LaCroix is currently relating accelerometer-measured activity to incident cardiovascular disease and mortality among female Seattle residents aged 80 years Several investigators have also related objective data on habitual physical activity to the metabolic syndrome and cardiovascular risk factors [87].

1.7.3.3 Ceilings of Benefit

A few questionnaire-based studies have suggested that there may be not only a ceiling to the benefits of increased physical activity, but that excessive physical activity may lead to a worsening of prognosis. Again, this issue is more readily explored using objective monitors; if a given health benefit is plotted against the recorded activity, a plateauing should be seen, with a substantial quadratic function or a negative exponential function limiting benefits. We have certainly seen a plateauing of response in terms of bone health, muscle mass, and health-related quality of life, although no negative effects of excessive activity within the limits of our data.

Bone Health In a comparison of activity patterns between those with osteopenia and those with normal bone health in 92 post-menopausal women, Jana Pelclová (Fig. 1.10) and her associates [88, 89] found the largest (although non-significant) inter-group difference was in the time allocated to light activity (430 vs. 537 minutes/week).

Fig. 1.10 Jana Pelclová and her colleagues have studied the relationship between actigraph measurements of habitual activity and bone health

We measured the bone health of seniors in terms of an osteosonic index [90]. The mathematically-fitted curves showed benefit approaching a plateau in the most active people, with negative exponential terms both for step counts ($M = -1.23 - 2.73e^{-x/2884}$; $F = -1.21 - 1.72e^{-x/6990}$) and for the duration of moderate activity ($M = -1.03 - 1.21e^{-x/17.1}$; $F = -1.43 - 1.08e^{-x/20.8}$). Moreover, when the sample was divided into physical activity quartiles, the osteosonic index was not significantly enhanced in those exceeding the activity of the second quartile (in men, averaging 6589 steps/day, and 13.0 minutes/day at an intensity >3 METs, and in women, averaging 6165 steps/minute, and 11.9 minutes/day at an intensity >3 METs). A longitudinal study in the same population yielded essentially similar results (Table 1.9). The osteosonic index showed a trend to a significantly increased risk of fractures (a T score of 1.5 below the population norm) in those with the lowest levels of habitual activity. Kitagawa and associates reported similar findings in a cross-sectional comparison of 7-day pedometer records with ultrasound measurements of bone health in women aged 61–87 years; their fitted graph was quadratic, with no further increase of bone density in those individuals taking more than 12,000 steps/day [92].

The optimal dose of physical activity may differ between bones. Thus, Vainionpää et al. [93] classified the intensity of activity of women aged 35–40 years in terms of acceleration bands; in the case of the femoral neck, the trochanter and the calcaneus, bone density was similar with in those with daily accelerations of 3.9–5.3 g and 5.4–9.2 g, but for the lumbar spine, significantly higher densities were associated with the highest accelerations (5.4–9.2 g). In 5-year-old children [94], the largest increase in bone mineral content was seen in moving from the third to the fourth (most active) quartile, the main effect being associated with minutes of what was described as "vigorous" activity per day (>2972 counts/minute).

Muscle Mass We evaluated the risk of sarcopenia in the same population, estimating appendicular muscle mass by an impedance device [95]. A ceiling of response was again apparent. The fitted curves showed clear evidence of plateauing in relation both to daily step count ($M = 7.90 - 2.96e^{-x/2423}$; $F = 6.27 - 1.99e^{-x/2522}$) and the

Table 1.9 Longitudinal data for Japanese seniors (M = males, F = females), showing relative risk and 95 % confidence limits of osteosonic index dropping to fracture range (T value of −1.5) in relation to objectively measured habitual physical activity (steps/day and minutes of activity at an intensity >3 METs) for males (M) and females (F) [91]

	Men		Women	
Activity quartile	Step count/day	Moderate activity (minutes/day)	Step count/day	Moderate activity (minutes/day)
M 3512 steps/day, 3.8 minutes/day F 3473 steps/day, 4.0 minutes/day	2.69 (1.77–4.96)	2.99 (1.48–5.91)	3.87 (2.53–6.02)	3.94 (2.37–6.41)
M 5973 steps/day, 11.2 minutes/day F 5909 steps/day, 9.9 minutes/day	1.51 (0.94–3.88)	1.43 (0.92–3.08)	2.66 (1.63–4.31)	1.85 (1.03–3.47)
M 7451 steps/day, 19.7 minutes/day F 7601 steps/day, 18.2 minutes/day	1.24 (0.86–3.61)	1.20 (0.64–2.24)	1.14 (0.65–1.95)	1.08 (0.76–2.24)
M 10650 steps/day, 32.4 minutes/day F 10334 steps/day, 30.9 minutes/day	1	1	1	1

duration of activity at an intensity >3 METs ($M = 7.93 - 0.92e^{-x/13.5}$; $F = 6.23 - 1.08e^{-x/5.9}$). When data were sorted by quartiles, the odds ratio for sarcopenia (adjusted for age, current smoking and alcohol intake) showed a gradient of relative risk for both step count (M 1.00, 0.79, 1.20, 2.00; F 1.00, 1.02, 1.57, 2.66) and for minutes of moderate activity (M 1.00, 1.05, 2.03, 3.39; F 1.00, 1.23, 3.15, 4.55), with the main and highly significant effect on moving from the first quartile (M averaging 3427 steps/day, 6.7 minutes/day of moderate exercise; F averaging 3049 steps/day, 5.9 minutes/day of moderate exercise) to the next more active quartile (M 6171 steps/day, 14.7 minutes/day; F 4999 steps/day, 10.1 minutes/day).

A 5-year longitudinal study of the same population [96] examined the risk of muscle mass falling below an arbitrary sarcopenia threshold in relation to habitual physical activity (Table 1.10). Again, there was a trend of risk between the four activity quartiles, with the greatest protection being seen on moving from the lowest to the next more active quartile, and little difference of risk between the two most active quartiles. It should be emphasized that no measure of possible involvement in resistance exercise was made, although the likelihood of such activity probably bore a moderate correlation with involvement in aerobic activity.

As with bone health, there was some evidence of site-specificity of response. Abe and associates [97, 98] found that in subjects aged 52–83 years, muscle mass in the lower leg was correlated with both moderate (3–6 METs) and vigorous

Table 1.10 Relative risk and 95 % confidence limits of muscle mass falling below an arbitrary sarcopaenia threshold in a sample of Japanese seniors (M = males, F = females), in relation to objectively measured habitual physical activity (steps/day and minutes of moderate activity)

	Men		Women	
Activity quartile	Step count/day	Moderate activity (minutes/day)	Step count/day	Moderate activity (minutes/day)
M 3512 steps/day, 3.8 minutes/day F 3473 steps/day, 4.0 minutes/day	2.33 (1.43–4.51)	3.01 (2.02–5.99)	2.99 (1.91–3.42)	3.49 (2.11–6.32)
M 5973 steps/day, 11.2 minutes/day F 5909 steps/day, 9.9 minutes/day	1.97 (1.00–2.86)	1.78 (1.32–4.17)	2.01 (1.01–3.03)	2.21 (1.03–3.61)
M 7451 steps/day, 19.7 minutes/day F 7601 steps/day, 18.2 minutes/day	0.95 (0.43–2.01)	1.13 (0.57–2.15)	1.03 (0.58–2.25)	0.91 (0.37–2.51)
M 10650 steps/day, 32.4 minutes/day F 10334 steps/day, 30.9 minutes/day	1	1	1	1

Odds ratios adjusted for initial lean mass, age, smoking status and alcohol consumption [96]

(6 METs) habitual physical activity, but this was not true for muscle mass in the upper leg.

Health-Related Quality of Life The health-related quality of life (HRQOL) of Japanese seniors was assessed using the SF-36 questionnaire. When objectively measured physical activity was divided into quartiles, the HRQOL was greater for individuals in the second than those in the first quartile, but there was no additional advantage in the third and fourth quartiles [99]. However, the intensity of effort also seemed important in that the HRQOL was greater in those taking more than 25 % of their physical activity at an intensity >3 METs [100].

In a study of colon cancer survivors [101], Kerry Courneya (Fig. 1.11) and his colleagues found that quality of life was positively associated with both light and moderately vigorous physical activity, and in those taking moderate activity, the HRQOL increased progressively through to the quartile exercising for the longest daily time (>40 minutes/day).

1.7.3.4 Form of Physical Activity/Health Relationship

Observers using questionnaires concluded that in general the main benefit from greater physical activity and fitness was seen at the lower end of the population distribution [80]. Objective monitoring can provide further detail on the shape of this relationship, although in order to gain such information, it is important to

Fig. 1.11 Kerry Courneya is encouraging cancer survivors to increase their objectively measured daily physical activity

Table 1.11 Odds ratios (and 95 % confidence limits) for the risk of the metabolic syndrome in Japanese seniors in relation to habitual physical activity [104]

Activity quartile	Steps/day	Activity >3 METs (minutes/day)
3427 steps/day, 4.4 minutes/day	4.55 (1.81–11.41)	3.67 (1.50–8.97)
5581 steps/day, 12.1 minutes/day	3.10 (1.22–7.88)	2.29 (0.92–5.71)
7420 steps/day, 19.4 minutes/day	2.63 (1.02–6.75)	2.10 (0.83–5.27)
10,129 steps/day, 33.5 minutes/day	1	1

recruit adequate subject numbers, including a substantial number of individuals who are engaging in voluntary physical activity. The problems that can arise if this precaution is neglected are illustrated by a study of Gerdhem and associates [102]. Accelerometry measurements of physical activity showed no significant correlations with balance, muscle strength or bone density in a sample of 57 eighty-year-old women, but only 8 of the 57 subjects were engaging in moderate or vigorous physical activity. This negative conclusion was quickly reversed in a larger study by some of the same authors where 152 men and 206 women aged 50–80 were followed for 10 years; in this group, annual bone loss was 0.6 % less in those who were classified as active relative to those who were inactive [103]. In the larger group, benefits were also seen in terms of balance, although there was no impact upon muscle strength or gait velocity.

For many benefits, including increased bone health, greater vascular distensibility, and a larger lean tissue mass, objective monitoring confirms the impression gained from questionnaire data that the biggest improvement of health status is seen on moving from a completely sedentary status to a modest level of physical activity. However, this is not true of all conditions; in particular, the risk of showing manifestations of the metabolic syndrome [104] decreases across each of the four quartiles of habitual physical activity (Table 1.11).

1.8 Inference of Causality

Given the practical impossibility of conducting a classical randomized controlled experiment to test the benefits of habitual physical activity in preventing various forms of chronic disease, the epidemiologist is faced with the task of excluding spurious and indirect associations, and then of weighing the likelihood that observed associations are causal rather than casual.

1.8.1 Spurious Associations

The commonest cause of a spurious association is the type I error of the statistical method. If the usual criterion—a probability of 0.05—is accepted, then there is 1 chance in 20 that an apparent association is no more than a statistical artifact. If (as in many of the associations with physical activity), the association is reported repeatedly, the likelihood of such an error is diminished, although it remains necessary to watch for the possibility that perceptions have been biased by a tendency for journals to publish papers with positive conclusions.

Problems may arise from failure to allow for particular characteristics of the active people within a population—they may be younger, or come from a higher socio-economic level, and this may account for their better health. The modern tendency is to introduce co-variates to make a statistical adjustment for such issues, but before the advent of powerful computers, some epidemiologists attempted to overcome this issue by investigating a single age-group of comparable social status (for example, the bus drivers and conductors studied by Morris, Chap. 2).

Bias may also arise in prospective trials, since the subjects selected for both experimental and control groups in such trials tend to be more active and health-conscious than the general population. This tendency weakens the potential contrast between active and inactive individuals.

1.8.2 Indirect Associations

A further issue is a potential relationship between both habitual physical activity and health. For instance, both inactivity and cardiovascular disease are associated through other variables such as cigarette smoking, obesity and a mesomorphic body build [105, 106]. Again, provided that such extraneous factors are recognized, their influence can largely be eliminated by the incorporation of co-factors into multi-variate analyses.

1.8.3 Criteria Suggesting a Causal Association

The big challenge for the epidemiologist is to move beyond the demonstration of associations between habitual physical activity and health to causal inferences that can be used in prevention and treatment of various medical conditions. The English statistician and epidemiologist Bradford Hill [107] (Fig. 1.12) enunciated nine criteria that pointed towards a causal association between epidemiological findings (Table 1.12). Although commonly termed "criteria," Hill seems to have viewed these nine items as guidelines, and they did not all have to be satisfied before making a causal inference. In the case of associations between physical activity and health, the shift from subjective to objective monitoring has strengthened several pillars of the causal inference, although in some studies the advantages of the objective monitor have been offset by smaller subject numbers.

1.8.3.1 Strength of the Association

The a priori assumption is that a causal assumption is a strong one. Thus, the relative risk of developing lung cancer is increased 20-fold in a pack-a-day smoker

Fig. 1.12 Sir Austin Bradford Hill proposed nine criteria for testing the causality of an epidemiological association

Table 1.12 Criteria proposed by Bradford Hill for testing the causality of an association

• Strength of the association
• Consistency of the association
• Temporally correct association
• Specificity of the association
• Biological gradient
• Biological plausibility
• Coherence
• Experimental verification
• Analogy

[108]. Unfortunately, the impact of inactivity upon health is generally small; questionnaire estimates suggest that in terms of cardiovascular disease, the relative risk is about 1.28 in men and 1.3 [109], and for colon cancer, it is around 1.8 in men and 1.1–1.2 in women [110]. However, Bradford Hill recognized that a factor could be causal without necessarily exerting a strong effect; much depends upon the care that has been taken to eliminate sources of bias. It is difficult to measure small differences in risk using questionnaires, and the more precise gradation of habitual activity obtained from objective monitors should be helpful in this regard.

1.8.3.2 Consistency of the Association

A causal association should be consistently observed in a wide variety of situations, using a variety of measuring techniques, different subjects, places, circumstances and times. Objective monitoring is useful in providing an alternative method of grading physical activity, and because of its greater precision, consistent associations are more likely to be seen among those using this measurement technique.

1.8.3.3 Temporally Correct Association

Exposure to physical activity must ante-date the improvement of health by a period commensurate with the likely mechanism of benefit. This is an important issue to explore in connection with the possible mental health benefits of greater physical activity; if a lesser anxiety or depression preceded the increase of physical activity, one might be looking at reverse causality. It is very difficult to obtain a detailed history of past physical activity from questionnaires. To date, there have been few long-term prospective studies using objective monitors, but in future such investigations should provide a much clearer picture of when an individual initiated an active lifestyle.

1.8.3.4 Specificity of the Association

Ideally, a history should always include evidence of a risk factor for the disease in question, and the existence of that risk factor should always give rise to manifestation of the condition. However, in conditions such as cardiovascular disease, there is some lack of specificity—the risk of developing the disease is increased not only by physical inactivity, but also by genetic abnormalities of metabolism and by other risk factors such as cigarette smoking and obesity. Multivariate analysis may help in elucidating an apparent lack of specificity of associations.

1.8.3.5 Biological Gradient

A causal relationship commonly has a graded dose-response relationship, and the objective monitoring of physical activity is helpful in demonstrating both the existence of a relationship and its format. Hill appeared to be anticipating a linear-dose response relationship, and a more complex dose-response relationship might imply the intervention of other factors.

1.8.3.6 Biological Plausibility

If the relationship is causal, it should be possible to identify at least one plausible physiological explanation of the association. An increase of habitual physical activity has quite a wide range of known effects that could enhance various aspects of health. However, as Cox and Wermuth have pointed out [111], a biological mechanism has greater plausibility if it is postulated before the association is demonstrated rather than if it is sought as a post-hoc explanation of findings.

1.8.3.7 Coherence

The requirement of coherence is closely related to that of plausibility. If the association is causal, then the hypothesis that is advanced should provide a coherent explanation of all available data, and should not contradict current knowledge.

1.8.3.8 Experimental Verification

Ideally, a causal explanation of an association should be validated by direct experiment. Unfortunately, it is difficult if not impossible to test the long-term effects of an increase of physical activity in humans through a randomized double-blind controlled trial. Many aspects of the exercise hypothesis can be evaluated in animal experiments, although it is sometimes difficult to translate findings in animals to the human experience.

1.8.3.9 Analogy

The causal hypothesis gains credence if components of the postulated mechanisms of benefit can be shown to operate at the cellular level (for instance, if the tars in cigarettes can be shown to exert carcinogenic effects when applied to isolated tissues or cells). Further, if physical activity produces demonstrated benefits in one area of ill-health, it becomes more reasonable to assume it will be of benefit in the prevention or treatment of a somewhat similar ailment.

1.8.4 Critique of the Bradford Hill Criteria

Several investigators have sought to revise Bradford Hill's criteria in the light of the needs of evidence-based medicine [112]. One proposal is to group information as direct, mechanistic and parallel evidence. Direct evidence is drawn from randomized or non-randomized trials, mechanistic evidence points to processes that link the cause with the outcome, and parallel evidence looks at related studies that have yielded similar results. We will consider these revised criteria briefly, with particular reference to physical activity; in general, the changes in thinking have not modified the value of objective monitoring of physical activity to the establishment of cause and effect.

1.8.4.1 Direct Evidence

Direct evidence should either exclude plausible confounders of an experiment such as outcome expectations by blinding both subjects and observers, or at least ensuring that the magnitude of the effect is substantially greater than the likely influence of confounders. The size of effects that are of interest thus needs to be larger in observational trials than in randomized experiments.

In the new approach, the criterion of temporality is expanded to encompass temporal and spatial proximity: is the time interval to the observed effect appropriate to a causal hypothesis, and is there reason to seek a response in a particular part of the body?

In terms of dose-response, the strongest inference can be drawn if the process is reversible. This is generally true of physical activity—an adequate dose of exercise generally enhances health, but gains are quickly lost if the activity ceases.

The criterion of specificity is now commonly omitted from the analysis, since it is recognized that most clinical conditions have multiple antecedents and multiple effects.

1.8.4.2 Mechanistic Evidence

Bradford Hill thought simply in terms of biological plausibility, but it is now recognized that some of the responses to increased physical activity can have a mechanical or chemical explanation, rather than a simple physiological basis.

1.8.4.3 Parallel Evidence

Observations should be replicable (Bradford Hill's "consistency"), with the anticipation of a similar response when others have applied a similar intervention to a similar population, using the same outcome measures.

1.9 Conclusions

Epidemiologists seek associations between environmental factors, lifestyle influences and human health. An up-date of principles first enunciated by Bradford Hill is used to test whether observed associations are causal in nature. There is a long list of medical conditions where physical activity has been suggested as having a beneficial influence. Until recently, evaluation of the benefits of physical activity in these conditions has been hampered by the limited reliability and validity of physical activity questionnaires. The availability of pedometer/accelerometers and other types of objective monitor now facilitates this task, allowing investigators to detail the effects of various intensities, frequencies and durations of effort, although further improvement of monitoring devices is needed to capture the full range of activities likely to be encountered in children and younger adults. Objective monitoring does not support the hypothesis that a minimum intensity of effort is needed for health benefit; indeed, for many conditions, the largest effects are associated with quite small increases of habitual activity. For many conditions, there also seems a ceiling of benefit, with no additional gains from a further increase of habitual activity beyond this ceiling. Causality can never be totally proven, but the inference that moderate physical activity confers multiple health benefits is sufficiently strong to urge people of all ages to incorporate regular physical activity into the pattern of daily living.

References

1. Caspersen CJ, Powell KE, Christenson GM. Physical activity, exercise and physical fitness: definitions and distinctions for health-related research. Public Health Rep. 1985;100 (2):126–31.
2. Bouchard C, Shephard RJ, Stephens T, et al. Exercise, fitness and health. Champaign, IL: Human Kinetics; 1990.
3. Bouchard C, Shephard RJ, Stephens T. Physical activity, fitness and health. Champaign, IL: Human Kinetics; 1994.
4. Brown JR, Crowden GP. Energy expenditure ranges and muscular work grades. Br J Ind Med. 1963;20:277–83.
5. McIntosh PC. "Sport for all" programs around the world. Paris, France: UNESCO; 1980.
6. Hanke H. Freizeit in der DDR (Leisure time in East Germany). Berlin, Germany: Dietz Verlag; 1979.
7. Stundl H. Freizeit und Erholungsport in der DDR (Free time and recreational sport in East Germany). Schorndorf, Germany: Karl Hofmann Verlag; 1977.
8. U.S. Bureau of Labor. American time use survey—2013 results. Washington, DC: U.S. Bureau of Labor; 2014.
9. Passmore R, Durnin JVGA. Human energy expenditure. Physiol Rev. 1955;35:801–40.
10. World Health Organisation. Global recommendations on physical activity for health. Geneva, Switzerland: World Health Organisation; 2010.
11. Godin G, Shephard RJ. A simple method to assess exercise behaviour in the community. Can J Appl Sport Sci. 1985;10:141–6.
12. Shephard RJ. Endurance fitness, (2nd ed.) Toronto, ON: University of Toronto Press; 1977.

13. Clemes SA, Matchett N, Wane SL. Reactivity: an issue for short-term pedometer studies? Br J Sports Med. 2008;42:68–70.
14. Clemes SA, Deans NK. The presence and duration of reactivity to pedometers in adults. Med Sci Sports Exerc. 2012;44(6):1097–101.
15. Matthews CE, Ainsworth BE, Thompson RW, et al. Sources of variance in daily physical activity levels as measured by an accelerometer. Med Sci Sports Exerc. 2002;34(8):1376–81.
16. Weiner JS, Lourie JA. Human biology: a guide to field methods. Oxford: Blackwell; 1969.
17. Blair SN, Haskell WL, Ho P, et al. Assessment of habitual physical activity by a seven day recall in a community survey and controlled experiments. Am J Epidemiol. 1985;122 (5):795–804.
18. Cain KL, Geremia CM. Accelerometer data collection and scoring manual. San Diego, CA: James Sallis Active Living Laboratory, San Diego State University; 2011.
19. Tudor-Locke C, Burkett L, Reis JP, et al. How many days of pedometer monitoring predict weekly physical activity in adults? Prev Med. 2005;40:293–8.
20. Kubota A, Nagata J, Sugiyama M, et al. How many days of pedometer monitoring predict weekly physical activity in Japanese adults? Nihon Koshu Eisei Zasshi (Jap J Publ Health). 2009;56:805–10.
21. Trost SG, Pate RR, Freedson PS, et al. Using objective physical activity measures with youth: how many days of monitoring are needed? Med Sci Sports Exerc. 2000;32(2):426–31.
22. Tudor-Locke C, Bassett DR, Swartz AM, et al. A preliminary study of one year of pedometer self-monitoring. Ann Behav Med. 2004;3:158–62.
23. Yasunaga A, Togo F, Watanabe E, et al. Sex, age, season, and habitual physical activity of older Japanese: the Nakanojo study. J Aging Phys Act. 2008;16:3–13.
24. Togo F, Watanabe E, Park H, et al. Meteorology and the physical activity of the elderly: the Nakanojo study. Int J Biometeorol. 2006;50(2):83–9.
25. Hjorth MF, Chaput J-P, Michaelsen K, et al. Seasonal variation in objectively measured physical activity, sedentary time, cardio-respiratory fitness and sleep duration among 8–11 year-old Danish children: a repeated-measures study. BMC Public Health. 2013;13:8098.
26. Basiotis PP, Welsh SO, Cronin FJ, et al. Number of days of food intake records required to estimate individual and group nutrient intakes with defined confidence. J Nutr. 1987;117 (9):1638–41.
27. Matthews CE, Hebert J, Freedson PS. Sources of variance in daily physical activity levels in the seasonal variation of blood cholesterol study. Am J Epidemiol. 2001;153:987–95.
28. Togo F, Watanabe E, Park H, et al. How many days of pedometer use predict the annual activity of the elderly reliably? Med Sci Sports Exerc. 2008;40(6):1058–64.
29. Levin S, Jacobs DR, Ainsworth BE, et al. Intraindividual variation and estimates of usual physical activity. Ann Epidemiol. 1989;9(8):481–8.
30. Hatano Y. Use of the pedometer for promoting daily walking exercise. ICHPER. 1993;29:4–8.
31. Welk GJ, Differding J, Thompson RW, et al. The utility of the Digi-walker step counter to assess daily physical activity patterns. Med Sci Sports Exerc. 2000;32(9):S481–8.
32. Wilde BE, Sidman CL, Corbin CB. A 10,000 step count as a physical activity target for sedentary women. Res Q Exerc Sport. 2001;72(4):411–4.
33. Moreau KL, Degarmo R, Langley J, et al. Increasing walking lowers blood pressure in postmenopausal women. Med Sci Sports Exerc. 2001;33(11):1825–31.
34. Swartz AM, Thompson DL. Increasing daily walking improves glucose tolerance in overweight women. Res Q Exerc Sport. 2002;73(Suppl):A16.
35. Iwane M, Arita M, Tomimoto S, et al. Walking 10,000 steps/day or more reduces blood pressure and sympathetic nerve activity in mild essential hypertension. Hypertens Res. 2000;23:573–80.
36. Tudor-Locke C, Bassett DR. How many steps/day are enough? Preliminary pedometer indices for public health. Sports Med. 2004;34(1):1–8.

37. Rowlands AV, Eston RG, Ingledew DK. Relationship between activity levels, aerobic fitness, and body fat in 8- to 10-yr-old children. J Appl Physiol. 1989;86(4):1428–35.
38. Public Health Agency of Canada. Physical activity tips for adults (18–64 years). 2014. http://www.phac-aspcgcca/hp-ps/hl-mvs/pa-ap/07paap-engphp.
39. American Heart Association. American Heart Association recommendations for physical activity in adults. 2014. http://www.heart.org/HEARTORG/GettingHealthy/PhysicalActivity/FitnessBasics/American-Heart-Association-Recommendations-for-Physical-Activity-in-Adults_UCM_307976_Articlejsp.
40. Tudor-Locke C, Jones GR, Myers AM, et al. Contribution of structured exercise class participation and informal walking to daily physical activity in community-dwelling older adults. Res Q Exerc Sport. 2002;73(3):350–6.
41. Aoyagi Y, Shephard RJ. Steps per day: the road to senior health? Sports Med. 2009;39:423–38.
42. Public Health Agency of Canada. Physical activity tips for children (5–11 years). 2014. http://www.phac-aspcgcca/hp-ps/hl-mvs/pa-ap/05paap-engphp.
43. American Heart Association. AHAs recommendations for physical activity in children. 2014. http://www.heart.org/HEARTORG/GettingHealthy/HealthierKids/ActivitiesforKids/The-AHAs-Recommendations-for-Physical-Activity-in-Children_UCM_304053_Articlejsp.
44. Bassett Jr DR, Cureton AL, Ainsworth BE. Measurement of daily walking distance: questionnaire versus pedometer. Med Sci Sports Exerc. 2000;32(5):1018–23.
45. Mizuno C, Yoshida T, Udo M. Estimation of energy expenditure during walking and jogging by using an electro-pedometer. Ann Physiol Anthropol. 1990;9:283–9.
46. Shephard RJ, Aoyagi Y. Objective monitoring of physical activity in older adults: clinical and practical implications. Phys Ther Rev. 2010;15:170–82.
47. Tudor-Locke C, Myers AM. Methodological considerations for researchers and practitioners using pedometers to measure physical (ambulatory) activity. Res Q Exerc Sport. 2001;72:1–12.
48. Freedson P, Melanson E, Sirard J. Calibration of the computer science and applications accelerometer. Med Sci Sports Exerc. 1998;30(5):777–81.
49. Matthews CE, Chen KY, Freedson PS, et al. Amount of time spent in sedentary behaviors in the United States, 2003–2004. Am J Epidemiol. 2008;167(7):875–81.
50. Kozey-Keadle S, Libertine A, Lyden K, et al. Validation of wearable monitors for assessing sedentary behavior. Med Sci Sports Exerc. 2011;43:1561–7.
51. Brage S, Wedderkopp N, Franks PW, et al. Reexamination of validity and reliability of the CSA monitor in walking and running. Med Sci Sports Exerc. 2003;35:1447–54.
52. Leenders NY, Nelson TE, Sherman WM. Ability of different physical activity monitors to detect movement during treadmill walking. Int J Sports Med. 2003;24:43–50.
53. Metzger JS, Catellier DJ, Evenson KR, et al. Patterns of objectively measured physical activity in the United States. Med Sci Sports Exerc. 2008;40:630–8.
54. Yngve A, Nilsson A, Sjostrom M, et al. Effect of monitor placement and of activity setting on the MTI accelerometer. Med Sci Sports Exerc. 2003;35(2):320–6.
55. Abel MG, Hannon JC, Sell K, et al. Validation of the Kenz Lifecorder EX and Actigraph GT1M accelerometers for walking and running in adults. Appl Physiol Nutr Metab. 2008;33:1155–64.
56. Schneider PL, Crouter SE, Lukajic O, et al. Accuracy and reliability of 10 pedometers for measuring steps over a 400-m walk. Med Sci Sports Exerc. 2003;35:1779–84.
57. Bassett DR, Ainsworth BE, Leggett SR, et al. Accuracy of five electronic pedometers for measuring distance walked. Med Sci Sports Exerc. 1996;28:1071–7.
58. Albright C, Hultquist CN, Thompson DL. Validation of the Lifecorder EX activity monitor. Med Sci Sports Exerc. 2006;35 Suppl 5:S500.
59. Bassey EJ, Dalloso HM, Fentem PH, et al. Validation of a simple mechanical accelerometer (pedometer) for the estimation of walking activity. Eur J Appl Physiol. 1987;56:323–30.

60. Yokoyama Y, Kawamura T, Tamakoshi A, et al. Comparison of accelerometry and oxymetry for measuring daily physical activity. Circulation. 2002;66:751–4.
61. Kumahara H, Yoshioka M, Yoshitake Y, et al. Validity assessment of daily expenditure in a respiration chamber by accelerometry located on the waist vs the wrist or in combination. Med Sci Sports Exerc. 2002;34 Suppl 5:S140.
62. Cyarto EV, Myers AM, Tudor-Locke C. Pedometer accuracy in nursing home and community dwelling older adults. Med Sci Sports Exerc. 2004;36:205–9.
63. Sirard JR, Pate RR. Physical activity assessment in children and adolescents. Sports Med. 2001;31:439–54.
64. Le Masurier GC, Tudor-Locke C. Comparison of pedometer and accelerometer accuracy under controlled conditions. Med Sci Sports Exerc. 2003;35:867–71.
65. McClain JJ, Craig CL, Sisson BB, et al. Comparison of Lifecorder EX and Actigraph accelerometers under free-living conditions. Appl Physiol Nutr Metab. 2007;32(4):753–61.
66. Dondzila CJ, Swartz AM, Miller NL, et al. Accuracy of uploadable pedometers in laboratory, overground, and free-living conditions in young and older adults. Int J Behav Nutr Phys Act. 2012;9:143.
67. Eston RG, Rowlands AV, Ingledew DK. Validity of heart rate, pedometry, and accelerometry for predicting the energy cost of children's activities. J Appl Physiol. 1998;84(1):362–71.
68. Choquette S, Chuin A, LaLancette R-A, et al. Predicting energy expenditure in elders with the metabolic cost of activities. Med Sci Sports Exerc. 2009;41:1915–20.
69. Colbert L, Matthews CE, Havighurst TC, et al. Comparative validity of physical activity measures in older adults. Med Sci Sports Exerc. 2011;43:867–76.
70. Fogelholm M, Hiilloskorpi H, Laukkanen R, et al. Assessment of energy expenditure in overweight women. Med Sci Sports Exerc. 1998;30:1191–7.
71. Gardner AW, Poehlman ET. Assessment of free-living daily activity in older claudicants: validation against the double labeled water technique. J Gerontol A Biol Sci Med Sci. 1998;53A:M275–80.
72. Rafamantanantsoa HH, Ebine N, Yoshioka M, et al. Validation of three alternative methods to measure total energy expenditure against the doubly labeled water method for older Japanese men. J Nutr Sci Vitaminol. 2002;48:517–23.
73. Starling RD, Matthews DE, Ades PA, et al. Assessment of physical activity in older individuals: a doubly labeled water study. J Appl Physiol. 1999;86:2090–6.
74. Leenders NYJM, Sherman WM, Nagaraja HN, et al. Evaluation of methods to assess physical activity in free-living conditions. Med Sci Sports Exerc. 2001;33:1233–40.
75. Kesaniemi YK, Danforth E, Jensen PJ, et al. Dose-response issues concerning physical activity and health: an evidence-based symposium. Med Sci Sports Exerc. 2001;33:S351–8.
76. Lamonte MJ, Ainsworth BE. Quantifying energy expenditure and physical activity in the context of dose response. Med Sci Sports Exerc. 2001;33(6 Suppl):S370–8.
77. Haskell WL, Yee MC, Evans A, et al. Simultaneous measurement of heart rate and body motion to quantitate physical activity. Med Sci Sports Exerc. 1993;25:109–15.
78. Ainsworth BE, Bsssett DR, Strath SJ, et al. Comparison of three methods for measuring the time spent in physical activity. Med Sci Sports Exerc. 2000;32:S457–64.
79. Welk GJ, Blair SN, Wood K, et al. A comparative evaluation of three accelerometry-based physical activity monitors. Med Sci Sports Exerc. 2000;32:S489–97.
80. Blair SN, Cheng Y, Holder JS. Is physical activity or physical fitness more important in defining health benefits? Med Sci Sports Exerc. 2001;33 Suppl 2:S379–99.
81. Shephard RJ, Aoyagi Y. Associations of activity monitor output and an estimate of aerobic fitness with pulse wave velocities: the Nakanojo study. J Phys Activ Health. 2014 (in press).
82. Morris JN, Clayton DG, Everitt MG. Exercise in leisure time: coronary attack and death rates. Br Heart J. 1990;63:325–34.
83. Aoyagi Y, Park H, Kakiyama T, et al. Yearlong physical activity and regional stiffness of arterial segments in older adults: the Nakanojo study. Eur J Appl Physiol. 2010;109 (3):455–64.

84. Sugawara J, Otsuki T, Tanabe T, et al. Physical activity duration, intensity and arterial stiffening in post-menopausal women. Am J Hypertens. 2006;19:1032–6.
85. Gando Y, Yamamoto K, Murakami H, et al. Longer time spent in light physical activity is associated with reduced arterial stiffness in older adults. Hypertension. 2010;56:540–6.
86. Hawkins M, Gabriel KP, Cooper J, et al. The impact of change in physical activity on change in arterial stiffness in overweight or obese sedentary young adults. Vasc Med. 2014;19(4):257–63.
87. Alhassan S, Robinson TM. Objectively measured physical activity and cardiovascular disease risk factors in African American girls. Ethn Dis. 2008;18(4):421–8.
88. Pelclová J, Gába A, Kapuš S. Bone mineral density and accelerometer-determined habitual physical activity in postmenopausal women. Acta Univ Palacki Olomuc Gymn. 2011;41(3):47–53.
89. Gába A, Kapuš S, Pelclová J, et al. The relationship between accelerometer-determined physical activity (PA) and body composition and bone mineral density (BMD) in postmenopausal women. Arch Gerontol Geriatr. 2012;54(3):e315–22.
90. Park H, Togo F, Watanabe E, et al. Relationship of bone health to yearlong physical activity in older Japanese adults: cross-sectional data from the Nakanojo study. Osteoporosis Int. 2007;18:285–93.
91. Shephard RJ, Park H, Park S et al. Objectively measured physical activity and calcaneal bone health in older Japanese adults: dose/response relationships in longitudinal data from the Nakanojo study. J Am Geriatr Soc. 2016 (in press).
92. Kitagawa J, Omasu F, Nakahara Y. Effect of daily walking steps on ultrasound parameters of the calcaneus in elderly Japanese women. Osteoporosis Int. 2003;14(3):219–24.
93. Vainionpää A, Korpelainen R, Vihriälä E, et al. Intensity of exercise is associated with bone density change in premenopausal women. Osteoporosis Int. 2006;17:455–63.
94. Janz KF, Burns TL, Torner JC, et al. Physical activity and bone measures in young children: the Iowa bone development study. Pediatrics. 2001;107:1387–93.
95. Park H, Park S, Shephard RJ, et al. Year-long physical activity and sarcopenia in older adults: the Nakanojo study. Eur J Appl Physiol. 2010;109(5):953–61.
96. Shephard RJ, Park H, Park S, et al. Objectively measured physical activity and progressive loss of lean tissue in older Japanese adults: Longitudinal data from the Nakanojo study. J Am Geriatr Soc. 2013;61(11):1887–93.
97. Abe T, Mitsukawa N, Thiebaud RS, et al. Lower body site-specific sarcopenia and accelerometer-determined moderate and vigorous physical activity: the Hiregasaki study. Aging Clin Exp Res. 2012;24(6):657–62.
98. Ogawa M, Mitsuwaka N, Loftin M, et al. Association of vigorous physical activity with age-related, site-specific loss of thigh muscle in women: the HIREGASAKI study. J Trainol. 2012;1:6–9.
99. Yasunaga A, Togo F, Watanabe E, et al. Year-long physical activity and health-related quality of life in older Japanese adults: the Nakanojo study. J Aging Phys Act. 2006;14:288–301.
100. Yasunaga A, Togo F, Park H, et al. Interactive effects of the intensity and volume of physical activity on health-related quality of life in older adults: the Nakanojo study. J Aging Phys Act. 2008;20:S184.
101. Vallance JK, Boyle T, Courneya KS, et al. Associations of objectively assessed physical activity and sedentary time with health-related quality of life among colon cancer survivors. Cancer. 2014;120:2919–26.
102. Gerdhem P, Dencker M, Ringsberg K, et al. Accelerometer-measured daily physical activity among octogenerians: results and associations to other indices of physical performance and bone density. Eur J Appl Physiol. 2008;102(2):173–80.
103. Daly RM, Ahlkborg HG, Ringsberg K, et al. Association between changes in habitual physical activity and changes in bone density, muscle strength, and functional performance in elderly men and women. J Am Geriatr Soc. 2008;56(12):2252–60.

104. Park S, Park H, Togo F, et al. Yearlong physical activity and metabolic syndrome in older Japanese adults: cross-sectional data from the Nakanojo study. J Gerontol A Biol Sci Med Sci. 2008;63:1119–23.
105. Morgan P, Gildiner M, Wright GR. Smoking reduction in adults who take up exercise: a survey of a running club. CAHPER J. 1976;42(5):39–43.
106. Yamaji K, Shephard RJ. Longevity and causes of death of athletes: a review of the literature. J Hum Ergol. 1977;6:13–25.
107. Hill AB. Principles of medical statistics. 9th ed. New York, NY: University Press; 1971.
108. Pope CA, Burnett RT, Turner MC, et al. Lung cancer and cardiovascular disease mortality associated with ambient air pollution and cigarette smoke: shape of the exposure-response relationship. Environ Health Perspect. 2011;119:1616–21.
109. Schnor P, Jensen JS, Scharling H, et al. Coronary heart disease risk factors ranked by importance for the individual and the community. Eur Heart J. 2002;23:620–6.
110. Ballard-Barbash R, Schatzkin A, Albanes D, et al. Physical activity and risk of large bowel cancer in the Framingham study. Cancer Res. 1990;50:3610–3.
111. Cox DR, Wermuth N. Multivariate dependencies: models, analysis and interpretation. London: Chapman Hall; 1996.
112. Howick J, Glasziou P, Aronson JK. The evolution of evidence hierarchies: what can Bradford Hill's "guidelines for causation" contribute? J R Soc Med. 2009;102(5):186–94.

Chapter 2
A History of Physical Activity Measurement in Epidemiology

Roy J. Shephard

Abstract Although Hippocrates is often considered the father of epidemiology, John Snow also played an important role with his studies of cholera epidemics in Victorian London. A detailed study of relationships between physical activity and the prevention of chronic disease did not begin until the mid-twentieth century, with Jeremy Morris in London, and Henry Taylor and Ralph Paffenbarger in the U.S. leading investigations of the epidemic of ischaemic heart disease. Occupation or athletic status was initially used to classify the habitual physical activity of study participants, but as daily energy expenditures diminished at most work sites, interest shifted to questionnaire and diary assessments of leisure activity. Other options to classify the habitual activity of subjects included occasional quasi-experimental assignments to exercise programmes, determinations of aerobic fitness, and a study of "natural experiments" where community activity patterns were known to have diminished. Such initiatives generally distinguished active from inactive individuals, but attempts to determine the intensity and volume of exercise that was undertaken often yielded unrealistically large values. The introduction of modern pedometer/accelerometers at first seemed to promise accurate, objective assessments of habitual activity. Although quite successful in assessing standardized activities such as steady walking, the newer monitors have shown much less consistency in measuring the wide range of activities encountered in normal daily living. Future research may focus upon some combination of activity monitoring with global position-sensing and posture detecting devices.

Roy J. Shephard (✉)
Faculty of Kinesiology & Physical Education, University of Toronto, Toronto, ON, Canada
e-mail: royjshep@shaw.ca

R.J. Shephard, C. Tudor-Locke (eds.), *The Objective Monitoring of Physical Activity: Contributions of Accelerometry to Epidemiology, Exercise Science and Rehabilitation*, Springer Series on Epidemiology and Public Health,
DOI 10.1007/978-3-319-29577-0_2

2.1 Introduction

The term epidemiology means literally "*what is upon the people*." From the classical period of ancient Greece, physicians such as Hippocrates (460–370 BCE) have sought to find a logic underlying human illness and disease. Rather than blaming the random act of an angry God, the more discerning physicians have sought to explain and treat illness in terms of adverse external influences. For Hippocrates, the answer lay in restoring the internal balance of the four humours he saw as underlying the structure of our universe (air, fire, water, and earth) [1, 2]. Exposure to an unusual climate, an excess of moisture or noxious vapours were seen as external factors that could upset this delicate balance. Thus, Hippocrates noted that malaria and yellow fever were prevalent in swampy areas, although it was not until 1900 that Walter Reed linked these diseases to the breeding grounds of the mosquito rather than to unhealthy miasmata that were emerging from the swamps [1].

Even in Classical times, a small proportion of physicians, philosophers and other opinion-makers had formed a vague idea that moderate exercise was good for the health of the average citizen [2], but until about 70 years ago, there was no attempt to explore the epidemiology of physical inactivity; no one had defined the health benefits of a physically active lifestyle, and no one had attempted to specify the minimal amount of daily exercise that was needed for good health. In the mid-twentieth century, epidemiologists such as Jeremy Morris in England, and Ralph Paffenbarger and Henry Taylor in the United States began to address this challenge.

The present chapter first considers the apparent epidemic of cardiovascular disease that stimulated this new interest in the epidemiology of physical activity. It relates early ground-breaking initiatives, and traces through to the present day how changing methods of classifying an individual's habitual physical activity have helped to elucidate both the likely range of health benefits associated with regular exercise and the optimal weekly dose of physical activity.

2.2 The Cardiac Epidemic

Much early epidemiology was concerned with the causes of acute disease, as exemplified by the observations of Hippocrates on malaria and yellow fever, and more recently by John Snow's (1813–1858 CE) study of a cholera epidemic in the area served by the Broad Street pump in Central London [3]. However, there had been occasional ventures into the epidemiology of chronic disease. Paracelsus (1493–1541 CE) had demonstrated a geographic relationship between the mineral content of drinking water and the development of goiter, and Snow had himself proposed a chronic epidemiology project. The latter author had written an article in the Lancet suggesting that the adulteration of commercial bread was a probable cause of rickets [4], and he proposed testing this by a comparison of bone health between families baking their own bread and those purchasing it from bakers.

One new feature of epidemiology in the mid-twentieth century was its exploration of the idea that personal lifestyle could be among the adverse external influences predisposing to chronic disease. This concept was first applied to ischaemic heart disease, and subsequently to lung cancer, obesity, diabetes mellitus and other chronic conditions. The issue that persuaded "Jerry" Morris to begin his study of the epidemiology of physical activity was his perception that during the 1940s, an epidemic of cardiovascular disease affecting middle-aged adults was sweeping across England and Wales.

Morris cited the impression of many clinicians [5–8] that coronary heart disease was becoming ever more common and was being seen in younger patients. He thus examined the statistics of the Registrar General for England and Wales for the period 1931–1948 [9]. The statistics for older men (aged 50–54 and 60–64 years) showed a progressive increase in the number of deaths attributed to acute ischaemic heart disease (including coronary atheroma or atherosclerosis, and coronary, ischaemic or atherosclerotic heart disease). In contrast, deaths attributed to chronic myocardial disease showed little change over this same period (Fig. 2.1). Among women of comparable age, deaths attributed to coronary vascular disease were much less frequent, but nevertheless they also showed an increase in acute cardiovascular mortality. Morris [9] concluded cautiously "*If the increase is real, in whole or in part....causes may be discoverable in changing ways of living, in people who are changing in a changing environment.*" He also cautioned "*It is extremely difficult to determine how much of the apparent increase of coronary heart disease*

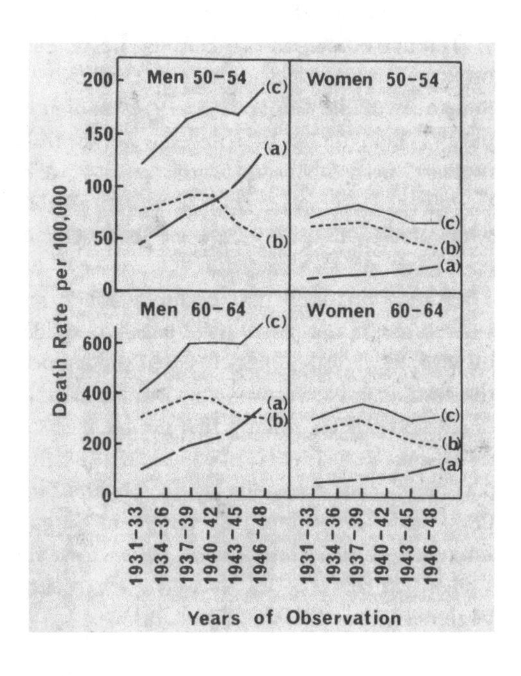

Fig. 2.1 Annual death rates for age and sex specific groups of the English and Welsh populations, as recorded by the Registrar General for the year 1931–1948 CE. (a): Acute ischaemic heart disease (coronary arterial disease plus angina, including coronary atheroma or atheroslerosis and coronaryt, ischaemic or arteriosclerotic hear disease); (b): chronic myocardial disease (including late deaths from myocardial infarction); (c): sum of (a) + (b). Based on the analysis of Morris [9]

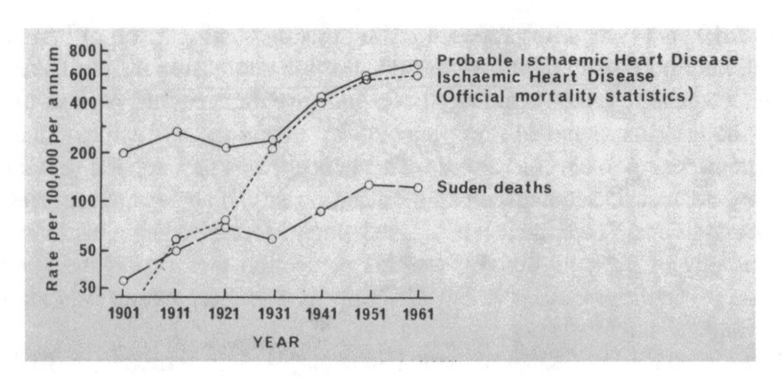

Fig. 2.2 Deaths from ischaemic heart disease in the province of Ontario, as analyzed by Anderson and LeRiche [10].The figure illustrates the official mortality data for ischaemic heart disease (comprising angina pectoris, coronary thrombosis, myocardial infarction and arteriosclerotic heart disease) in men aged 45–64 years, the probable incidence of ischaemic heart disease (based upon a reinterpretation of the original death certificates), and the incidence of sudden death (adjusted for certificates providing no information on the duration of illness). Note that death rates are expressed on a logarithmic scale

represents greater prevalence and how much is due to better diagnosis or changing fashions in diagnosis." Nevertheless, the Registrar General's statistics were supported by a doubling in the number of sudden deaths due to ventricular rupture from 1932 to 1949, and data for recent coronary thrombosis and acute myocardial infarction from the necropsy room of the London Hospital from 1907–1914 to 1944–1949 showed a six to sevenfold overall increase, with the steepest increase occurring around the time of the first World War.

Terence Anderson and Harding LeRiche, working in the Department of Epidemiology and Biometrics at the University of Toronto, decided to explore the generality of the epidemic, using Canadian mortality data [10]. Vital statistics for the Province of Ontario showed a similar trend to that observed by Morris, with an increase in deaths from cardiovascular disease from 1931 to 1961 (Fig. 2.2). However, Anderson and LeRiche recognized the problem that in the early 1900s, some practitioners may have ascribed ischaemic heart disease deaths to such vague causes as "*acute indigestion,*" "*apoplexy,*" or "*chronic myocardial degeneration.*" They reasoned that there could be much less room for error in the reporting of sudden death, and since most cases of sudden death were due to ischaemic heart disease, the substantial increase of sudden deaths from 1931 to 1961 suggested that the cardiac epidemic was a real phenomenon. They next re-examined the original death certificates, finding that no deaths were in fact recorded as "acute indigestion," and very few had been classed as "apoplexy." Sex ratios provided a final argument against explaining the apparent epidemic in terms of misdiagnoses. If there had been simply a change in diagnostic labels, sex ratios should have remained close to unity throughout. However, statistics for Canada, England and Wales and the U.S. all showed a sharp surge in cardiovascular disease in men, beginning around 1920 (Fig. 2.3) [11].

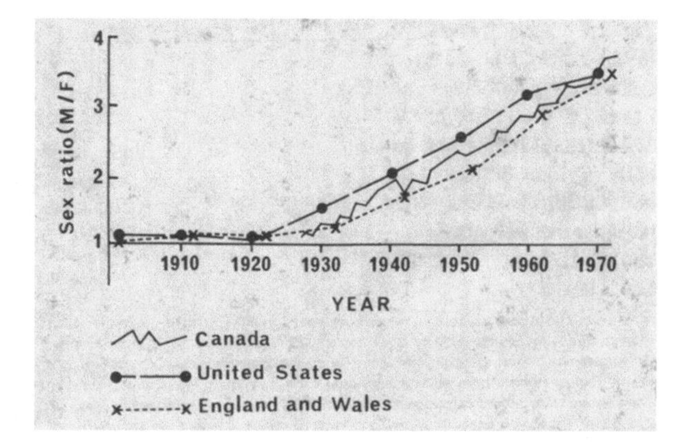

Fig. 2.3 The sex ratio (male/female) for all forms of heart disease in adults aged 45–64 years. Data for Canada, England and Wales and the United States. Based on an analysis of Anderson [11]

The reality of the cardiac epidemic thus seems established. However, its etiology is less certain. Shephard [12] made a detailed analysis of changes in ways of living over the twentieth century, looking specifically at such factors as hypertension, occupational stress, cigarette smoking, habitual physical activity, cholesterol levels, consumption of refined carbohydrates, diabetes, and obesity. Associations were sought between changes in each of these factors and the course of the epidemic. A growing prevalence of cigarette smoking during this era was one undesirable aspect of personal lifestyle; heavy smoking doubled the risk of a fatal heart attack [13], but it did not seem a strong enough influence to account for the epidemic in its entirety. A decline of habitual physical activity during the 1940s did not seem a major contributing factor, at least in England and Wales. Relatively few Britishers owned cars prior to World War II, and during the war most people had no access to petrol for the operation of private vehicles. Moreover, in the Britain of Morris's era, there had been little attempt to modernize industry or to reduce the energy expenditures of employees. Leisure habits also had shown little change; although public television broadcasting had officially begun in 1936, the majority of the population was unable to purchase a television receiver until the mid 1950s.

Given that the secular trend to a decline in habitual physical activity had yet to occur in Britain, it is a little surprising that Morris and his fellow epidemiologists developed an interest in potential relationships between habitual physical activity and cardiovascular disease. Nevertheless, over the next decade their efforts were rewarded by a clear demonstration that physical inactivity was a substantial risk factor for cardiovascular disease.

Fig. 2.4 "Jerry" Morris (1910–2009 CE), was an epidemiologist at the London School of Hygiene and Tropical Health and an early explorer of the relationships between physical inactivity and chronic ill-health. *Source*: LSHTH Blog

2.3 The Occupational Epidemiology of Physical Activity and Health

Although a proportion of physicians had long commended moderate physical activity as good for health, a serious examination of relationships between habitual physical activity and the prevention of chronic disease was not initiated until the late 1940s, with the classical epidemiological studies of Jerry Morris [14], Ralph Paffenbarger [15, 16] and Henry Taylor [17] on the role of regular physical activity in the prevention of cardiovascular disease.

Their early investigations were based upon comparisons of health experience between groups of workers with presumed differences of occupational energy expenditure. Statistics that were examined included total death rates, sudden death rates, deaths from ischaemic heart disease, the annual incidence of myocardial infarction, the annual incidence of angina, and combinations of these statistics.

2.3.1 Jeremiah Noah Morris

"Jerry" Morris (1910–2009 CE) was the son of Jewish refugees from Poland (Fig. 2.4). He had grown up in poverty in the slums of Glasgow, and this experience shaped his lifelong interest in the social determinants of health. After graduating in Medicine from University College London, he joined the Medical Research Council's Social Medicine Research Unit, which he brought to the London School of Hygiene and Tropical Medicine. He became a keen proponent of incorporating what he termed "vigorous getting about" into daily living.

He initiated studies on the epidemiology of physical activity by comparing cardiovascular data for London bus-drivers (who sat in their driver's seats for at least 90 % of their shifts) with similar figures for the bus conductors, who in that era had to collect fares from passengers on both decks of their vehicle every few minutes; the conductors climbed 500–750 steps every working day [18]. In a

Fig. 2.5 Ralph Paffenbarger (1922–2007 CE) was a U.S. epidemiologist who carried out landmark studies on the relationships between activity and cardiovascular health on San Francisco long-shore workers and Harvard Alumni. Photograph reproduced by kind permission of Stanford School of Medicine's Office of Communication & Public Affairs

massive sample of 31,000 male workers aged 35–65 years, Morris found that the respective annual incidences of coronary arterial disease were 2.7/1000 for the drivers and 1.9/1000 for the conductors. Moreover, in the conductors who did eventually develop chronic cardiovascular disease, it occurred at a later age than in the drivers, often presenting as angina rather than a heart attack, and it was less likely to be fatal (30 vs. 50 %).

Morris next examined an even larger sample of 110,000 employees of the British post-office. He again observed a gradation of cardiovascular disease incidence with the presumed physical demands of employment. The lowest rates were seen in postal carriers who cycled or walked to deliver the mail; rates were higher in workers who engaged in less demanding physical activity (counter-hands, postal supervisors, and higher grade postal staff) and the highest rates were seen in totally sedentary workers (telephone operators, clerks and executives). Respective annual rates for the incidence of coronary arterial disease were 1.8/1000, 2.0/1000, and 2.4/1000, and annual case fatality rates were 0.6/1000, 0.9/1000 and 1.2/1000 in the three groups of workers. But as with the London Transport employees, the annual incidence of angina pectoris was highest in the most active group (0.7/1000), with rates of 0.4/1000 and 0.5/1000 for the other two categories of employees [18].

2.3.2 Ralph Paffenbarger

Ralph Paffenbarger (1922–2007 CE) (Fig. 2.5) was the son of a physician. He completed medical training at Northwestern University, and then took a Ph.D. degree in Epidemiology at Johns Hopkins University in Baltimore. After spending time teaching at Harvard and the University of California at Berkeley, he joined Stanford University, where he taught health research and policy. He was himself a committed exerciser, and engaged in many marathon and ultramarathon events.

His first study on the epidemiology of physical activity involved a group of 3263 San Francisco longshore workers who had undergone multiple health screenings over a 16-year period of employment. He classified these workers in terms of their presumed energy expenditure per shift, distinguishing a group of workers who each day expended 4 MJ more than their peers. The "heavy" workers, thus identified, had lower coronary artery disease death rates than other dock workers (59 vs. 80 incidents per 10,000 person-years of work) [15]; moreover, these differences persisted after statistical adjustment of the data for other coronary risk factors (smoking habits, weight for height and blood pressure).

A further analysis of data for the same population distinguished heavy workers (energy expenditures averaging 7.8 MJ/day), moderately heavy workers (expenditures of 6.1 MJ/day) and light workers (expenditures of 3.6 MJ/day). Again, the data showed a gradation of cardiac risk between the three categories of worker, with age-adjusted coronary arterial disease death rates of 26.9 vs. 46.3 vs. 49.0 deaths per 10,000 person-years of work; protection of the heavy workers was particularly marked in terms of sudden deaths (5.6 vs. 19.9 vs. 15.7 deaths per 10,000 person-years) [16].

A final report on the longshore workers looked at the 22-year health experience of this population. It distinguished men who engaged in bursts of high intensity activity (29 kJ/minute) with those with the lowest energy expenditures on the waterfront (4 kJ/minute). Advances in computer technology now allowed adjustment of the data for multiple co-variates (age, race, systolic blood pressure, smoking, glucose intolerance and ECG status). After making such adjustments, the sedentary group had twice the risk of sustaining a fatal myocardial infarction when compared with the highly active group [19].

2.3.3 Henry Longstreet Taylor

Henry Taylor (1912–1983 CE) (Fig. 2.6) was one of a group of distinguished physiologist-epidemiologists who spent much of his career in Ancel Key's

Fig. 2.6 Henry Longstreet Taylor (1912–1983 CE) was one of a group of distinguished physiologist-epidemiologists who spent much of his career in Ancel Key's *Laboratory of Physiological Hygiene* at the University of Minnesota. He studied the relationship between occupational physical activity and cardiovascular health in a large sample of railroad workers. *Source*: The Physiologist, 27(1), p. 22, 1984, used with courtesy from the American Physiological Society

Laboratory of Physiological Hygiene at the University of Minnesota. He gave serious consideration to the possibility of conducting a randomized controlled trial of exercise in the primary prevention of heart disease, but he reluctantly concluded that such an experiment was not practicable because of its complexity, magnitude and the number of confounding factors.

Like Morris and Paffenbarger, Taylor thus approached the epidemiology of physical activity in terms of workers with differing occupational energy expenditures. He chose as his subjects railroad employees, classifying them as clerks (85,112 person-years), switchmen (61,630 person-years) and section workers (44,867 person-years) [17]. Although the clerks were clearly sedentary, and greater energy expenditures were required by the other two groups of workers, no precise estimates of occupational energy demands were made. All subjects were white males aged 40–64 years who had worked for the railroad company for at least 10 years. Age-adjusted death rates for the three categories of employee were, respectively, 11.8/1000, 10.3/100 and 7.6/1000 for all-cause mortality, and 5.7/1000, 3.9/1000, an 2.8/1000 for deaths ascribed to arteriosclerotic heart disease. Sub-maximal treadmill tests were subsequently carried out on many of the study participants, using as a laboratory a converted Pullman car that travelled around the U.S. [20]; unfortunately, these findings were not related to the occupational categories identified by Henry Taylor.

2.3.4 Conclusions from Occupational Comparisons

Morris was convinced the evidence from his occupational comparisons pointed towards a beneficial effect of physical activity, advancing the hypothesis the "*men doing physically active work have a lower mortality from coronary heart disease in middle age than men in less active work*" [18]. However, the job-classification approach initially encountered strong skepticism from the medical establishment.

One obvious issue was the self-selection of active vs. sedentary employment. A follow-up analysis at London Transport demonstrated that bus drivers began their employment with a larger trouser and jacket size than the conductors, implying that they had a greater initial body fat content [21]. Nevertheless, at a 10-year follow-up, rates of sudden death in the drivers were twice as high as those in the conductors, irrespective of their initial physique [22]. A subsequent multivariate analysis examined the contribution of other risk factors to the development of cardiac disease. It found that systolic blood pressure and serum cholesterol levels accounted for 75 % of new cases. Blood pressure and lipid levels were lower in the conductors than in the drivers, but for any given systemic blood pressure, the risk of cardiovascular disease was almost twice as great in the drivers as in the conductors [23].

Paffenbarger's occupational comparisons were complicated somewhat by job reclassifications secondary to cardiac symptoms; those with angina frequently transferred out of heavy work, and he noted that the risk of a heart attack was greatest in those whose occupational energy expenditure had decreased over a

4-year observation period [19]. However, Taylor suggested that there was little problem of reclassification amongst his railroad workers, because sedentary employees belonged to a different union that did not welcome track workers. Despite the potential challenges of self-selection of active employment, reclassification of active workers as cardiac disease develops, and differences of social class and leisure activities between sedentary and active workers, the occupational studies point strongly towards benefit from engaging in physically demanding employment. However, they do not define the intensity and volume of effort needed for better health.

The interest of epidemiologists shifted from occupational classifications to assessments of leisure-time activity during the latter part of the twentieth century, in part because mechanization was eroding the physical demands of much traditional heavy industry, in part because there seemed the greatest scope for public health intervention in terms of augmenting the voluntary leisure activity of sedentary individuals, and in part because analyses of leisure behaviour could be made on groups having a similar socio-economic status. Options for assigning an individual to an active leisure category became reports of athletic status, assignment to the experimental group in quasi-experimental studies, questionnaire self-reports of vigorous leisure activity, and demonstrations of a high level of aerobic fitness. In some instances, it was also possible to examine associations between changes in reported physical activity or fitness and health status.

2.4 Athletic Status and Health

2.4.1 Comparisons Between Athletes and the General Population

Many Victorian physicians had fears that participation in athletic competition would create what they regarded as the dangerous condition of "athlete's heart" [2]. Epidemiological studies intended to disprove this view (Table 2.1) commonly compared the overall and/or cardiovascular death rates of former athletes with the experience of the general population [48–50], or the death rates that Actuaries established for people purchasing life insurance. Almost all of such comparisons have favoured the athletes. A recent meta-analysis based upon a total of 2807 competitors (707 of whom were women) showed an all-cause standardized mortality ratio of 0.67 for the athletes, with no evidence of publication bias; six studies reported data for cardiovascular disease, with a mortality ratio of 0.73 favouring the athletes [48].

During much of the era under consideration, athletes (as university graduates or high-earning professional competitors) had access to better health-care and a considerable socio-economic advantage over the general population. Furthermore, they were usually selected initially in terms of physique, often showing differences

Table 2.1 Comparison of overall and/or cardiovascular (CV) death rates between athletes and the general population

Author	Populations	Athletes	Comment
Anderson [24]	808 Yale University athletes 1855–1905 vs. Actuarial tables	Athletes favoured	Death rate 52 % of actuarial table, 46 % of American table
Baron et al. [25]	3439 National football league players playing from 1959 to 1988 vs. general population	Athletes favoured	All-cause mortality 0.53, CV mortality 0.68
Belli et al. [26]	24,000 professional Italian soccer players, 1960–1996 vs. general population	Athletes favoured	CV mortality 0.83 (all-cause mortality 1.00)
Cooper et al. [27]	100 Australian oarsmen vs. Actuarial tables	Athletes favoured	Mortality ratio 75.4 %
Dublin [28]	4976 Athletes from 10 U.S. Colleges, 1890–1905 vs. actuarial table	Athletes favoured	Mortality 93.2 %, 91.5 % relative to two Actuarial tables; BUT greater advantage in non-athletic honours graduates
Gaines and Hunter [29]	808 Yale athletes vs. actuarial tables	Athletes favoured	Death rate 49 % of expected by actuarial table, 44 % relative to American table
Gajewski and Poznanska [30]	2113 Polish Olympic competitors (414 F) during twentieth century vs. gender adjusted Polish urban population	Athletes favoured	All-cause mortality 0.51
Hartley and Llewellyn [31]	767 Oxford and Cambridge oarsmen 1829–1928 vs. Actuarial tables	Athletes favoured	Mortality ratio 76.7–93.5 % in four comparisons
Hill [32]	3424 British County or University cricketers, 1800–1888 vs. English Life Tables	Athletes favoured	Significant advantage in terms of number dying before given age in all comparisons
Kalist and Peng [33]	2641 S. Major league baseball players born 1945–1964	Athletes favoured	All-cause mortality 0.31
Karvonen et al. [34]	396 Champion skiers vs. general Finnish population	Athletes favoured	3–4 year advantage of longevity
Kujala et al. [35]	2009 Finnish international competitors 1920–1965 vs. general population	Athletes favoured	All-cause mortality 0.74, CV mortality 0.72
Marijon et al. [36]	786 Tour de France competitors, 1947–2012 vs. general French male population	Athletes favoured	All-cause mortality 0.59, CV mortality 0.67
Menotti et al. [37]	700 M, 283 F Italian track and field athletes vs. Italian life tables	Athletes favoured	All-cause mortality 0.73 (M), 0. 48 (F)
Metropolitan Life [38]	6753 Baseball players 1876–1973 vs. insured population	Athletes favoured	Mortality ratio 1876–1900, 72 %
Meylan [39]	152 university rowers 1852–1892 vs. standard mortality tables	Athletes favoured	2.9 year advantage of life expectancy

(continued)

Table 2.1 (continued)

Author	Populations	Athletes	Comment
Morgan [40]	251 Oxford and Cambridge rowers 1829–1859 vs. Farr's English life tables	Athletes favoured	Death rate 64 % of expected, 2.0 year advantage of life expectancy
Reed and Love [41]	4991 West Point military academy officers 1901–1916 vs. Actuarial tables	Officers favoured	0.25–1.25 year advantage of longevity
Schmid [42]	400 Czechoslovak athletes, 1861–1900 vs. general population of non-athletes	Athletes favoured	Longevity advantage of 8.66–1.44 years
Schnohr [43]	297 Danish athletes vs. general population	Athletes favoured	To age 50 year mortality 0.69 of expected; CV deaths 34.7 %, expected 36.7 %
Schnohr et al. [44]	4658 Joggers vs. other Copenhagen residents	Athletes favoured	Relative risk of death in persistent joggers 0.37
Taioli [45]	5389 Italian professional soccer players 1975–2003 vs. population statistics	Athletes favoured	All-cause mortality 0.68, CV disease 0.41
Wakefield [46]	State high school basketball champions vs. general male population of Indiana	Athletes favoured;	16.3 % of deaths due to cardiac or renal disease, BUT mortality 69 % of general population; 13.3 % of age-matched deaths due to cardiac or renal disease
Waterbor et al. [47]	985 baseball players 1911–1915 vs. general population	Athletes favoured	All-cause mortality 0.94 (0.88–1.00)

of body build relative to the average person, and at least in the case of endurance athletes they were generally non-smokers.

2.4.2 *University Athletes and Their Academic Peers*

Some studies of athletic status and health have drawn comparisons between those winning "letters," as representatives of their university, and other individuals who were attending the same institution in the same era, but did not distinguish themselves as athletes (Table 2.2). This approach largely overcame the issue of a differing socio-economic status in the comparison group, and perhaps in consequence the advantage of the athletes was less clearly demonstrated.

Rook [58] compared the mortality experience of members of Cambridge University athletic teams with a random sample of students and with "intellectuals" marked by academic distinction in their final examinations. He found no significant inter-group difference in average age at death, although the percentage of cardiovascular deaths was marginally lower in the athletes than in the other students. Rook commented that those involved in muscular sports ("heavy" sportsmen) had a

Table 2.2 Comparison of deaths from cardiovascular disease between athletes and other similar populations

Author	Population	Athletes	Non-athletes
Abel and Kruger [51]	Baseball players debuting between 1900 and 1939 vs. other baseball players	Top athletes favoured	4.8 year advantage of life expectancy
Dublin [28]	4976 Athletes from 10 U.S. Colleges, 1890–1905 vs. non-lettermen	Intellectuals favoured	2 year greater life expectancy than athletes or other graduates
Greenway and Hiscock [52]	686 students earning athletic letters vs. other graduates	Controls favoured	Mortality 93 % of expected in athletes, 83 % of expected in other graduates
Montoye et al. [53, 54]	628 Pre-1938 University of Michigan lettermen vs. other graduates	Non-athletes tended to be favoured	2 year difference in age at death (non-significant)
Paffenbarger et al. [55]	63 Univ. of Pennsylvania and Harvard lettermen 1921–1950 vs. other graduates	Athletes favoured	Coronary death ratio 0.6
Polednak and Damon [56]	Harvard major athletes (letter winners) vs. minor athletes, vs. non-participating students	Minor athletes and non-athletes tended to be favoured relative to major athletes	40.0 % of deaths due to cardiovascular disease, no significant intergroup difference in age at death
Prout [57]	172 Harvard and Yale crews 1882–1902 vs. other graduates	Athletes favoured	6.3 year advantage in age at death
Rook [58]	Cambridge University athletes vs. "intellectuals" and random Cambridge students	36.4 % of deaths due to cardiovascular disease, NO significant difference in average age at death	39.9 % of "intellectuals," 41.4 % of random students died of cardiovascular disease

slight disadvantage relative to endurance competitors ("light" sportsmen), the latter living some 1.7 years longer. He further noted that the literature provided little support for the idea that University athletes gained a large amount of weight when their competitive career was over.

Other studies from Scandinavia found a several year advantage of longevity in endurance athletes relative to the general population [34]. However, it remained unclear whether this was because they continued to undertake a large volume of habitual physical activity, or whether their advantage was attributable to other characteristics of the endurance competitor such as an ectomorphic body build and abstinence from cigarettes. The potential influence of initial selection was seen in a comparison between Baseball Hall of Famers and other baseball players [51],

Fig. 2.7 Henry J. Montoye of the University of Michigan, Ann Arbor, MI, conducted a major longitudinal study of cardiovascular disease in athletes, and made major contributions to questionnaire analyses of habitual physical activity

with the top baseball players having a substantial health advantage over other less successful league players.

An extensive study conducted by Henry J. Montoye (Fig. 2.7) and his colleagues at the University of Michigan showed that athletic "*Letter Holders*" tended to have a shorter lifespan than their peers who had attended the same university, but had not participated in inter-collegiate athletics [53, 54]. This could suggest a negative effect of sport participation upon cardiovascular health. However, further investigation of this data suggested two alternative explanations. Many of the *Letter-Holders* at the University of Michigan had been football players, a sport where a mesomorphic body build not only gave them a physical edge over lighter opponents, but also predisposed them to cardiovascular disease in middle age. Moreover, by the time that they had reached middle age, many of the former *Letter-Holders* were engaging in no more physical activity than their peers and had gained some 10 kg of body mass since leaving the university.

2.4.3 Conclusions from Studies Based on Athletes

Unfortunately, a variety of factors conspire to limit the conclusions about physical activity and health that can be drawn from comparisons between former athletes and their supposedly sedentary peers. The initial selection of the athlete is based upon a specific body build and outstanding sport-specific fitness. Most studies have obtained little information on the exercise habits of the athletes after they ceased competition, and by middle age some former competitors have allowed themselves to become more sedentary and obese than their non-athletic peers. Finally, top athletes differ in temperament from the general population, and thus have an above average death rate from such causes as motor vehicle collisions, homicides and suicides.

Fig. 2.8 In 1969–1970 CE, the late Hugues Lavallée initiated a quasi-experimental study that examined the impact of five additional hours of physical education per week upon the fitness and health of 546 primary school students in Trois Rivières, Québec

2.5 Exercise Group Assignment in Quasi-experimental Studies of Fitness and Health

No one has disputed the assertion of Henry Taylor that it is impractical to carry out a true randomized controlled study looking at the long-term effects of physical activity upon various facets of fitness and health. However, several quasi-experimental studies have attempted to explore this question.

2.5.1 The Trois Rivières Regional Study

In terms of experimental design, perhaps the most convincing of the quasi-experimental studies was initiated by the late Hugues Lavallée (Fig. 2.8). Between 1970 and 1977, 546 children attending two state primary schools (one urban, and one rural) in the Trois Rivières region of Quebec were assigned to either an experimental programme (an additional 1 hour per day of vigorous physical education, taught by an enthusiastic physical education professional), or the standard control programme (where students received only minimal physical education, taught by their home-room teacher) [59]. Student assignment was based simply on the year of school enrolment; alternate years followed the experimental or the control programme at the same school, with the same teaching staff; the only inter-group difference in treatment was that 14 % of the academic teaching time was replaced by physical activity instruction for students who were assigned to the experimental group. Students continued in their assigned group throughout their 6 years in primary school, the only exceptions being a few pupils who skipped one grade, or who failed a year. By the age of 10–12 years, children in the experimental group showed the expected gains of aerobic fitness, muscular strength and physical performance relative to control students, but there was little inter-group difference of health experience while students were attending the primary school.

An attempt was made to re-evaluate participants 25–35 years later. At this stage, there were still a few inter-group differences of lifestyle (the experimental group still tended to greater habitual physical activity and had a smaller proportion of smokers); however, bone density was similar for the two groups [60]. Possibly, differences in the incidence of cardiovascular disease or cancer might have been seen if the study could have continued further, until the subjects were aged 50 or 60 years. However, even in a city such as Trois Rivières, where the population has a relatively limited mobility, it would be difficult to maintain contact with an adequate proportion of the original subjects in order to establish significant lifetime differences in health experience. In most North American cities, mobility is much greater, and few of the original sample would be available for testing after an interval of 40–50 years. Moreover, although the experimental programme was effective in enhancing the immediate fitness of the students in Trois Rivières, it would be necessary to extend modification of the academic programme throughout the critical adolescent years in order to ensure that there was still a substantial inter-group difference of lifestyle when the subjects were adults.

2.5.2 *Quasi-experimental Assignment to a Work-Site Fitness Programme*

A second potential option for a quasi-experimental study is to introduce an effective work-site fitness programme to one major company, and to select as a control a closely comparable company at a nearby location, where similar evaluations can be made, with the promise of an equivalent fitness programme at a later date. The control group are unlikely to accept a delay of more than a year in development of their fitness programme, and anticipation of this event could compromise their control status. Further, most companies are unlikely to agree to a programme that focuses uniquely upon exercise; typically, advice is given on other health issues such as obesity, stress, and substance abuse. Further, programme attendance is voluntary, and despite attractive incentives, less than a third of employees may be following the programme at the end of the first year of its operation. In one quasi-experimental work-site fitness study in Toronto, we were able to demonstrate that there was a small reduction in health care costs over the first year of programme operation [61], but a fitness programme was provided to the control company at the end of the year, so it would not have been possible to evaluate long-term effects upon health.

2.5.3 *Conclusions*

Although quasi-experimental studies allow an examination of short and medium term influences of increased physical activity upon fitness and health, population mobility and poor long-term compliance with the assigned regimen have to date precluded use of this approach in examining the impact of increased physical activity upon the incidence of chronic ill-health.

2.6 Questionnaire and Diary Assessments of Leisure Activity

The physical activity questionnaire has long been the main tool of the chronic disease epidemiologist. Its main attraction is that it can easily be applied to large population samples. However, the limitations of such methodology have also long been appreciated [62, 63].

2.6.1 *Types of Instrument*

The questionnaires available to epidemiologists have ranged from very simple forms, requiring only two or three responses from users [64–66], to complex instruments 20 or more pages in length encompassing every conceivable type of physical activity [67–69]. The latter have sometimes been completed using cue sheets to remind subjects of less common sources of energy expenditure, and often a trained assistant has been needed to make nuanced interpretations of responses. The latter requirement immediately negates the primary virtues of the questionnaire approach: simplicity and low cost, and the act of interpretation could bias the data. One alternative or complement to the questionnaire is an activity diary. Morris [70] had his sample of civil servants complete sheets noting their activities every 5 minutes during a Friday and a Saturday. The need to complete such a record can in itself modify activity patterns, particularly by restricting the number of changes in activity over the course of a day [70].

Most investigators have recognized the limitations of human memory, particularly in older individuals. Thus questionnaires have generally asked about activity performed during the previous week or the previous month. This is problematic in countries with a continental climate, since outdoor activities are decreased by rainfall and extremes of warm or cold weather [71–73]. Ideally, responses should be obtained over the various seasons of the year, but few observers have taken this precaution.

In terms of some chronic conditions such as cancers, the relevant time-frame of preventive physical activity is much longer than a month or even a year. A few questionnaires have covered the entire life span [68, 74–76]. Friedenrich et al. [75] used trained interviewers and an anchored calendar as a memory aid; with this approach, they found that the 6–8 week reliability when estimating lifetime hours per week of physical activity was relatively good (intra-class correlation coefficients for occupational activity, 0.87; for domestic activity, 0.77; for exercise or sports participation, 0.72; and for all forms of activity, 0.74). A 1-year re-evaluation of female U.S. College graduates [77] achieved similar intra-class correlation coefficients (vigorous activities, 0.86; moderate activities, 0.80; recreational activities, 0.87; domestic activities, 0.78; total activities, 0.82). Both of these studies were conducted on well-educated subjects, and it is less clear that a comparable reproducibility could be achieved with the poorly educated, or with recent immigrants who have a poor understanding of the English language and differing cultural norms of physical activity. Indeed, even with Harvard Alumni, the reliability coefficient for the questionnaire used by Paffenbarger and his associates was 0.72 over a 1-month interval, dropping to 0.3–0.4 when the recall period was 8–12 months [78].

2.6.2 *Types of Information Collected*

There is little argument that leisure questionnaires can draw a relatively accurate binary distinction between active and inactive individuals [70, 79–83]. However, many investigators have attempted to collect a much greater range of information from their questionnaires, including the intensity and the total volume of physical activity performed in a typical week.

2.6.2.1 Intensity of Physical Activity

Many questionnaires have ignored brief or low intensity bursts of physical activity, although in the context of modern sedentary society, the cumulative impact of short periods of activity could conceivably be important to health. Thus, Blair et al. [84] took no account of activities if they were less intense than brisk walking, or lasted for less than 10 minutes. Likewise, a questionnaire developed by Gaston Godin (Fig. 2.9) [64] asked "how many times on the average do you do the following types of exercise for more than 15 minutes..."

Physical activity compendia [85, 86] have been used to convert reported activities such as walking, jogging or cycling to approximate energy expenditures, although unfortunately secular changes in technology such as the introduction of ultra-lightweight bicycles have progressively reduced the actual energy costs of activities relative to the values indicated in compendia [87]. Accurate information on pace is also critical to any estimate of the intensity of energy expenditures, and many of the activities listed in the compendia were assessed on fit young adults rather than on sedentary older people who may be heavier, but tend to move much

Fig. 2.9 Gaston Godin is a exercise psychologist and physiologist at Laval University, Quebéc. He developed a simple physical activity questionnaire with scores having a good correlation with the individual's level of attained aerobic fitness

more slowly. In terms of health, the relative intensity of effort is also more important than absolute energy expenditures [88]. Subjective descriptions of the intensity of effort (light, moderate or hard) depend on age, physical fitness, culture and the duration of activity [89], the stoicism of the subject [90], and sometimes on the social desirability of the response [91]. It may thus be best to tie descriptors of intensity to physiological anchors such as breathlessness [92] or the onset of sweating [64].

2.6.2.2 Duration of Physical Activity

It remains unclear whether a single exercise session must have a minimum duration in order to induce health benefits. Objective monitoring has allowed the analysis of the effects of the total volume of physical activity and the effects of its segmentation. Very few people of the general population engage in the recommended 30 minutes of daily aerobic exercise [93, 94], but it has been suggested that they might obtain similar benefit by taking three 10 minute sessions of physical activity per day [95]; possibly, every step counts! Sixteen studies of fractionated activity [94] have as yet failed to resolve this issue, although it remains important to the design and interpretation of physical activity questionnaires.

2.6.2.3 Aspects of Exercise Other than Aerobic Activity

With a few exceptions [96, 97], physical activity questionnaires have focussed uniquely upon participation in aerobic exercise. However, current public health recommendations also call for resistance and flexibility exercises, plus activities to enhance bone structure and improve balance [93, 94]. Most questionnaires have also neglected the environmental context in which activity is performed, although it is now recognized that high temperatures can increase the energy cost of activities [89, 98, 99], and that extremes of temperature reduce programme compliance. Further, favourable psychological reactions such as a reduction of stress are more likely if activity is taken in a pleasant natural environment [100].

2.6.3 Reliability and Validity of Information Obtained from Questionnaires

2.6.3.1 Reliability

The reliability of questionnaires has commonly been expressed as intra-class test-retest correlation coefficients. Results have been most reliable in groups marked by intelligence and/or a high socio-economic status (for example, senior civil servants [101]; Harvard alumnae [102]; members of an exclusive fitness club [103]; and nursing graduates [104]). The consistency of reported activity was also greatest if a single stereotyped activity such as walking was the predominant form of physical activity. Test-retest correlations were higher for reports of strenuous physical activity (>0.7) than for reports of moderate or low intensity activity [105]. Data from the Canada Fitness Survey of 1982 also suggested that intensity was reported less reliably than the frequency or duration of physical activity [106].

2.6.3.2 Validity

Although many questionnaires can sort a population into active and inactive categories, doubt has been cast upon the validity of more detailed and quantitative interpretations. The lack of validity is highlighted by discrepancies between the substantial proportions of Canadian National and Provincial populations reporting activity patterns that meet current physical activity guidelines, and the very low proportion of active individuals that are seen by casual observers in most North American cities [69, 71, 107–111]. Serious systematic errors seem inherent in many questionnaire studies [112, 113]. Thus, three surveys conducted by the U.S. National Center for Health Statistics provided widely divergent estimates of the prevalence of limited physical activity among women of child-bearing age (from 3.9 to 39 % [114]). Likewise, estimates of the prevalence of moderate physical activity have varied widely from one survey to another, depending on such issues as the use of cue cards and the exact form of the questions asked [115, 116]. The problem reflects in part the ever-diminishing average level of physical activity in most populations. Questionnaires that were appropriate to discern the occupational and leisure activities of earlier generations lack the sensitivity to quantitate the very limited energy expenditures required by modern society [62].

Response options in even the most elaborate questionnaires have frequently omitted important component activities of a typical day, such as care for an ailing parent or grandchildren. Memory has also been a problem (particularly for older people). On the other hand, subjects have commonly exaggerated their activity patterns, partly to satisfy observer expectations (particularly if a physical activity programme has been prescribed for them [71, 117–119]). Because of over-estimation of times spent in various activities, some respondents have indicated total day lengths much greater than 24 hours, and it has then been necessary to introduce an arbitrary downward scaling in the duration of reported activities [71, 120].

Direct comparisons between questionnaire estimates and external criteria have shown discouragingly large errors, even when assessing something as simple as the daily walking distance [78, 91, 121]. Thus, the College Alumnus Questionnaire data suggested a daily walking distance of 1.4 km vs. the 4.2 km that was scored on a pedometer [121]. Even more alarming, for each km of walking distance that was reported, a 48-hour activity diary suggested that subjects had actually walked 5.3 km [78]. A recent study of elderly Japanese (where walking was the predominant form of physical activity) found moderate and statistically significant correlations of questionnaire responses with pedometer/accelerometer measures of daily step count, minutes of activity <3 METs and minutes >3 METS ($r = 0.41$, 0.28 and 0.53, respectively), but in this community the reported total energy expenditures relative to the objective pedometer/accelerometer measurements were over-estimated by a factor of three [122]. One reason for over-estimating the duration of activities is that subjects often include such items as travel to a gymnasium, periods of passive instruction, changing, showering, and socializing in their estimates of the time that they have devoted to physical activity [123].

Formal assessments of the validity of questionnaires have been made relative to doubly-labelled water (DLW) estimates of overall energy expenditures, pedometer/accelerometer data, and physiological indices such as aerobic fitness and obesity. All of these objective measures have their limitations. DLW measurements are costly, and thus can be applied only to small samples; moreover, they only provide an average of energy expenditures over a 2-week period, which is necessarily heavily weighted by low intensity activities. Correlation coefficients with DLW were 0.68 for the Baecke index, 0.64 for the Tecumseh questionnaire, and 0.68 (women) and 0.79 (men) for the physical activity scale for the elderly [124, 125]. Coefficients of correlation between pedometer/accelerometer scores or physiological status and reports of strenuous activity are typically in the range 0.4–0.5 [105, 126]. Among 10 questionnaires [105], the closest correlations with the directly measured maximal oxygen intake (0.56) were for the Minnesota [82] and Godin/Shephard [64] questionnaires.

Relatively simple questionnaires [64, 127–130] seem at least as valid as complex forms, although there is often a poor correlation between the data obtained from simple and complex questionnaires (0.14–0.41) [131, 132]. The questionnaire of Godin and Shephard [64] required three-option responses to two simple questions; in well-educated individuals, the most satisfactory univariate item was the report of performing strenuous physical activity over the previous 7 days (2-week reliability coefficient of 0.94, a correlation of 0.98 with an age and sex-adjusted step test prediction of maximal oxygen intake, and an 0.74 correlation with an age and sex-adjusted skinfold prediction of body fat content).

2.6.3.3 Conclusions on the Reliability and Validity of Questionnaires

The general direction of relationships between physical activity and health was probably indicated correctly in most questionnaire-based surveys. However, the demonstration of substantial systematic errors in most questionnaire responses has inevitably cast a pall of doubt upon the quantitative epidemiological research carried out over the past 50 years. Moreover, the lack of precision in questionnaire data may well have attenuated the apparent strength of relationships with health, making correlations statistically insignificant [133, 134].

2.6.4 Population Studies of Leisure Activity Using Questionnaires and Activity Diaries

As examples of analyses of leisure activity using questionnaires and diaries, we may instance in Britain a study of executive-class civil servants [70] and a regional analysis drawn from 25 representative medical practices across Britain [135]. From the U.S., we will consider the experience of Seventh-Day Adventists Harvard alumni [81], and U.S. railroad workers. Many of these studies not only identified active individuals, but also attempted to assess the intensity, duration and volume of habitual physical activity. In particular, attempts were made to predict the minimum intensity [70, 101, 136], the relative intensity [137] and the total weekly volume of effort needed [138] to prevent heart attacks in various age groups.

2.6.4.1 Executive-Class Civil Servants

Between 1968 and 1970, Morris and his associates [70] undertook a study of 16,682 male executive-class civil servants aged 40–64 years. Participants recorded on a Monday their recollection of leisure activities undertaken at 5-minute intervals on the preceding Friday and Saturday. Without knowledge of clinical histories, three observers then noted for each participant periods >5 minutes per day that had been ascribed to active recreation, keeping fit, or vigorous getting about (activities demanding near maximal effort), bouts of heavy work reported as lasting >15 minutes (gardening, building or moving heavy objects) and stair climbing. All of these activities were thought to demand at least peak energy outputs >31 kJ/minute, and thus could be considered as heavy work. Reported vigorous physical activity in each of the first 214 of the sample who sustained a clinical attack of ischaemic heart disease was compared with that of two matched peers who had not sustained a heart attack (Table 2.3). Those sustaining the heart attack were less likely to have been active than their peers. Protection was also seen for fatal heart attacks in all of the categories of physical activity that were analyzed. However, no such differences were seen with respect to reports of moderate activity, or for total

Table 2.3 Frequency of vigorous physical activity reported by 214 executive-class civil servants sustaining a first clinical attack of myocardial infarction relative to the frequency of attacks observed in matched controls [70]

Type of physical activity	Matched controls (n = 428)	Men sustaining heart attack (n = 214)	Expected number in those sustaining attack
Active recreation	15	5	7.5
Keeping fit	15	3	7.5
Vigorous getting about	18	1	9
Heavy work	73	17	36.5
Climbing >500 stairs/day	8	0	4
Total reporting vigorous activity	111	23	55.5

activity (which included much light activity). When participants were grouped by age, the benefit of habitual physical activity seemed similar for those aged 40–49 years and for those aged 60–64 years. Further, the data provided no evidence that early symptoms had previously impeded physical activity in those sustaining a first heart attack. Those engaging in much vigorous exercise seemed to gain more benefit (risk ratio 0.18) than those doing some vigorous exercise (risk ratio 0.42–0.55). The identification of those engaging in any vigorous physical activity also proved relatively consistent from March or September 1969 (38 individuals) to November 1971 (31 individuals). The one issue that could not be excluded was that participation in vigorous physical activity was serving as a marker of some other unidentified personal characteristic.

This study thus confirmed the association of regular vigorous physical activity with a low risk of cardiac disease, as seen in the occupational studies, and it made the first tentative steps towards identifying the required intensity and volume of activity. By 1980, participants in the study had sustained 1138 first clinical episodes of coronary heart disease, but the general conclusion that vigorous habitual activity protected against heart attacks, particularly fatal incidents, remained unchanged [139].

A second and somewhat similar study was launched in 1976 [101]; this followed 9376 healthy male executive civil servants initially aged 45–64 years for an average of 9 years. During this period, there were 474 heart attacks. Some 9 % of employees reported that over the preceding 4 weeks that they had often participated in vigorous sports, undertaken considerable amounts of cycling, and/or rated their walking pace as >6.4 km/hour, they experienced less than a half as many non-fatal and fatal heart attacks as their peers. However, no protection against cardiac disease was observed unless the sport or exercise was reported as vigorous (Table 2.4), and no protection was obtained from sports participation at a younger age.

Table 2.4 Relationship of participation of male executive class civil servants in sport to age-adjusted risk of coronary heart disease, based on data of Morris and associates [101]

	Vigorous sports (>31 kJ/minute)		Non-vigorous sports (<31 kJ/minute)	
Episodes of sports play in past 4 weeks	Cases/1000 man-year	Relative risk	Cases/1000 man-year	Relative risk
None	5.8	1.00	5.4	1.00
1–3	4.5	0.78	5.9	1.09
4–7	4.1	0.71	5.9	1.09
8–11	2.1	0.36	3.5	0.65
>12			6.8	1.26

2.6.4.2 Coronary Disease and Physical Activity in Regional Medical Practice

Shaper and Wannamethee [135] followed 7735 men initially aged 40–59 years for an average of 8 years. Their sample was drawn from 24 medical practices that were considered as representative of socio-economic conditions across Britain. Nurses administered to all participants a questionnaire that included questions on physical activity (walking, recreational activity, cycling and sport activity).

A six-level gradation of physical activity was calculated arbitrarily from the reported frequency and intensity of each of these forms of activity. The relative risk of coronary heart disease over the 8 years was adjusted for age, body mass index, social class and smoking status. Setting the risk of inactive subjects at 1.00, values for patients placed in the other five categories were: occasional physical activity, 0.8; light activity, 0.8; moderate activity, 0.4; moderately vigorous activity, 0.4, and vigorous activity 0.8. Although there was considerable statistical overlap between groups, in contrast to the views of Morris, there was a suggestion that vigorous activity might give a poorer prognosis than moderate or moderately vigorous activity. However, when a similar analysis was made for the relative risk of stroke among men initially aged 45–59 years, the best prognosis was seen in those men who reported undertaking vigorous physical activity [140].

2.6.4.3 Seventh-Day Adventist Men

Linsted and his colleagues [141] followed a group of 9484 male Seventh-Day Adventist men for an average of 26 years. There were 4000 deaths over this period. A self-administered lifestyle questionnaire completed at entry to the study allowed the categorization of participants as inactive, moderately active, or highly active. The all-cause mortality rate for the three groups was calculated for each of five decades (Table 2.5). In late middle-age, the moderate and highly active groups had a substantial advantage, but as age advanced, there was a crossover, with the highly active men having a poorer prognosis after the age of 78 years, and the moderately

Table 2.5 All-cause mortality (deaths/1000 man-years) of Seventh-Day Adventist males, classified by age group and reported activity category (based on the data of Linsted et al. [141])

Age category (years)	Inactive	Moderately active	Very active
50–59	4.0	2.4	2.5
60–69	11.2	8.4	9.1
70–79	36.6	27.4	33.5
80–89	85.1	81.9	94.1
90–99	169.6	152.5	156.5

Table 2.6 Relationships between questionnaire estimates of weekly leisure energy expenditures in U.S. male railroad workers, coronary heart disease mortality and all-cause mortality, as seen over 17–20 year follow-up (based on the data of Slattery et al. [142])

Reported leisure energy expenditure (MJ/week)	Relative risk of CHD mortality	Relative risk of all-cause death
15.1	1.00	1.00
5.7	1.05	1.04
2.3	1.11	1.08
0.2	1.28	1.21

active faring worse than the inactive after 95 years of age. As in the study of Shaper and Wannamethee, there was little suggestion that the very active had a better prognosis than those who were moderately active in any of the age decades that were analyzed.

2.6.4.4 U.S. Railroad Workers

The U.S. railroad study [142] followed the population recruited by Henry Taylor, a sample of 3032 white male employees. Subjects completed the very detailed Minnesota Leisure-time activity questionnaire in order to determine each individual's participation in more than 50 activities; the reported frequency, duration and hours per week of involvement in each of these activities were noted, and an estimate of the person's weekly leisure activity energy expenditures was calculated from responses (Table 2.6). Over a 17–20 year follow-up, the data showed a small trend towards an increased risk of death both from coronary heart disease and from all-causes among those who were taking less leisure-time exercise than the most active group, after statistical adjustment of the data for age, cigarette smoking, blood pressure and serum cholesterol level (but surprisingly, not for differences of occupational activity); the adverse effect of physical inactivity was statistically significant for all-cause but not for coronary deaths.

2.6.4.5 Harvard Alumni

Paffenbarger and his associates [138] questioned 11,864 initially healthy male Harvard alumni on their exercise habits, noting the number of city blocks walked, the number of stairs climbed, and the intensity and duration of any sport involvement during a typical week. Arbitrary rates of energy expenditure were ascribed to each of these activities to provide an estimate of the individual's gross leisure energy expenditure. Over an 11- to 15-year follow up, there were 1413 deaths, 45 % of these being due to cardiovascular disease and 32 % to cancer. Setting the risk of cardiovascular disease in the least active group as 1.00, an age-adjusted benefit was seen from walking (<5 vs. >15 km/week, 0.67), stair-climbing (<20 vs. >55 flights/week, 0.75), sports (none vs. moderate sports play >4.5 METs, 0.63), and a large total energy expenditure (<2 to 8 vs. 8 to >14 MJ/week, 0.70). The data were interpreted as showing a need to spend more than 8 MJ/week to enhance cardiovascular prognosis, although in fact risk ratios did not differ greatly between participants with expenditures of 2–4 MJ/week (0.63) and 12–14 MJ/week (0.68).

A further 8-year follow-up of the same population was undertaken in 1985, examining health outcomes in relation to changes in moderate sport involvement and weekly energy expenditures >8 MJ. The baseline risk was set as failing to meet the 8 MJ/week standard in either 1962–1966 or in 1977. The risk tended to an insignificant increase if activity had diminished over the 8 years, but in those whose activity had risen, the cardiac risk had dropped to approximately the same level as seen in those who had remained active throughout the 8 years (for moderate sports, 0.71 vs. 0.69, and for weekly energy expenditures >8 MJ, 0.74 vs. 0.79).

2.6.4.6 Relative Risk Assessment

I-M Lee and associates [137] noted that whereas earlier analyses of Harvard alumni data had called for a weekly leisure energy expenditure of >8 MJ, the usual public health recommendation of 30 minutes of moderate physical activity (>3 METS) on most days generated a weekly energy expenditure of no more than 4 MJ. They highlighted the further problem that specifying an absolute energy expenditure took little account of the relative stress that such a requirement imposed upon a fit young adult relative to an unfit elderly individual. They thus investigated the usefulness of assessing the relative intensity of effort, even in people who were currently failing to meet minimum public health recommendations for weekly physical activity.

A sample of 7337 men with an initial age averaging 66 years used the Borg scale to report their perceived intensity of effort when exercising. Over a 7-year follow-up, the risks of coronary heart disease relative to men who perceived their efforts as weak were 0.86, 0.69 and 0.72 for those rating their effort as moderate, somewhat strong and strong respectively. Moreover, this statistically significant trend was shown even among individuals with total energy expenditures of less than 4 MJ/week, and among those never engaging in activities >3 METS.

2.6.4.7 Conclusions

The questionnaire study of leisure activities provides strong support for the concept that regular exercise protects against both coronary vascular disease and all-cause deaths. However, disagreement remains concerning the optimal intensity and volume of effort. While some authors argue that activity must be vigorous to improve cardiac health, other studies suggest that particularly in older people moderate activity may be preferable to vigorous effort. Likewise, although some studies have argued the need for a weekly leisure energy expenditure as large as 8 MJ, others have found benefit from much smaller amounts of vigorous activity, particularly in the elderly.

2.7 Aerobic Fitness Assessments

Given the substantial uncertainties and errors inherent in questionnaire assessments of habitual physical activity, Steve Blair (Fig. 2.10) and his associates at the Aerobics Fitness Center in Dallas [143] suggested that it might be better to classify the activity behaviour of subjects in terms of an objectively measurable outcome, such as a high level of maximal oxygen intake. Certainly, physiological data have greater precision than questionnaire reports, although a part of a high maximal oxygen intake may reflect body build and genetic endowment rather than participation in vigorous physical activity. Moreover, the type of fitness assessment made in many laboratories (including the Cooper Aerobics Center) has been the endurance time during a progressive treadmill test, and this value is strongly influenced not only by habitual physical activity, but also by obesity.

Fig. 2.10 Steve Blair for many years directed epidemiological research for the Cooper Aerobics Center in Dallas Texas. He conducted many noteworthy studies of relationships between the fitness of clients and their subsequent health

2.7.1 Review Evidence

A review of 67 articles published between 1990 and 2000 showed consistent evidence of greater longevity and a reduced risk of coronary heart disease, cardiovascular disease, stroke and colon cancer in the more active members of a subject-sample, whether judged from questionnaire reports or measurements of aerobic fitness [143]. Commonly, much of the benefit appeared to occur at low levels of physical activity. It was not entirely clear from this review whether questionnaire assessments of physical activity or fitness assessments provided the better indication of prognosis. Nevertheless, gradients of health benefit were generally steeper for fitness than for questionnaire data, and if habitual activity was included in a multivariate analysis of the treadmill data, there appeared to be a residual effect of aerobic fitness.

2.7.2 Cooper Aerobics Center Data

Blair and his colleagues made treadmill assessments of aerobic fitness levels in 10,244 men and 3120 women, subsequently following them for an average of 8 years. The age-adjusted relative risk of death over this period was expressed relative to the experience of the group in the lowest quintile of fitness. For men, the values for the four higher quintiles were 0.40, 0.42, 0.34 and 0.29, respectively, and for the women, the corresponding figures were 0.52, 0.31, 0.15 and 0.22 [103].

A further analysis of 9777 men from this sample looked at the impact of a change in fitness level over an average interval of 4.9 years in terms of health experience in the following 5.1 years [144]. The highest age-adjusted death rate was seen in those who were unfit at both examinations (12.2/1000 man-years), and the lowest rate was found in men who were fit at both examinations (4.0/1000 man-years). The men whose fitness improved between the two evaluations showed a 44 % decrease in their death rate, to 6.8/1000 man-years; the risk decreased by some 7.9 % for each minute increase in their treadmill endurance time.

2.7.3 Canada Fitness Survey Data

In addition to asking subjects to complete a detailed physical activity questionnaire which yielded rather questionable results, participants in the original Canada Fitness Survey undertook a simple step test in their own homes, under the supervision of a health professional. A 7-year follow-up of 2174 male and female participants distinguished individuals who reached recommended, minimum acceptable and unacceptable age and sex-linked standards of performance on the step test [145]. In terms of all-cause deaths, the crude risk-ratios for the two groups with less than recommended fitness levels were 1.3 and 3.0, and after adjusting data for age, sex, body mass index and smoking status, the corresponding ratios were 1.6 and 2.7.

Subjects were also classed into four categories in terms of their habitual physical activity (very active, active, moderately active and inactive), as reported in the questionnaires, but both crude and adjusted risk ratios for the three lower categories of physical activity were too variable to discern any trends.

2.7.4 Conclusions

Classification of subjects by fitness level rather than reported physical activity gives a steeper gradient of subsequent health experience, suggesting that fitness data have greater accuracy than questionnaire responses. This is particularly evident in the findings of the Canadian Fitness Survey, where a simple step test showed a clear gradient of subsequent mortality, but this gradient was not seen in analysis of responses to a lengthy questionnaire.

2.8 Natural Experiments

In a few instances, there is good evidence that secular change has altered the energy expenditures of a community (generally in a downward direction), thus allowing comparisons of health experience when activity was high with the current experience, when activity is much less frequent and less intensive. Striking instances of a secular reduction of habitual physical activity have occurred over the past several decades among indigenous populations such as the Inuit [146] and North American Indians [147], in association with epidemic increases in the prevalence of obesity and type 2 diabetes mellitus. However, the etiology of these epidemics is complex, as there has also been a shift from country foods such as fish and caribou to store products with a high carbohydrate content, and the diagnosis of illness has also been more complete with greater access to medical care.

2.9 The Objective Monitoring of Human Activity in Epidemiology

During the past decade, the development of inexpensive and relatively accurate objective monitors has challenged the accuracy of what seemed sophisticated questionnaire-based estimates of habitual physical activity. Questionnaires appear to have been under- or over-estimating energy expenditures by factors of two or even three [121, 122, 148, 149]. After brief comments on studies based upon pulse monitors and the Kofranyi-Michaelis respirometer, we will look at the potential information yield from the objective monitoring of body movement patterns, using odometers, pedometers and accelerometers.

2.9.1 Pulse Monitoring

Potential methods for the field monitoring of heart rate have included ECG telemetry or tape-recording, ear-lobe photo cells, and electrochemical integration of pulse signals.

The latest type of telemetric pulse-rate recorders [150, 151] can transmit an ECG signal from a chest band to a wrist-watch type recorder or a nearby laboratory; although useful in athletic training, the need to maintain effective chest electrode contact has limited use of this approach in epidemiology.

Ear-lobe photo cells [152] have also been used mainly in short-term industrial studies rather than in epidemiology; slippage of the earpiece was a strong argument against epidemiological use.

In contrast, Wolff's electrochemical integrator [153] was developed specifically for epidemiological studies, including the International Biological Programme and Morris's studies of bus drivers and conductors. The ECG was used to drive a chemical reaction, and the average heart rate for the day could be estimated in the laboratory by passing a current in the reverse direction. Evolution of this instrument introduced multiple electrochemical cells, so that heart rates could be counted in selected target ranges [154], but problems remained from poor electrode contacts and erroneous counts [155]; in the study of more than 5000 London Transport workers (above), successful heart rate recordings were reported on only 10 subjects. Other wrist-band devices that accumulated ECG signals [156, 157] suffered from similar problems.

Even if these technical difficulties could be overcome, the theoretical basis for use of heart rate monitoring to monitor the intensity of physical activity is dubious. Data interpretation rests upon the existence of a relatively linear relationship between heart rate and oxygen consumption from 50 to 100 % of an individual's maximal aerobic effort [158]. There are many limitations to exploitation of this relationship: the slope of the line varies with the age, sex, physical fitness and posture of the subject, and it differs radically between arm and leg work. Moreover, the slope is increased by static effort, anxiety, and exposure to a hot environment or high altitudes [159]. Finally, in the context of population studies of physical activity, most of modern daily activity demands less than 50 % of maximal aerobic effort, an intensity of exercise where the relationship is non-linear.

2.9.2 Kofranyi-Michaelis Respirometer

The Kofranyi-Michaelis respirometer measures a subject's respiratory minute volume by a mechanical gas meter, and small aliquots of expired gas are collected in a balloon for subsequent chemical analysis [160]. The device has proven useful in determining the average energy cost of specified types of physical activity, in publishing compendia that show the energy demands of various common activities [161], and in demonstrating that Inuit engaged in various types of traditional hunting developed high energy expenditures on the days when they were hunting [146]. However, only a limited number of observations are possible with such a respirometer, and the information must be combined with direct observation of the subjects, diary records or questionnaire responses to form a picture of total energy expenditures over a typical week.

2.9.3 *Odometers*

The odometer may have been invented by Archimedes of Syracuse (287–212 BCE), although its use to measure walking distances was first described by Vitruvius in Book X of *de Architectura*, around 27 BCE. The device allowed the accurate placement of milestones on Roman roads, ensuring that armies covered an appropriate distance on their daily marches.

More recently, car odometers have been used to measure walking distances on city streets, thus allowing cardiologists to prescribe physical activity for their patients [162]. For example, a patient might be advised to walk a measured distance of two level miles in 40 minutes at least five times a week. This would equate to a pace of 4.8 km/hour, equivalent to an oxygen consumption of about 1.25 L/minute, or to a net weekly increase in energy expenditure of about 4 MJ. If a person's main source of leisure activity was deliberate walking or active commuting, a similar measurement of times and distances could be exploited to determine his or her active energy expenditures.

2.9.4 *Pedometers*

2.9.4.1 Early History

Invention of the pedometer has been ascribed to Leonardo da Vinci (1452–1519). One beautiful early instrument in Southern Germany has been dated to 1590 CE (Fig. 2.11). da Vinci's underlying purpose in designing the pedometer was military—he suggested that it should be used to make accurate maps of contested territory [163].

Reintroduction of the pedometer has been traced to the Swiss watchmaker Abraham-Louis Perrelet (1729–1826). He invented a watch that rewound itself from the impulses generated by 15 minutes of brisk walking, and a logical extension of this device was an instrument that measured the number of steps taken, allowing the distance walked to be estimated from the average stride length. Thomas

Fig. 2.11 An early pedometer from Southern Germany (1590 CE). *Source*: http://en.wikipedia.org/wiki/Pedometer#History

Fig. 2.12 Han Kemper evaluated the traditional type of pedometer, based on a pocket watch mechanism. He found that sometimes it counted two steps rather than one during jogging, and that it failed to record at all during slow walking

Jefferson (US President, 1801–1809) [164] brought the pedometer to North America, using it to monitor the length of his 2-hour daily walks.

Some mechanical instruments that were produced in the 1950s followed the concept of Perrelet, using the step impulse to activate a watch escapement, advancing the dial reading by a single unit. Designers have nevertheless experienced difficulty in choosing an appropriate loading of the escapement, so that only one "escape" is allowed per step. Han Kemper (Fig. 2.12) and Verschuur demonstrated that the greater force associated with jogging sometimes advanced the dial by two units rather than one, and steps might not be counted at all during very slow walking [165]. An accurate response to slow gaits and minor movements is plainly important in our sedentary and aging society.

Miscounts tended to be exacerbated if a belt-worn pedometer was tilted from the vertical, as when assessing an obese person [166]. There have thus been claims that a recently developed "Step-watch Activity Monitor" detects movements more accurately if it is worn on the ankle rather than a waist belt [167], and some investigators regard this as the current criterion instrument for assessing the activity of free-living individuals.

2.9.4.2 Pedometer/Accelerometers

There is a growing consensus that the errors inherent in the original watch-type pedometer are too large for accurate epidemiological work [168, 169]. The Seiko and Epson companies began developing a digital timepiece prior to the Tokyo Olympic Games (1964), and a liquid crystal display (LCD) was adapted to display accumulated step-impulses as the Yamasa Tokei (Keiki) company marketed the manpo-kei, or "10,000 steps meter" in 1965 CE. In modern electronic pedometers, a lever arm moves with each stride, making an electrical contact or compressing a piezo-electric crystal, and the electrical impulse thus generated is recorded as a step. The user can input an estimated stride length, so that the instrument displays a figure for the distance travelled; however, a person's stride length is quite variable, and it is probably better to report simply the number of steps taken.

The difficulty of setting an appropriate sensitivity to record one impulse per step has also not been entirely overcome. However, devices such as the Kenz Lifecorder incorporate a filter that limits the range of recorded accelerations, thus reducing false counts from incidental movements. Sophisticated types of equipment also allow electronic storage of information for up to 200 days, and some instruments can estimate the minute-by-minute energy expenditures from the force of impulses [170, 171].

2.9.4.3 Reliability and Validity

Most reports suggest that modern electronic pedometer/accelerometers are more reliable than the older mechanical watch-type pedometers. Reliability has been assessed by fitting two instruments to the left and right-hand sides of a person's waist belt [172, 173]; good reliability has been seen with devices such as Yamax, Kenz, Omron, New Lifestyle and Digiwalker recorders [173, 174].

When one instrument was shaken on a test rig, the correlation with the actual number of oscillations was 0.996; the threshold acceleration for this particular device was 2 m/s^2, with a coefficient of variation of 1.5 % [172]. Pedometer/accelerometers also respond accurately to a consistent movement pattern such as level walking. The 24-hour step count determined with one pedometer/accelerometer had an intramodal reliability of 0.998 and the error relative to 500 actual paces taken on a level 400 m track was only -0.2 ± 1.5 steps [175]. Of five instruments tested, the Yamax device gave the best estimates for both moderate and slow walking speeds; the average systematic error over a distance of 4.8 km being about 2 % [173]. At a walking speed of 4.8 km/hour, most of 10 commercially available pedometer/accelerometers were able to estimate treadmill walking distances to within ±10 % and gross energy expenditures to within ±30 % of the actual value [175, 176]. As with the older mechanical pedometers, activity tended to be under-estimated at speeds below 4.8 km/hour [177]; at 3.2 km/hour, readings from the Kenz Lifecorder and the Actigraph were, respectively, 92 ± 6 % and 64 ± 15 % of the true values [178]. Other validation assessments have compared step counts with impulses recorded by a heel-mounted resistance pad (an error of 460 ± 1080 steps/day [172]), with the directly measured oxygen consumption during treadmill walking at speeds of 3–9 km/hour (a Pearson correlation coefficient of 0.97 and a mean difference in estimated energy expenditures ranging from −3.2 to +0.1 kJ/minute [179]), and with 24-hour metabolic chamber data that included some treadmill walking and running (more substantial errors of 9 % for total energy expenditure and 8 % for physical activity energy expenditure [180]).

Although many of these trials suggest that modern pedometer/accelerometers have a reasonable validity, inaccuracies are introduced by slow walking speeds, a short stride length and abnormalities of gait [181]. Moreover, errors are increased on moving from the laboratory to free-living conditions [182], as subjects choose their own activity patterns rather than walking over a fixed course at a fixed pace. Both pedometers and accelerometers respond poorly to cycling, skating, load-carrying, household chores, and other non-standard activities [183]. Further, they take no account of energy expended when climbing hills or making movements against external resistance, and artifacts may arise when traveling over bumpy ground in a car [184].

In general, the current generation of pedometer/accelerometers functions reasonably well when assessing old people, whose main source of physical activity is moderate walking. However, attempts to garner information on younger adults remains much more questionable, and sometimes the conversion of step counts to energy expenditures using dubious and secret equations has under-estimated doubly-labelled water measurements by as much as 30–60 % [185].

2.9.4.4 Uni-axial Accelerometers

Uni-axial accelerometers measure the acceleration forces exerted upon small weights rather than summing a series of electrical contacts. But given the similarity of principle to the modern pedometer/accelerometer, it is not surprising that the performance of accelerometers is generally comparable [175]. Moreover, the interpretation of accelerometer data remains controversial. Different accelerometers integrate movements over intervals ranging from 1 to 15 seconds. Also, complex (and sometimes secret) equations are often used to interpret the data. Thus, the Actical instrument includes a three-part algorithm: (1) below an inactivity threshold, the individual is credited with an energy expenditure of 1 MET; (2) a walk/run regression equation is used when the inactivity threshold is exceeded but the coefficient of variation (CV) of impulses over four consecutive 15-second epochs is <13 %; and (3) a third, lifestyle regression equation is used when the inactivity threshold is exceeded and the CV of impulses is >13 % [186].

2.9.4.5 Tri-axial Accelerometers

Tri-axial accelerometers measure forces in three planes, and thus can theoretically capture a wider range of body movements than uni-axial devices (Chap. 3). Park et al. [187] found differences of score between the two types of device even during walking, with uni-axial devices underestimating step counts and/or metabolic equivalents, particularly during slow walking (55 m/minute). Further, the accuracy of both types of accelerometer was affected by step frequency at any given walking speed.

One type of tri-axial accelerometer (the DynaPort Move Monitor) includes an algorithm to evaluate gaits and postures [188]. A quadratic discriminant analysis and a Hidden-Markov model are used to infer activity patterns in another tri-axial accelerometer (the MTI Actigraph) [189]. However, these algorithms were established on small samples, and their generality remains to be established. Moreover, the interpretation of tri-axial data under free-living conditions is as yet too complex for most epidemiological surveys.

2.9.4.6 Conclusions

To date, most pedometer and accelerometer studies have used relatively small samples, and have not been continued long enough to provide statistically significant information on clinical outcomes such as the onset of cardiovascular disease or sudden death, although this situation is now changing. As yet, it has been necessary to examine relationships between the recorded activity patterns and precursors of disease such as attained fitness levels [190], metabolic risk factors [191–193] or a deterioration of arterial elasticity [194]. Interestingly, the outcome of one study of arterial elasticity shows a greater correlation with measured habitual activity than with attained fitness. This is in contrast with the conclusion from questionnaire data, probably because the pedometer/accelerometer provides a measure of habitual physical activity which is at least as valid and accurate as the estimate of aerobic fitness [195].

2.10 New Directions in Objective Monitoring

Given problems in using objective monitors under free-living conditions, some investigators have looked at the benefits of altering the placement of accelerometers. Information has also been paired with data collected from global positioning systems (GPS), and multi-phasic devices have recorded audio signals, light intensity, barometric pressure, humidity and environmental temperatures. As yet, most of these devices are too costly and complicated for epidemiological use.

2.10.1 Accelerometer Placement

Accelerometers have sometimes been mounted on places other than the subject's waist belt. One commercial accelerometer known as the "footpod" has been attached to the foot, recording the impact associated with each stride. Another innovation has been the attachment of a small inertia-sensing device to the ear lobe. This monitor records information on both posture and linear acceleration that is introduced into an algorithm predicting energy expenditures. An initial report has claimed a good correlation with the directly measured oxygen consumption when performing 11 daily activities; the investigators also claimed success in identifying the nature of the activities that were being undertaken [196].

2.10.2 GPS Devices

Some recent studies have combined GPS and accelerometer data [197]. The GPS is helpful in detecting artifacts arising from travel in vehicles. However, the sampling rate is insufficient to detect very rapid movements. Moreover, the quality of the device is critical when recording data in urban areas with tall buildings [198], and the signal is lost when travelling on underground trains.

2.10.3 Multiphasic Devices

Multiphasic devices seek to enhance the interpretation of objective monitors when examining non-standard movements [199]. Corder et al. [200] claimed some improvement in the estimation of children's energy expenditures when they supplemented accelerometry with ECG data. Haskell et al. [159] combined 3-lead ECG data with information captured by uni-axial accelerometers mounted on the wrist and the thigh. When using this arrangement, the correlation with directly measured oxygen consumption across a range of activities (0.73) was not particularly impressive, and the absolute error in measurements of oxygen consumption remained a substantial 5.2 mL/(kg minute).

One recent multi-phasic device (the Multi-sensor Board) combines a tri-axial accelerometer with GPS, audio, light, barometric pressure, temperature and humidity recording [134]. Supposedly, this device can determine the proportion of activity taken out of doors, local versus distant travel, the vibrations resulting from car journeys, and the possible influence of weather conditions upon movement patterns.

The Sensewear arm band is another multi-phasic device. It incorporates information from two accelerometers with heat flux, skin temperature, and galvanic skin response data [201–204]. The manufacturers claim that this instrument can distinguish activities performed by the upper and lower limbs, and that it takes account of hill climbing, load carriage and non-ambulatory activity. However, one trial found a 24–56 % under-estimation of directly measured energy expenditures during skating, and another trial found significant under-estimation of energy expenditures during high-speed running (40 %) and cycling (25–50 %) [205]. Other studies of the Sensewear arm band have found 16–43 % over-estimates of energy expenditure in children [206] and large individual errors relative to indirect calorimetry in adults [207]. Much seems to depend on the algorithm that is used [208], and it remains a challenge to find formulae appropriate for the interpretation of multi-phasic data on all subjects and all circumstances.

2.11 Conclusions

Estimation of the intensity and total volume of weekly energy expenditures has become practicable for epidemiologists with the replacement of questionnaires by relatively low cost objective monitoring devices. Step counting has become progressively more sophisticated, with an ability to classify the intensity of impulses and accumulate activity data over long periods. Pedometer/accelerometers yield quite precise data for standard laboratory exercise, and in groups where steady, moderately paced walking is the main form of energy expenditure they can provide very useful epidemiological data. Nevertheless, such instruments remain vulnerable to external vibration and they fail to reflect adequately the energy expenditures incurred in hill climbing and isometric activity, as well as many of the everyday activities of children and younger adults. Multi-phasic devices hold promise as a means of assessing atypical activities, but appropriate and universally applicable algorithms have as yet to be developed. Moreover, the equipment is at present too costly and complex for epidemiological use.

References

1. Merrill R. Introduction to epidemiology. Sudbury, MA: Jones and Bartlett Learning; 2010.
2. Shephard RJ. An illustrated history of health and fitness: from prehistory to our post-modern world. New York, NY: Springer; 2015.
3. Johnston S. The ghost map: the story of London's most terrifying epidemic—and how it changed science, cities and the modern world. New York, NY: Riverhead; 2006.
4. Snow J. On the adulteration of bread as a cause of rickets. Lancet. 1857;70 Suppl 766:4–5.
5. Cassidy M. Coronary disease; the Harveian oration of 1946. Lancet. 1946;2(6426):587–90.

6. Hutchinson R. Medicine today and yesterday. Br Med J. 1950;1(4644):72–3.
7. Perring RJ. Family doctor. The impact of modern therapeutic ideas on general practice. Lancet. 1949;254(6591):1163–7.
8. Yater WM, Traum AH, Brown WG, et al. Coronary artery disease in men eighteen to thirtynine years of age: report of eight hundred sixty-six cases, four hundred fifty with necropsy examination. Am Heart J. 1948;36(3):334–72.
9. Morris JN. Recent history of coronary disease. Lancet. 1951;257(6645):1–7.
10. Anderson TW, Le Riche WH. Ischemic heart disease and sudden death, 1901–1961. Brit J Prev Soc Med. 1970;24:1–9.
11. Anderson TW. The myocardium in coronary heart disease. In: Kavanagh T, editor. Proceedings of International Symposium on exercise and coronary artery disease. Toronto, ON: Toronto Rehabilitation Centre; 1976. p. 32–44.
12. Shephard RJ. Ischemic heart disease and exercise. London: Croom Helm; 1981.
13. Paffenbarger RS. Physical activity and fatal heart attack: protection or selection? In: Amsterdam EA, Wilmore JH, deMaria AN, editors. Exercise in cardiovascular health and disease. New York, NY: Yorke Medical Books; 1977. p. 35–49.
14. Morris JN, Crawford MD. Coronary heart disease and physical activity of work. Br Med J. 1958;2(5111):1485–96.
15. Paffenbarger RS, Laughlin ME, Gima AS, et al. Work activity of longshoremen as related to death from coronary heart disease and stroke. N Engl J Med. 1970;20:1109–14.
16. Paffenbarger RS, Hale WE. Work activity and coronary heart mortality. N Engl Med J. 1975;292:545–50.
17. Taylor HL, Klepetar E, Keys A, et al. Death rates among physically active and sedentary employees of the railroad industry. Am J Publ Health. 1962;52:1697–707.
18. Morris JN, Heady J, Raffle P, et al. Coronary heart disease and physical activity of work. Lancet. 1953;262:1053–7. 1111–20.
19. Brand RJ, Paffenbarger RS, Sholtz RJ, et al. Work activity and fatal heart attack studied by multiple logistic regression risk analysis. Am J Epidemiol. 1979;110:52–62.
20. Slattery ML, Jacobs DR. Physical fitness and cardiovascular disease mortality. The US railroad study. Am J Epidemiol. 1988;127(3):571–80.
21. Morris JN, Heady J, Raffle P. Physique of London busmen: epidemiology of uniforms. Lancet. 1956;271(6942):569–70.
22. Paffenbarger RS, Blair SN, Lee I-M, et al. A history of physical activity, cardiovascular health and longevity: the scientific contributions of Jeremy N. Morris. Int J Epidemiol. 2001;30:1184–92.
23. Morris JN, Kagan A, Pattison DC, et al. Incidence and prediction of ischaemic heart disease in London busmen. Lancet. 1966;288(7463):553–9.
24. Anderson WG. Further studies in longevity of Yale athletes. Med Times. 1916;44:75–7.
25. Baron SL, Hein MJ, Lehman E, et al. Body mass index, playing position, race, and the cardiovascular mortality of retired professional football players. S Am J Cardiol. 2012;109 (6):889–96.
26. Belli S, Vanacore N. Proportionate mortality of Italian soccer players: is amyotrophic lateral sclerosis an occupational disease? Eur J Epidemiol. 2005;20(3):237–42.
27. Cooper EL, Sullivan J, Hughes E. Athletes and the heart: an electrocardiologic study of the responses of healthy and diseased hearts to exercise. Med J Austr. 1937;1:569.
28. Dublin LI. College honor men long lived. Stat Bull Metropol Life. 1932;13:5–7.
29. Gaines JM, Hunter AN. Mortality among athletes and other graduates of Yale University. Trans Actuar Soc Am. 1905;IX(33):47–52.
30. Gajewski AK, Poznanska A. Mortality of top athletes, actors and clergy in Poland: 1924-2000 follow-up study of the long term effect of physical activity. Eur J Epidemiol. 2008;23 (5):335–40.
31. Hartley PHS, Llewellyn GF. The longevity of oarsmen. Br Med J. 1939;1(4082):657–9.
32. Hill AB. Cricket and its relation to the duration of life. Lancet. 1927;210(5435):949–50.

33. Kalist DE, Peng Y. Does education matter? Major league baseball players and longevity. Death Stud. 2007;31(7):653–70.
34. Karvonen MJ, Klemola H, Virkajarvi J, et al. Longevity of endurance skiers. Med Sci Sports. 1974;6:49–61.
35. Kujala UM, Tikkanen HO, Sarna S, et al. Disease-specific mortality among elite athletes. JAMA. 2001;285(1):44–5.
36. Marijon E, Tafflet M, Antero-Jacquemin J, et al. Mortality of French participants in the Tour de France (1947–2012). Eur Heart J. 2013;34(40):3145–50.
37. Menotti A, Amici E, Gambelli GC, et al. Life expectancy in Italian track and field athletes. Eur J Epidemiol. 1990;6(3):257–60.
38. Metropolitan Life Insurance. Characteristics of major league baseball players. Stat Bull Metrop Life Ins Co. 1975;56:2–4.
39. Meylan G. Harvard University oarsmen. Harvard Grad Mag. 1904;9:362–76.
40. Morgan JE. University Oars, being a critical enquiry into the after-health of the men who rowed in the Oxford and Cambridge boat race from the year 1829–1859. London: Macmillan; 1873.
41. Reed LJ, Love AG. Biometric studies on U.S. army officers. Longevity of army officers in relation to physical fitness. Mil Surgeon. 1931;69:397.
42. Schmid VL. Contributions to the study of the causes of death of sportsmen. Dtsch Z Sportmed. 1967;18(10):411–5.
43. Schnohr P. Longevity and cause of death in male athletic champions. Lancet. 1971;298 (7738):1364–5.
44. Schnohr P, Parner J, Lange P. Mortality in joggers: population based study of 46589 men. Br Med J. 2000;321(7261):602–3.
45. Taioli E. All causes of mortality in male professional soccer players. Eur J Publ Health. 2007;17(6):600–4.
46. Wakefield MC. A study of mortality amongst the men who have played in the Indiana High School State final basketball tournament. Res Quart. 1944;15:3–11.
47. Waterbor J, Cole P, Delzell E, et al. The mortality experience of major league baseball players. N Engl J Med. 1988;318:1278–80.
48. Garatachea N, Santos-Lanzano A, Sanchis-Gomar F, et al. Elite athletes live longer than the general population: a meta-analysis. Mayo Clin Proc. 2014;89(9):1195–200.
49. Sarna S, Kaprio J. Life expectancy of former elite athletes. Sports Med. 1994;17:149–51.
50. Yamaji K, Shephard RJ. Longevity and causes of death in athletes: a review of the literature. J Hum Ergol. 1977;6:13–25.
51. Abel E, Kruger ML. The longevity of baseball hall of famers compared to other players. Death Stud. 2005;29:959–63.
52. Greenway JC, Hiscock IV. Mortality among Yale men. Yale Alum Wkly. 1926;35:1806–8.
53. Montoye HJ, Van Huss WD, Olson H, et al. Study of the longevity and morbidity of college athletes. Ann Arbor, MI: Phi Epsilon Kappa Fraternity, Michigan State University; 1957.
54. Olson HW, Teitelbaum H, Van Hus WD, et al. Years of sport participation and mortality in college athletes. J Sports Med Phys Fitness. 1977;17:321–6.
55. Paffenbarger RS, Natkin J, Krueger D, et al. Chronic disease in former students. II. Methods of study and observations on mortality from coronary heart disease. Am J Publ Health. 1966;56:962–71.
56. Polednak AP, Damon A. College athletics, longevity and cause of death. Hum Biol. 1970;42:28–46.
57. Prout C. Life expectancy of college oarsmen. JAMA. 1972;220:1709–11.
58. Rook A. An investigation into the longevity of Cambridge sportsmen. Br Med J. 1954;1 (4865):773–7.
59. Trudeau F, Shephard RJ. Lessons learned from the Trois-Rivières physical education study: a retrospective. Pediatr Exerc Sci. 2005;17(2):2–25.
60. Trudeau F, Shephard RJ. Is there a long-term legacy of required physical education? Sports Med. 2008;38(54):265–70.

61. Shephard RJ, Corey P, Renzland P, et al. The influence of an industrial fitness programme upon medical care costs. Can J Publ Health. 1982;73:259–63.
62. Shephard RJ. Limits to the measurement of habitual physical activity by questionnaires. Br J Sports Med. 2003;37:197–206.
63. Sternfeld B, Goldman-Rosas L. A systematic approach to selecting an appropriate measure of self-reported physical activity or sedentary behavior. J Phys Act Health. 2012;10 Suppl 1: S19–28.
64. Godin G, Shephard RJ. A simple method to assess exercise behaviour in the community. Can J Appl Sport Sci. 1985;10:141–6.
65. Haskell WL, Taylor HL, Wood PD, et al. Strenuous physical activity treadmill exercise test performance and plasma lipoprotein cholesterol. The lipid clinics research program. Circulation. 1980;62:53–61.
66. Mundal R, Eriksson J, Rodahl K. Assessment of physical activity by questionnaire and by personal interview with particular reference to fitness and coronary mortality. Eur J Appl Physiol. 1987;54:245–52.
67. Montoye HJ, Kemper HCG, Saris WH, et al. Measuring human energy expenditure. Champaign, IL: Human Kinetics; 1996.
68. Vuillemin A. Revue des questionnaires d'évaluation de l'activité physique (A review of evaluation questionnaires for physical activity). Rev Epidemiol Santé Publique. 1998;46:49–55.
69. Stephens T, Craig CL. The well-being of Canadians: the 1988 Campbell's survey. Ottawa, ON: Canadian Fitness & Lifestyle Research Institute; 1990.
70. Morris JN, Chave SP, Adam C, et al. Vigorous exercise in leisure time and the incidence of coronary heart disease. Lancet. 1973;1(7799):333–9.
71. Shephard RJ. Fitness of a nation: lessons from the Canada fitness survey. Basel, Switzerland: Karger; 1986.
72. Uitenbroeck DG. Seasonal variation in leisure time physical activity. Med Sci Sports Exerc. 1993;25:753–60.
73. Yasunaga A, Togo F, Watanabe E, et al. Sex, age, season, and habitual physical activity of older Japanese: the Nakanojo study. J Aging Phys Activ. 2008;16:3–13.
74. Dan AJ, Wilbur JE, Hedricks C, et al. Lifelong physical activity and older women. Psychol Women Quart. 1990;14:531–42.
75. Friedenreich CM, Corneya KS, Bryant HE. The lifetime total physical activity questionnaire: development and reliability. Med Sci Sports Exerc. 1998;30:266–74.
76. Kriska AM, Black-Sandler R, Cauley JA, et al. The assessment of historical physical activity and its relation to adult bone parameters. Am J Epidemiol. 1988;127:1053–63.
77. Chasan-Taber L, Erickson JB, McBride JW, et al. Reproducibility of a self-administered lifetime physical activity questionnaire among female college alumnae. Am J Epidemiol. 2002;155(3):282–91.
78. Ainsworth BE, Leon AS, Richardson MT, et al. Accuracy of the college alumnus physical activity questionnaire. J Clin Epidemiol. 1993;46:1403–11.
79. Magnus K, Matroos A, Strackee J. Walking, cycling, or gardening, with or without seasonal interruption, in relation to acute coronary events. Am J Epidemiol. 1979;110:724–33.
80. Montoye HJ. Estimation of habitual activity by questionnaire and interview. Am J Clin Nutr. 1971;24:1113–8.
81. Paffenbarger RS, Wing AL, Hyde RT. Physical activity as an index of heart attack risk in college alumni. Am J Epidemiol. 1978;108:161–75.
82. Taylor HL, Jacobs DR, Schucker B, et al. A questionnaire for the assessment of leisure time physical activities. J Chron Dis. 1978;31:741–5.
83. Williams CL, Carter BJ, Eng A. The "know your body" program: a developmental approach to health education and disease prevention. Prev Med. 1980;9:371–83.
84. Blair SN, Haskell WL, Ho P, et al. Assessment of physical activity by a seven-day recall in a community survey and controlled experiments. Am J Epidemiol. 1985;122:794–804.

85. Ainsworth BE, Haskell WL, Leon AS, et al. Compendium of physical activities: classification of energy costs of human physical activities. Med Sci Sports Exerc. 1993;25:71–80.
86. Passmore R, Durnin JVGA. Human energy expenditure. Physiol Rev. 1955;35:801–40.
87. Ainsworth BE, Haskell WL, Herrmann S, et al. 2011 compendium of physical activities: a second update of codes and MET values. Med Sci Sports Exerc. 2011;43:1575–81.
88. Shephard RJ. Intensity, duration and frequency of exercise as determinants of the response to a training regimen. Int Z Angew Physiol. 1968;26:272–8.
89. Bouchard C, Shephard RJ, Stephens T. Physical activity, fitness and health. Champaign, IL: Human Kinetics; 1994.
90. Sallis JF, Saelens BE. Assessment of physical activity by self-report: status and limitations. Res Quart. 2000;71:1–14.
91. Klesges RC, Eck LH, Mellon MW, et al. The accuracy of self-reports of physical activity. Med Sci Sports Exerc. 1990;22:690–7.
92. Goode RC, Mertens R, Shaiman S, et al. Voice, breathing and the control of exercise intensity. Adv Exp Med Biol. 1998;450:223–9.
93. American College of Sports Medicine. ACSM's guidelines for exercise testing and prescription. 7th ed. Philadelphia, PA: Lippincott, Williams & Wilkins; 2007.
94. Warburton DER, Katzmarzyk PT, Rhodes RE, et al. Evidence informed physical activity guidelines for Canadian adults. Appl Physiol Nutr Metab. 2007;32(Suppl 2E):S16–68.
95. Hardman AE. Issues of fractionalization of exercise (short vs. long bouts). Med Sci Sports Exerc. 2001;33 Suppl 6:S421–8.
96. Lamonte MJ, Ainsworth BE. Quantifying energy expenditures and physical activity in the context of dose response. Med Sci Sports Exerc. 2001;33 Suppl 6:S370–80.
97. Washburn RA, Heath GW, Jackson AW. Reliability and validity issues concerning large-scale surveillance of physical activity. Res Quart. 2000;71 Suppl 2:S104–13.
98. Pandolf KB, Haisman MF, Goldman RF. Metabolic expenditure and terrain coefficients for walking on snow. Ergonomics. 1976;19:683–90.
99. Armstrong LE, Maresh CM. The induction and decay of heat acclimatisation in trained athletes. Sports Med. 1999;12:302–12.
100. Shephard RJ. Exercise and relaxation in health promotion. Sports Med. 1997;23:211–7.
101. Morris JN, Clayton DG, Everitt MG, et al. Exercise in leisure time. Coronary attack and death rates. Br Heart J. 1990;63:325–34.
102. Paffenbarger RS, Lee I-M. Physical activity and fitness for health and longevity. Res Quart. 1995;67 Suppl 3:S11–28.
103. Blair SN, Kohl HW, Paffenbarger RS, et al. Physical fitness and all-cause mortality: a prospective study of healthy men and women. JAMA. 1989;262:2395–401.
104. Baer HJ, Glynn RJ, Hu FB, et al. Risk factors for mortality in the nurses' health study: a competing risks analysis. Am J Epidemiol. 2011;173:319–29.
105. Jacobs DR, Ainsworth BE, Hartman D, et al. A simultaneous evaluation of 10 commonly used physical activity questionnaires. Med Sci Sports Exerc. 1993;25:81–91.
106. Weller IR, Corey PN. A study of the reliability of the Canada fitness survey questionnaire. Med Sci Sports Exerc. 1998;30:1530–6.
107. National Health & Nutrition Examination Survey. Questionnaires, data sets and related documentation. Atlanta, GA: Centers for Disease Control & Prevention; 2011.
108. Fine LJ, Philogene GS, Gramling R, et al. Prevalence of multiple chronic disease risk factors: 2001 National Health Interview Survey. Am J Prev Med. 2004;27 Suppl 2:18–24.
109. Fitness and Amateur Sport. Fitness & Lifestyle in Canada: The Canada Fitness Survey. Ottawa, ON: Fitness & Amateur Sport, Government of Canada; 1983.
110. Ontario F. Physical activity patterns in Ontario Part II. Toronto, ON: Fitness Ontario; 1983.
111. Stephens T. Fitness and activity measurements in the 1989 Canada Fitness Survey. In: Drury T, editor. Assessing physical fitness and physical activity. Hyattsville, MD: U.S. Department of Health & Human Services; 1989. p. 401–32.
112. Boon RM, Hamlin MJ, Steel GD, et al. Validation of the New Zealand Physical Activity Questionnaire (NZPAQ-LF) and the International Physical Activity Questionnaire (IPAQ-LF) with accelerometry. Br J Sports Med. 2010;44:741–6.

113. Prince SA, Adfamo KB, Hanmel ME, et al. A comparison of direct versus self-report measures for assessing physical activity in adults: a systematic review. Int J Behav Nutr Phys Activ. 2008;5:56.
114. Slater CH, Green LW, Vernon SW, et al. Problems in estimating the prevalence of physical activity from national surveys. Prev Med. 1987;16:107–19.
115. Brownson RC, Jones DA, Pratt M, et al. Measuring physical activity with the behavioral risk factor surveillance system. Med Sci Sports Exerc. 2000;32:1913–8.
116. Sarkin JA, Nichols JF, Sallis JF, et al. Self-report measures and scoring protocols affect prevalence estimates of meeting physical activity guidelines. Med Sci Sports Exeerc. 2000;32:149–56.
117. Sims J, Smith F, Duffy A, et al. The vagaries of self-reports of physical activity: a problem revisited and addressed in a study of exercise promotion in the over 65s in general practice. Fam Pract. 1999;16:152–7.
118. Moti RW, McAuley EA, Stefano C. Is social desirability associated with self-reported physical activity? Prev Med. 2005;40:735–9.
119. Rzewnicki R, Auweele YV, De Bourdeaudhuij I. Addressing the issue of over-reporting on the International Physical Activity Questionnaire (IPAQ) telephone survey with a population sample. Publ Health Nutr. 2003;6:299–305.
120. Adam CL, Mercer JG. Appetite regulation and seasonality: implications for obesity. Proc Nutr Soc. 2004;63:413–9.
121. Bassett DR, Cureton QL, Ainsworth BE. Measurement of daily walking distance questionnaire vs. pedometer. Med Sci Sports Exerc. 2000;32:1018–23.
122. Yasanuga A, Park H, Watanabe E, et al. Development and evaluation of the physical activity questionnaire for elderly Japanese: the Nakanojo study. J Aging Phys Activ. 2007;16:3–13.
123. Shephard RJ. Normal levels of activity in Canadian city dwellers. Can Med Assoc J. 1967;96:912–4.
124. Phillippaerts RM, Westerterp KM, Lefevre J. Doubly-labelled water validation of three physical activity questionnaires. Int J Sports Med. 1999;20:284–9.
125. Schuit AJ, Schouten EG, Westerterp KR, et al. Validity of the physical activity scale (PASE) for the elderly according to energy expenditure assessed by the double labeled water method. J Clin Epidemiol. 1997;50:541–6.
126. Katzmarzyk P, Tremblay MS. Limitations of Canada's physical activity data: implications for monitoring trends. Appl Physiol Nutr Metab. 2007;32(Suppl 2E):S185–94.
127. Bailey DA, Shephard RJ, Mirwald RL, et al. Current levels of Canadian cardio-respiratory fitness. Can Med Assoc J. 1974;111:25–30.
128. Philippaerts RM, Westerterp KR, Lefevre J. Comparison of two questionnaires with a triaxial accelerometer to assess physical activity patterns. Int J Sports Med. 2001;22:34–9.
129. Sallis JF, Buono MJ, Roby JJ, et al. Seven-day recall and other physical activity self-reports in children and adolescents. Med Sci Sports Exerc. 1993;25:99–108.
130. Shephard RJ, McClure RL. The prediction of cardiorespiratory fitness. Int Z Angew Physiol. 1965;21:212–23.
131. Buskirk ER, Harris D, Mendez J, et al. Comparison of two assessments of physical activity and a survey method for calorimetric intake. Am J Clin Nutr. 1971;24:1119–25.
132. Weiss RW, Slater CH, Green LW, et al. The validity of single-item, self-assessment questions as measures of adult physical activity. J Clin Epidemiol. 1990;43:1123–9.
133. Celis-Morales CA, Perez-Bravo F, Ibañez L, et al. Measurement method on relationships with risk biomarkers. PLoS One. 2012;7(5):e36345.
134. Warren JM, Ekelund U, Besson H, et al. Assessment of physical activity—a review of methodologies with reference to epidemiolgical research: a report of the exercise physiology section of the European Association of Cardiovascular Prevention & Rehabilitation. Eur J Cardiovasc Prev Rehabil. 2010;17:127–39.
135. Shaper AC, Wannamethee G. Physical activity and ischaemic heart disease in middle-aged British men. Br Heart J. 1991;66:384–94.

136. Sesso HD, Paffenbarger RS, Lee IM. Physical activity and coronary heart disease in men: the Harvard Alumni study. Circulation. 2000;102(9):975–80.
137. Lee IM, Sesso HD, Oguma Y, et al. Relative intensity of physical activity and risk of coronary heart disease. Circulation. 2003;107:1110–6.
138. Paffenbarger RS, Hyde RT, Wing AL, et al. Some interrelationships of physical activity, physiological fitness, health and longevity. In: Bouchard C, Shephard RJ, Stephens T, editors. Physical activity, fitness and health. Champaign, IL: Human Kinetics; 1994.
139. Morris JN, Everitt MG, Pollard R, et al. Vigorous exercise in leisure time. Protection against coronary heart disease. Lancet. 1980;316(8206):1207–10.
140. Wannamethee G, Shaper AC. Physical activity and stroke in British middle-aged men. Br Med J. 1992;304:597–601.
141. Linsted KD, Tonstad S, Kuzma JW. Self-report of physical activity and patterns of mortality in Seventh-Day Adventist men. J Clin Epidemiol. 1991;44:355–64.
142. Slattery ML, Jacobs DR, Nichaman NZ. Leisure time physical activity and coronary heart disease death. The US railroad study. Circulation. 1989;79:304–11.
143. Blair SN, Cheng Y, Holder JS. Is physical activity or physical fitness more important in defining health benefits? Med Sci Sports Exerc. 2001;33 Suppl 2:S379–99.
144. Blair SN, Kohl HW, Barlow CE, et al. Changes in physical fitness and all-cause mortality: a prospective study of healthy and unhealthy men. JAMA. 1995;273:1093–8.
145. Arraiz GA, Wigle DT, Mao Y. Risk assessment of physical activity and physical fitness in the Canada Health Survey mortality follow-up study. J Clin Epidemiol. 1992;45:419–28.
146. Shephard RJ, Rode A. The health consequences of ‘modernization.’ Evidence from circumpolar peoples. London, UK: Cambridge University Press; 1996.
147. Young TK, Reading J, O’Neil JD. Type 2 diabetes mellitus in Canada’s First Nations: status of an epidemic in progress. Can Med Assoc J. 2000;183(9):1132–41.
148. Troiano RP, Dodd KW. Differences between objective and self-report measures of physical activity: what do they mean? Korean J Meas Eval Phys Ed Sports Sci. 2008;10:31–42.
149. Tucker JM, Welk GJ, Beyler NK. Physical activity in U.S.: adults compliance with the Physical Activity Guidelines for Americans. Am J Prev Med. 2011;10:454–61.
150. Burke E. Precision heart rate training. Champaign, IL: Human Kinetics; 1984.
151. Karvonen J, Chwalbinska-Moneta J, Säynäjäkangas S. Comparison of heart rates by ECG and minicomputer. Phys Sportsmed. 1984;12(6):65–9.
152. Müller EA, Himmelmann W. Geräte zur continuierlichen fotoelektrischen Pulszahlung (Equipment for continuous photoelectric pulse counting). Int Z Angew Physiol. 1957;16:400–8.
153. Wolff HS. Physiological measurement of human subjects in the field, with special reference to a new approach to data storage. In: Yoshimura H, Weiner JS, editors. Human adaptability and its methodology. Tokyo, Japan: Japanese Society for the Promotion of Sciences; 1966.
154. Baker JA, Humphrey SGE, Wolff HS. Advances in the technique of using SAMIs (socially acceptable monitoring instruments). J Physiol. 1969;200:89.
155. Edholm OG, Humphrey SGE, Lourie JA, et al. VI. Energy expenditure and climatic exposure of Yemenite and Kurdish Jews in Israel. Philos Trans R Soc Lond B. 1973;266:127–40.
156. Glagov S, Rowley DA, Cramer DB, et al. Heart rates during 24 hours of usual activity for 100 normal men. J Appl Physiol. 1970;29:799–805.
157. Masironi R, Mansourian P. Determination of habitual physical activity by means of a portable R-R interval distribution recorder. Bull WHO. 1974;51:291–8.
158. Åstrand P-O, Ryhming I. A nomogram for calculation of aerobic capacity (physical fitness) from pulse rate during submaximal work. J Appl Physiol. 1954;7:218–21.
159. Haskell WL, Yee MC, Evans A, et al. Simultaneous measurement of heart rate and body motion to quantitate physical activity. Med Sci Sports Exerc. 1993;25:109–15.
160. Müller EA, Franz H. Energieverbrauchmessungen bei beruflicher Arbeit mit einer verbesserten Respirations-Gasuhr (Measurement of energy consumption during occupational activities with an improved respiratory gas meter). Arbeitsphysiol. 1952;14:499–504.

161. Durnin JVGA, Passmore R. Energy, work and leisure. London, UK: Heinemann; 1967.
162. Kavanagh T. Heart attack? Counter attack! Toronto, ON: Van Nostrand; 1976.
163. MacCurdy E. The notebooks of Leonardo da Vinci. New York, NY: Reynal & Hitchcock; 1938. p. 166.
164. Wolf ML. Thomas Jefferson, Abraham Lincoln, Louis Brandeis and the mystery of the universe. Boston Univ J Sci Tech Law. 1995;1:1–15.
165. Kemper HCG, Verschuur R. Validity and reliability of pedometers in research on habitual physical activity. In: Shephard RJ, Lavallée H, editors. Frontiers of activity and child health. Québec, QC: Editions du Pélican; 1977. p. 83–92.
166. Ewalt LA, Swartz AM, Strath SJ, et al. Validity of physical activity monitors in assessing energy expenditure in normal, overweight, and obese adults. Med Sci Sports Exerc. 2008;40 Suppl 5:S198.
167. Busse ME, van Deursen RW, Wiles CM. Real-life step and activity measurement: reliability and validity. J Med Eng Technol. 2009;33:33–41.
168. Meijer GAL, Westerterp KR, Verhoeven FMH, et al. Methods to assess physical activity with special reference to motion sensors and accelerometers. IEEE Trans Biomed Eng. 1991;38:221–8.
169. Tudor-Locke C, Sisson SB, Lee SM, et al. Evaluation of quality of commercial pedometers. Can J Publ Health. 2006;97 Suppl 1:S10–5. S10–6.
170. Mizuno C, Yoshida T, Udo M. Estimation of energy expenditure during walking and jogging by using an electro-pedometer. Ann Physiol Anthropol. 1990;9:283–9.
171. Tudor-Locke C, Myers AM. Methodological considerations for researchers and practitioners using pedometers to measure physical (ambulatory) activity. Res Quart. 2001;72:1–12.
172. Bassey EJ, Dalloso HM, Fentem PH, et al. Validation of a simple mechanical accelerometer (pedometer) for the estimation of walking activity. Eur J Appl Physiol. 1987;56:323–30.
173. Bassett DR, Ainsworth BE, Leggett SR, et al. Accuracy of five electronic pedometers for measuring distance walked. Med Sci Sports Exerc. 1996;28:1071–7.
174. Welk GJ, Almeida J, Morss G. Laboratory calibration and validation of the Biotrainer and Actitrac activity monitors. Med Sci Sports Exerc. 2003;35:1057–64.
175. Schneider PL, Crouter SE, Lukajic O, et al. Accuracy and reliability of 10 pedometers for measuring steps over a 400-m walk. Med Sci Sports Exerc. 2003;35:1779–84.
176. Crouter SE, Schneider PL, Karabulut M, et al. Validity of 10 electronic pedometers for measuring steps, distance, and energy cost. Med Sci Sports Exerc. 2003;35:1455–560.
177. Albright C, Hultquist CN, Thompson DL. Validation of the Lifecorder EX Activity Monitor. Med Sci Sports Exerc. 2006;35 Suppl 5:S500.
178. Abel MG, Hannon JC, Sell K, et al. Validation of the Kenz Lifecorder EX and ActiGraph GT1M accelerometers for walking and running in adults. Appl Physiol Nutr Metab. 2008;33:1155–64.
179. Yokoyama Y, Kawamura T, Tamakoshi A, et al. Comparison of accelerometry and oxymetry for measuring daily physical activity. Circulation. 2002;66:751–4.
180. Kumahara H, Schutz Y, Makoto A, et al. The use of uniaxial accelerometry for the assessment of physical-activity-related energy expenditure: a validation study against whole-body indirect calorimetry. Br J Nutr. 2004;91:235–43.
181. Cyarto EV, Myers AM, Tudor-Locke C. Pedometer accuracy in nursing home and community dwelling older adults. Med Sci Sports Exerc. 2004;36:205–9.
182. McClain JJ, Sisson BB, Tudor-Locke C. Actigraph accelerometer interinstrument reliability during free-living in adults. Med Sci Sports Exerc. 2007;39:1509–14.
183. Sirard JR, Pate RR. Physical activity assessment in children and adolescents. Sports Med. 2001;31:439–54.
184. Le Masurier GC, Tudor-Locke C. Comparison of pedometer and accelerometer accuracy under controlled conditions. Med Sci Sports Exerc. 2003;35:867–71.
185. Leenders JM, Nicole Y, Sherman M, et al. Evaluation of methods to assess physical activity in free-living conditions. Med Sci Sports Exerc. 2001;33:1233–40.

186. Crouter SE, Churilla JR, Bassett DR. Accuracy of the Actiheart for the assessment of energy expenditure in adults. Eur J Clin Nutr. 2008;62:704–11.
187. Park J, Ishikawa-Takata K, Tanaka S, et al. Effects of walking speed and step frequency on estimation of physical activity using accelerometers. J Physiol Anthropol. 2011;30:119–27.
188. Dijkstra B, Kamsma Y, Ziljstra W. Detection of gait and postures using a miniaturized triaxial accelerometer-based system: accuracy in community dwelling older adults. Age Ageing. 2010;39:259–62.
189. Pober DM, Staudenmeyer J, Raphael C, et al. Development of novel techniques to classify physical activity mode using accelerometers. Med Sci Sports Exerc. 2006;38:1626–34.
190. Aoyagi Y, Park H, Watanabe E, et al. Habitual physical activity and physical fitness in older Japanese adults: the Nakanojo study. Gerontol. 2009;55:523–31.
191. Freak-Poll R, Wolfe R, Backholer K, et al. Impact of a pedometer-based workplace health program on cardiovascular and diabetes risk profile. Prev Med. 2011;53(3):162–71.
192. Cook I, Alberts M, Lambert EV. Relationship between adiposity and pedometer-assessed ambulatory activity in adult, rural African women. Int J Obesity. 2008;32:1327–30.
193. Park S, Park H, Togo F, et al. Yearlong physical activity and metabolic syndrome in older Japanese adults: cross-sectional data from the Nakanojo study. J Gerontol (Biol Sci Med Sci). 2008;63:1119–23.
194. Aoyagi Y, Park H, Kakiyama T, et al. Year-long physical activity and arterial stiffness in older adults: the Nakanojo study. Eur J Appl Physiol. 2010;109(3):455–64.
195. Ayabe M, Park S, Shephard RJ, et al. Associations of activity monitor output and an estimate of aerobic fitness with pulse wave velocities: the Nakanojo study. J Phys Act Health. 2015;12 (1):139–44.
196. Atallah L, Leong JJ, Lo B, et al. Energy expenditure prediction using a miniaturized ear-worn sensor. Med Sci Sports Exerc. 2011;43:1369–777.
197. Troped PJ, Oliveira MS, Matthews CE, et al. Prediction of activity mode with global positioning system and accelerometer data. Med Sci Sports Exerc. 2008;40:972–8.
198. Krenn PJ, Mag DI, Titze S, et al. Use of global positioning systems to study physical activity and environment. Am J Prev Med. 2011;41(5):508–15.
199. King GA, Torres N, Potter C, et al. Comparison of activity monitors to estimate energy cost of treadmill exercise. Med Sci Sports Exerc. 2004;36:1244–51.
200. Corder K, Brage S, Mattocks C, et al. Comparison of two methods to assess PAEE during six activities in children. Med Sci Sports Exerc. 2007;39(12):2180–8.
201. Berntsen S, Hageberg R, Aandstad A, et al. Validity of physical activity monitors in adults participating in free-living activities. Br J Sports Med. 2010;44:657–64.
202. Malavolti M, Pietrobelli A, Dugoni M, et al. A new device for measuring resting energy expenditure (REE) in healthy subjects. Nutr Metab Cardiovasc Dis. 2007;17:338–43.
203. Soric M, Mikulic P, Misigoj-Durakovic M, et al. Validation of the Sensewear Armband during recreational in-line skating. Eur J Appl Physiol. 2012;112(3):1183–8.
204. Welk GJ, McClain JJ, Eisenmann JC, et al. Field validation of the MTI Actigraph and BodyMedia armband monitor using the IDEEA monitor. Obesity. 2007;15:918–28.
205. Koehler K, Braun H, de Marées M, et al. Assessing energy expenditure in male endurance athletes: validity of the SenseWear Armband. Med Sci Sports Exerc. 2011;43:1328–33.
206. Dorminy CA, Choi L, Akohoue SA, et al. Validity of a multisensor armband in estimating 24-h energy expenditure in children. Med Sci Sports Exerc. 2008;40(4):699–706.
207. Fruin ML, Rankin JW. Validity of a multi-sensor armband in estimating rest and exercise energy expenditure. Med Sci Sports Exerc. 2004;36(6):1063–9.
208. Jakicic JM, Marcus M, Gallagher KI, et al. Evaluation of the Sense-Wear Pro arm band to assess energy expenditure during exercise. Med Sci Sports Exerc. 2004;36(5):897–904.

Chapter 3
Outputs Available from Objective Monitors

Catrine Tudor-Locke

Abstract Physical activity epidemiologists are focused upon the prevalence and patterns of physical activity and sedentary behavior in a population, as well as the causes and effects of these behaviours. Of utmost importance to such research is the careful and precise measurement of physical activity, historically reliant on self-reported behaviours. However, with the introduction, development, and enhancement of various body-worn accelerometer devices, the objective monitoring of physical activity has progressed from the simple recording of step counts obtained via pedometry to the objective assessment of a much broader range of time-stamped free-living behaviors. Most public health physical activity guidelines have been framed in terms of the recommended minimum weekly time to be spent in moderate-to-vigorous physical activity (MVPA), and researchers thus focused initially on translating manufacturers' activity count outputs into a comparable measure. Initial calibration research sought to establish instrument-specific activity count cut-points that were associated with absolutely defined intensity levels (indicated by metabolic equivalents or METs). However, it became increasingly apparent that any established cut-points were sample- and instrument-specific, hampering the generalization of findings and limiting epidemiological studies to an assessment of relative rankings of physical activity within a data set. Potential solutions to this issue advanced to date have considered data obtained from additional physiological sensors (there is little evidence as yet that these yield improved measurements), and the use of increasingly more complex statistical models, focused on additional features of the raw accelerometry signals. Success realized in the laboratory has been somewhat illusive under free-living conditions, and the scale of technical data storage was initially daunting. The steady emergence of a wide variety of consumer wearable monitors has opened up new opportunities for wireless data collection, but it has also complicated the landscape of measurement choices for epidemiologists and other investigators. We may thus need to enter a phase of reflection, followed by a right-sizing of epidemiological

C. Tudor-Locke (✉)
Department of Kinesiology, University of Massachusetts Amherst, Amherst, MA, USA
e-mail: ctudorlocke@umass.edu

R.J. Shephard, C. Tudor-Locke (eds.), *The Objective Monitoring of Physical Activity: Contributions of Accelerometry to Epidemiology, Exercise Science and Rehabilitation*, Springer Series on Epidemiology and Public Health,
DOI 10.1007/978-3-319-29577-0_3

expectations and public health needs, in order to standardize more simple, translatable, and generalizable physical activity data.

3.1 Introduction

This chapter covers some basic measurement principles specifically in relation to the objective monitoring of physical activity. It presents a sampling of the very diverse range of outputs and definitions for some currently available objective monitors, discusses data treatment options including the derivation of additional variables post-data collection, and provides an overview of challenges and future directions in this area of investigation.

3.2 Validity

Epidemiologists and other researchers rely on the validity of their measurement tools to provide a credible foundation as they seek to make evidence-based conclusions. Validity is simply the extent to which a measurement tool, in this case an objective physical activity monitor, measures what it purports to measure accurately. Validity can take different forms (e.g., criterion, construct, or convergent validity), and it can also be studied in different contexts (e.g., laboratory, institutional, and free-living conditions).

The scientific literature is rife with validity studies of the various available objective physical activity monitors; to conduct an epidemiological study without prior evidence for the validity of the preferred objective monitor is thus ill conceived. Although researchers often include evidence of validity for a specific objective monitor in their description of study methods, and they may state that a particular device "has been validated" (as if this is dichotomous yes vs. no outcome), such a statement in itself provides no assurance of the absolute veracity of any measured variable. It is important to remain mindful that validity (and reliability, a related concept) are quite complex concepts and that their determinations are at best estimates, with some estimates better than others.

3.2.1 Criterion Validity

The criterion validity of an objective monitor is assessed relative to a criterion, or "gold standard," if such is available. For example, an obvious criterion for evaluating an objective monitor's step counting capabilities is direct observation (or video recording) of the number of steps taken. This is typically done in a laboratory, while the participants walk a short-distance course or, even more commonly, walk and run at various speeds on a treadmill. In 1996 Bassett et al. [1] published the first study of the validity of different electronic pedometers, establishing the Yamax brand pedometer as an early favorite for researchers. Bassett and his colleagues also pointed out that pedometer accuracy was reduced with slower walking speeds (i.e., when they were subjected to lower force accelerations).

Measurement of the time spent at various intensities of physical activity can also be validated using indirect calorimetry as the criterion as research participants engage in treadmill walking and running. In 1995, Melanson and Freedson [2] used this approach to demonstrate that CSA (an early version of the ActiGraph) accelerometers attached to the ankle, waist, and wrist produced activity counts that correlated closely with energy expenditure (r = 0.66–0.81), relative $\dot{V}O_2$ (r = 0.77–0.89), heart rate (0.80–0.73), and treadmill speed (0.83–0.97). As an individual moved faster, the accelerometer signal became greater, and this related to intensity (and energy expenditure). In the laboratory age, sex, and body mass are constant within a given individual so that the relationship between detected movement and energy expenditure is very clear.

Between individuals and under free-living conditions, estimates of energy expenditure derived from some objective monitors can be compared to a criterion standard of doubly labeled water-determined energy expenditure, an approach used extensively by Klaus Westerterp (Fig. 3.1), and reviewed by Plasqui and Westerterp in 2007 [3]. Since many more brands of objective monitors have subsequently entered the commercial market-place, Plasqui, Bonomi, and Westerterp undertook a further review in 2013 [4]. In the first review, this team of investigators pointed out that the positive correlation observed between accelerometer and doubly labeled water estimates of energy expenditure was largely driven by body mass (not movement). To be clear, energy expenditure estimates derived from accelerometers are based on calculations that include subject information that (typically) includes age, sex, and body mass in addition to the detected movement signal; the exact contribution of the detected movement signal to the final estimate is not quite clear. However, it has become increasingly apparent that body mass and/or fat free mass [5] comprise a large component of this estimate. In their more recent review, this investigative team pointed out that at least so far, "there is little evidence that adding other physiological measures such as heart rate significantly improves the estimation of energy expenditure."

Fig. 3.1 Klaus Westerterp has been a leader in the measurement of energy expenditure using doubly labeled water

3.2.2 Construct Validity

Construct validity is evaluated when the output of an objective monitor is considered in relation to a theoretically-related construct. Since physical activity is theoretically related to body mass index, for example, one would expect the output of an objective monitor to be related in a predictable way to body mass. Put another way, we generally believe that people with a high body mass move less and that people with a lower body mass move more. In 2004 Tudor-Locke et al. [6] published construct validity evidence (the median r-values for reported correlations) for pedometry, based on 29 relevant studies published after 1980; this data is summarized on the left side of Table 3.1. The directions and magnitudes for these correlations are considered appropriate given the expected relationships. Given the relationship between accelerometer-determined energy expenditure and that derived from doubly labeled water, the negative association between pedometer steps/day and body mass index may at first seem perplexing. It is important to emphasize, however, that body mass is the greatest inter-individual factor driving estimates of energy expenditure, and not movement [5], as measured for example as steps/day. Energy expenditures assessed by doubly labeled water in the free-living condition are generally higher in people with a higher body mass index or fat free mass [7], but their objectively monitored movement is known to be less [8]. This important but frequently overlooked nuance is the reason that energy expenditure and physical activity should not be considered synonyms [9], and that measures of energy expenditure should not serve as an absolute gold standard for the evaluation of physical activity measures [5].

Table 3.1 Summary of construct and convergent validity for pedometry, based on correlations with pedometer output

Construct validity		Convergent validity	
	Median r value with pedometer output		Median r value with pedometer output
Age	–0.21	Accelerometer output	0.86
Body mass index	–0.27	Time in observed activity	0.82
Percentage overweight	0.22	Time in observed in activity	–0.44
6-minute walk test	0.69	Different estimates of energy expenditure	0.68
Timed treadmill test	0.41	Self-reported physical activity	0.33
Estimated $\dot{V}O_{2max}$	0.22	Self-reported time spent sitting	–0.38

3.2.3 Convergent Validity

Convergent validity represents the degree to which different measuring tools that claim to measure the same (or similar) attributes actually do so. Although this may seem similar to criterion validity, it is important to remember that the convergence of different outputs indicates agreement, but does not directly convey accuracy explicitly in relation to an accepted criterion. Examining the degree of agreement between the output of different accelerometers is an example of convergent validity. James McClain (Fig. 3.2) and his associates [10] examined the convergent validity of three different low cost objective monitors (Omron HJ-151, New Lifestyles NL-1000, and Walk4Life W4L Pro; ranging in price from US$15 to 49) relative to the ActiGraph GT1M accelerometer under both laboratory and free-living conditions. Although all of the low cost objective monitors compared favorably with the ActiGraph accelerometer, the HJ-151 and the NL-1000 produced similar free-living estimates of MVPA to it.

Tudor-Locke et al. [11] summarized median r values for the convergent validity of pedometry, based on 25 relevant studies published between 1980 and 2002 (the data are presented in on the right side of Table 3.1). Again, the directions and magnitudes of these relationships are considered appropriate and in line with rational thinking. The collective evidence for construct and convergent validity presented in Table 3.1 lends credibility to the utility of even simple objective monitors focused on detecting the movement component of physical activity as accumulated steps.

Fig. 3.2 James McClain demonstrated that low cost objective monitors compared favourably to a more expensive accelerometer

3.3 Reliability

Reliability speaks to the repeatability of outputs measured under similar conditions. Technical variability focuses on the performance of objective monitors relative to mechanically delivered accelerations. Intra-brand reliability focuses on the performance of two or more objective monitors of the same brand, worn concurrently under laboratory or free-living conditions. Test/retest reliability assesses the repeated performance of the same objective monitors over time and is affected in part by the wearer's behavior patterns (i.e., behavioural stability). With the production of new generations of objective monitors by the same company, it is also possible to explore a new type of reliability, focused on the performance stability of a company's technology.

3.3.1 Technical Variability

The technical variability of objective physical activity monitors has typically been assessed using a variety of mechanical apparatus (including vibrating tables [12], shakers [13] and turntables [14]) and statistical approaches that have included intraclass correlation coefficients, coefficients of variation, and percentage agreement. Esliger et al. [15] compared the technical variability of the activity counts/min provided by three different accelerometers (ActiGraph, Actical, and RT3) on a hydraulic shaker. Although the RT3 performed poorly and the Actical performed best overall, the authors remarked that the observed discrepancies in variability (related to acceleration and/or frequencies) of the ActiGraph and the Actical precluded any conclusions regarding which one should be considered a superior instrument. In a follow-up study, Esliger et al. [16] re-examined the technical reliability of the Actical on a hydraulic shaker, but this time focused on its steps/min output. The authors reported that the results were "perfect" ($r = 1.0$).

3.3.2 Intra-brand Reliability

Intra-brand reliability might be described in terms of the reliability of measurements taken from differing body attachment locations. It has been assessed in controlled settings (e.g., laboratories, short distance track walks) and under free-living conditions. In the laboratory, Takacs et al. [17] reported excellent agreement (and accuracy relative to direct observation) in the step counts (intraclass correlation >0.95) obtained from three FitBit One objective monitors worn concurrently (two on either hip, and one in a shirt pocket) while research participants walked at five different treadmill speeds. They also reported that, although inaccurate, the estimates of distance obtained from the three objective monitors were comparable (intraclass correlation >0.95).

Outside of the laboratory, Holbrook et al. [18] examined the intra- and inter-model reliability of two Omron brand pedometer models (HJ-151 and HJ-720ITC). Research participants completed a self-paced 1-mile walk wearing three HJ-151 pedometers (positioned at the right hip, the left hip, and the midback) and three JH-720ITC pedometers (positioned in the right pocket, left pocket, and backpack). The authors reported that all coefficient of variation values were <2.1 %, a clear indication of the reliability of these pedometers worn at various body attachment sites.

McClain et al. [19] reported free-living reliability for two Actigraph 7164 accelerometers worn concurrently on the right and left hips for the waking hours of a 24-hour monitored period. Intraclass correlation coefficients were high (0.97–0.99) for step counts, activity counts, and the time spent in sedentary, light, moderate, vigorous and MVPA intensity physical activity. Only MPVA displayed a statistically significant (2 minutes) difference between the right and left hips.

3.3.3 Test Re-test Reliability

Russ Jago et al. [20] compared step counts obtained from the New Lifestyles Digiwalker SW-200 pedometer, worn by 78 Boy Scouts while they performed two successive bouts of externally paced moderate intensity track walking (10 minutes), fast walking (10 minutes), and running (5 minutes). Correlations between similarly paced bouts of activity were all statistically significant; however, the correlations for running were weaker (0.51–0.77) than those obtained during moderate intensity (0.75–0.89) and fast walking (0.61–0.92), likely due to differences of behaviour between bouts rather than to some mechanical flaw. Regardless of the explanation, the values were considered reproducible.

Gwen Felton et al. [21] reported the reliability of pedometer-determined steps/day in college women over two different weeks (separated by an interval of 12 weeks, representing the beginning and end of the semester). The intraclass correlation between weeks was moderate (0.72), indicating that the women's physical activity was stable across the semester; individual participants also held their rank order, at least to a moderate extent, over time.

As part of the Jackson Heart Study, Robert Newton Jr. (Fig. 3.3) and his associates [22] examined the repeatability of data from a 3-day pedometer monitoring period, assessed on two or three distinct occasions, each separated by approximately 1 month. Intraclass correlations ranged from 0.57 for measurements on two occasions to 0.76 for measurements on three occasions. Additionally, 85 % of those assessed on two occasions and 76 % of those assessed on three occasions either remained in the same steps/day quartile or at most changed their ranking by only one quartile over time.

Fig. 3.3 Robert Newton, Jr. demonstrate the repeatability of pedometer-determined physical activity in the Jackson Heart Study

3.3.4 Inter-generation Reliability

Objective physical activity monitors are all commercial products, and either by plan or happenstance, their features change over time, as new versions and other hardware/firmware/software updates are introduced. However, epidemiologists and other investigators require measurement tools whose performance is stable over time. New studies have thus addressed the reliability of objective monitors over successive commercial generations.

Reid-Larsen et al. [23] documented significant differences of output between an earlier version of the ActiGraph (7164) and more recent generations (GT1M, GT3X, and GT3X+) during both a mechanical evaluation and under free-living conditions. Enabling the low extension frequency filter (to increase sensitivity) in the GT3X appeared to attenuate differences in mean physical activity estimates relative to the earlier version of the apparatus, but it also biased the estimate of moderate intensity physical activity. Cain et al. [24] demonstrated that using the low extension filter improved agreement between the newer generation ActiGraph accelerometers and the 7164 with regards to the time spent at various intensities of physical activity, but neither the default nor the normal filter on the newer devices produced comparable step counts to the older generation instruments.

3.4 Sensitivity and Specificity

Objective physical activity monitors are measuring devices, and they are thus subject to the well-known tension between the sensitivity and the specificity of measurements. This issue exists whether we are discussing how to measure individual steps, activity counts, time spent in sedentary behavior, cut-points appropriate to identifying the time spent in MVPA, or an index of the number of steps/day that is associated with health benefits. To measure any parameter we must have a clear understanding of what should be included in the measurement (true positives, values that are classified correctly as positive) and what should be excluded (true negatives, values that are classified correctly as negative). For example, true positives would be those steps taken while walking that are correctly identified by an objective monitor. False positives would be any steps erroneously detected by an objective monitor while the wearer was travelling in a car and not actually accumulating any ambulatory movement [25]. We convey sensitivity when true positives are expressed as a proportion relative to the sum of both true positive and false negatives. We convey specificity when true negatives are expressed relative to the sum of both true negatives and false positives.

As an objective monitor is modified to increase its sensitivity, there is an inevitable decrease in its specificity, and vice versa. There is no consensus on the perfect trade-off between sensitivity and specificity. Some researchers may prefer an instrument that is rigorous in its determination of a step (or any other objectively monitored parameter) to spur intervention participants to meet a specific goal such as brisk walking, for example. Other investigators working with frail populations may prefer to detect even the lowest force movements, and they are thus more tolerant about accepting non-step movements as steps. To be clear, the two objective monitors described in these examples would likely yield very different real estimates of the ambulatory activity performed.

Researchers can also manipulate sensitivity and specificity for some derived variables (e.g., time in MVPA) during post-processing, at least for some types of objective monitor. For example, applying a lower ActiGraph cut-point when determining the time spent in MVPA would increase measurement sensitivity and decrease its specificity. As a result, more subjects would be counted as meeting public health recommendations for the duration of MVPA-based physical activity. Since earlier generations of the ActiGraph accelerometer were shown to be more sensitive to lower force accelerations than common research grade pedometers [25], it became challenging to interpret the NHANES ActiGraph 7164 accelerometer-based step data in terms relevant to the pedometers which are more likely to be used in clinical and practical applications. Tudor-Locke et al. [26] devised a post-processing data treatment plan to not count (i.e., to censor) any steps detected at a rate of less than 500 activity counts/min. The manufacturers of the ActiGraph accelerometers now also provide two different sensitivity filters to use during post-processing of their GTX generation accelerometers [27]. As noted above, neither filter provides step count estimates that are comparable to those obtained from the earlier 7164 version [24].

3.5 Currently Available Objective Monitors

The range of commercially available objective monitors has grown considerably in recent years. In addition, a number of the earlier devices are no longer produced (e.g., the Caltrac). The Yamax Digiwalker pedometer gained early credibility as a

Table 3.2 A sample of currently available objective monitors, their body attachment site, manufacturer/distributor, and cost per unit

Objective monitor	Body attachment site	Manufacturer/ distributer	Cost per unit (US$)	Validation reference
New Lifestyles Digi-Walker SW-200 pedometer	Waist	New Lifestyles	19.95	[1]
NL-1000 pedometer	Waist	New Lifestyles	54.95	[28]
Fitbit® One™	Waist	Fitbit, Inc.	99.95	[16]
Nike+ FuelBand SE	Wrist	Nike, Inc.	149.00	[29]
UP^{24}	Wrist	Jawbone	149.99	[30]
StepWatch Activity Monitor (SAM)	Ankle	Orthocare Innovations	525.00	[31]
SenseWear Armband	Upper arm	BodyMedia	500.00	[32]
GT3X	Waist/wrist	ActiGraph	225.00	[33]
GENEActiv	Waist/wrist	Activeinsights	225.00	[34]
Actical	Wrist	Respironics	670.00	[16]
ActivPal	Thigh	Pal Technologies	350.00	[35]
Kenz Lifecorder EX	Waist	New Lifestyles	245–300	[36]

"research grade pedometer" and it has been subsequently used in numerous research studies. Similarly, the ActiGraph accelerometer (previously known as the CSA, the MTI, and many subsequent ActiGraph versions) has been extensively researched and is widely considered the most popular accelerometer choice. Although not exhaustive, Table 3.2 presents a sample of the range of validated pedometers and accelerometer-based devices currently available.

3.5.1 Mechanisms of Measurement

The internal mechanisms shaping the measurement of objectively monitored outcomes are unique to each instrument. The traditional pedometer mechanism is a horizontal, spring-suspended lever arm that moves up and down with vertical accelerations at the hip during walking. For example, the Yamax pedometer records a step when it detects a vertical acceleration above the 0.35 g sensitivity threshold [37]. Different pedometer brands have hair springs, coiled springs or other type of mechanical feature that provide the counter-force tension that ultimately governs measurement sensitivity/specificity and determines whether or not a force is sufficient to close an electronic circuit and be counted as a step. Attachment of such devices with a "tilt" off the vertical plane is known to impair function (for example, in the case of abdominally obese individuals [38]) so attention may be required to adjust their attachment to optimize measurement characteristics [39].

Some pedometers, for example the New LifeStyles NL-2000, have a piezo-electric accelerometer mechanism. This is a horizontal cantilevered bean, weighted at one end to compress a piezo-electric crystal with acceleration, ultimately producing a proportional voltage that is then translated into accumulated steps, using an internal proprietary algorithm. This type of device is less sensitive to false negative errors associated with tilt off the vertical plane [40].

The electro-mechanical aspects of the accelerometer mechanism were similar for earlier versions of the popular ActiGraph accelerometer, whereas newer versions have a capacitive microelectromechanical system (i.e., capable of detecting variations in electric charge storage potential) that is sensitive to both static (e.g., gravitational forces while stationary) and dynamic accelerations [41]. The analog electronic charge is filtered, digitized, and full-wave-rectified before being converted to more readily accessible activity count data. Not all accelerometer brands share the same mechanism of measurement either. The Actigraph sensor 7164 used in 2003–2006 NHANES [42] was based on a seismic mass cantilever beam mechanism, whereas the Actical sensor (another common research-grade accelerometer used in Canadian surveillance systems [43]) is surfaced-mounted on a printed circuit board and does not have a seismic mass at one end of its piezo-electric element [43]. It goes without saying that the internal proprietary algorithms shaping the sensitivity/specificity of data outputs are also likely different, although such trade secrets are held closely and unfortunately are not readily available for scrutiny.

3.5.2 Memory and Data Access

Simple pedometers display readily accessed digital data and retain a record of accumulated steps only until they are manually reset. Accelerometer-based pedometers (like the NL-1000) keep a rolling record of daily steps and "active minutes" for 7 days and these can be accessed manually by scrolling through the device's on-board internal memory and digital display. Research-grade accelerometers like the ActiGraph have, over time, extended the on-board memory to 120 days, limited only by battery life (25 days). Data must be downloaded using the manufacturer's software; there is no on-board data display.

3.5.3 Technical Limitations

Objective motion sensing monitors, regardless of whether they are worn at the waist or wrist, do not accurately capture non-ambulatory physical activity, for example, bicycling, strength training and other forms of body conditioning, or carrying loads. They also underestimate the intensity of effort associated with climbing hills or stairs. Unless they are waterproof they cannot be worn during swimming (or other water-based activities), and there is as yet no known way to process such data. Miller et al. [44] demonstrated that non-ambulatory activity accounted for only a small proportion of the physical activity performed by one sample of Australians who were wearing pedometers. They thus argued that the effect on population estimates (i.e., epidemiological applications) was minimal and did not require correction to account for such non-ambulatory physical activity. This issue may be more significant in populations where participation in non-ambulatory physical activity is more prevalent, for example, commuting by bicycle in some European countries. Miller et al. did point out that the effect on individual estimates (i.e., clinical applications) was apparent if the individual self-reported regularly engaging in such activities.

3.5.4 The New Wave of Wearable Devices

Accelerometer-based movement-detecting technology is now much more readily available directly to consumers, embedded in smart-phones and a range of personal devices known commercially as personal trackers, wearable technology, or "wearables."

In particular, a number of mainstream commercial companies are producing personal activity monitors that are based on accelerometer technology but output their data simply, via a readily available digital display. This new wave of technology allows users to upload their recorded data to personalized accounts via the internet or mobile phone application (often wirelessly). This facilitates personal reflection on progress and offers an ability to connect and share data through social media. Such applications typically have at least some degree of built-in automated behavioural support mechanism. The primary target of such devices is the individual consumer who is interested in behavioural change; such monitors have less appeal for epidemiological study, which requires a more generalized sample and no motivational exposure potentially leading to reactivity. It may be possible in the future to work with the companies who hold the very large resource of data collected by users and held in their online accounts, but it is difficult to think that this information would be based on a representative population sample. Regardless of this issue, it is possible that interesting behavioural patterns may indeed emerge, although access to the data would require careful negotiation with corporations and equally careful consideration of potential ethical and privacy considerations.

3.6 Outputs and Their Definitions

3.6.1 Direct Versus Derived Outputs

Direct outputs from accelerometers are typically presented as a summarized digital display, either on-board the instrument or automatically uploaded to a computer interface. No additional data queries are required to obtain such outputs. In contrast, derived outputs are not direct or otherwise readily available, but instead must be generated by post-processing, using manufacturer's software, public-use computer programmes, or researcher-developed specialty programmes. This post-processing may be quite simple (built in software data query) or quite involved, requiring multiple steps and/or programming expertise. Direct outputs can be generally classified as volume (e.g., steps/day, total activity counts/day) or rate (e.g., cadence or steps/min, activity counts/min) indicators. Derived outputs can be generally categorized as time indicators (e.g., time in MVPA, sedentary time, time >100 steps/minute, etc.), peak effort indicators (e.g., peak 30-minute cadence), and event counts (e.g., the number of breaks/transitions in sedentary time). Categorization as direct or derived outputs is not rigid, however. Some objective monitors provide a direct output of time spent in MVPA without any additional post-processing, for example. In addition, some objective monitors readily display estimates of energy expenditure; others produce data that must be processed subsequently to distill such information. The universe of derived outputs is likely to expand, and is only limited by imagination, as researchers move towards using the raw data signals from accelerometers (described more below). A summary of commonly used ActiGraph objective monitoring outputs and their definitions is presented in Table 3.3.

3.6.2 Volume Indicators

3.6.2.1 Steps/Day

The simplest output available from most objective monitors is an accumulated step count. This can be presented as steps/day, if accumulated over an entire day, or as a count relative to some other preferred unit of time, for example, the number of steps

Table 3.3 A summary of ActiGraph objective monitoring outputs and their definitions

Variable	Definitions
	Volume indicators
Steps/day	Sum of steps/min accumulated in a day
Total activity counts/day	Sum of total activity counts/min accumulated in a day
	Rate indicators
Cadence or steps/minute	Count of steps accumulated in a minute [45]
Activity counts/minute	Counts of activity counts accumulated in a minute
	Peak effort indicators
Peak 1-minute cadence	Steps/minute recorded for the highest single minute in a day [46]
Peak 30-minute cadence	Average steps/minute recorded for the 30 highest, but not necessarily consecutive, minutes in a day [46]
Peak 60-minute cadence	Average steps/minute recorded for the 60 highest, but not necessarily consecutive, minutes in a day [47]
	Time indicators (cadence determined)
Non-movement	Total minutes at 0 steps/minute during valid wear time [48]
Incidental movement	Total daily minutes at 1–19 steps/minute [48]
Sporadic movement	[48]
Purposeful steps	Total daily minutes at 40–59 steps/minute [48]
Slow walking	Total daily minutes at 60–79 steps/minute [48]
Medium walking	Total daily minutes at 80–99 steps/minute [48]
Brisk walking	Total daily minutes at 100–119 steps/minute [48]
Faster locomotion	Total daily minutes ≥120 steps/minute [48]
Any movement	Total daily minutes >0 steps/minute [49]
Non-incidental movement	Total daily minutes >19 steps/minute [49]
	Time indicators (activity count determined)
Sedentary time	Total daily minutes <100 activity counts/minute [50]
Low intensity	Total daily minutes at 100–499 activity counts/minute [8]
Light intensity	Total daily minutes at 500–2019 activity counts/minute [26]
Lifestyle intensity	Total daily minutes at 760–2019 activity counts/minute [51]
Moderate intensity	Total daily minutes at 2020–5998 activity counts/minute [42]
Vigorous intensity	Total daily minutes ≥5999 activity counts/minute [42]
Moderate-to-vigorous intensity	Total daily minutes ≥2020 activity counts/minute [42]
Moderate-to-vigorous intensity in 10 min bouts	Total daily minutes ≥2020 activity counts/minute accumulated in modified 10 minute bouts (modified 10 minutes bout defined as 10 or more consecutive minutes ≥2020 activity counts/minute, with allowance for 1–2 minutes <2020 activity counts/minute) [42]
	Event counts
Breaks in sedentary time (transitions/day)	Count of daily occurrences where activity counts rose from <100 activity counts in 1 minute to ≥100 activity counts in the subsequent minute [52]

Table adapted from: Schuna JM, Jr., Johnson WD, Tudor-Locke C. Adult self-reported and objectively monitored physical activity and sedentary behavior: NHANES 2005–2006. Int J Behav Nutr Phys Act 2013; 10(1):126

Fig. 3.4 Kong Chen has provided leadership in understanding the technology underlying today's accelerometer-based objective monitors

accumulated over children's recess period at school [53]. The measurement of accumulated steps has been described as "a simple, raw or pure measure of ambulatory activity" [54]. Although early accelerometers (e.g., Caltrac) did not yield step counts, and early versions of the ActiGraph did not provide step counts as a default option (to conserve battery life), most consumer and research grade objective monitors now do provide at least some estimate of accumulated steps. Hatano [55] reported that in Japan, industrial standards set by the Ministry of Industry and Trading regulate a maximum 3 % miscounting rate during normal walking for manufactured pedometers. Outside of Japan there is no such regulatory body. The accuracy of Japanese pedometers has been noted by others [30].

3.6.2.2 Total Activity Counts/Day

Most accelerometer-based objective monitors present physical activity data in counts per unit of time (or epoch), for example, activity counts/min, based on detected, filtered, amplified, and digitized electronic signals. Single axis accelerometers typically output activity counts from the vertical axis, and therefore are theoretically more sensitivity to ambulatory movements. Tri-axial accelerometers detect outputs in three planes (vertical, antero-posterior and medio-lateral) and their summed activity count data are often presented as "vector magnitudes." As Chen (Fig. 3.4) and Bassett have acknowledged, however, "it is often unclear what an activity count truly means, physically or physiologically" [56]. Further, presenting activity counts relative to different epochs can affect data interpretation; shorter durations permit higher resolution, but reduce physiological value [56]. What is more, activity counts are not translatable across accelerometer devices because the technology underlying their creation is proprietary to each device's manufacturer. Regardless of this problem, the summed total of daily activity counts is considered a condensed single indicator of physical activity frequency, intensity, and duration [57].

3.6.3 *Rate Indicators*

3.6.3.1 Cadence or Steps/Min

Steps/day is widely accepted as a simple indicator of the daily volume of ambulatory physical activity. A common criticism of this simple metric, however, is its inability to convey the intensity of movement. Since cadence and intensity (assessed with indirect calorimetry and expressed as metabolic equivalents or METs) are strongly correlated ($r = 0.94$) in laboratory-based studies [58], Tudor-Locke and Rowe [45] urged researchers to consider the usefulness of capturing and describing free-living ambulatory behavior in terms of cadence, or step accumulation patterns, expressed as steps/min. Accelerometer-based devices possess the time-stamping capability necessary to track cadence under free-living conditions. Replicating Freedson's [59] well-known accelerometer calibration study, Tudor-Locke et al. [60] established the first step-based cut-point of 100 steps/minute demarking time spent in MVPA. Since that time, four additional studies [61–64] have directly measured cadence and verified absolutely-defined moderate intensity activity in adults. All agree that despite inter-individual variation, 100 steps/minute represents a reasonable heuristic value associated with at least moderate intensity walking. This level of consistency and consensus is in stark contrast to the on-going debate surrounding cut-points set using activity counts/min, as described in more detail below.

3.6.3.2 Activity Counts/Min

Early accelerometers provided "black box" outputs such as energy expenditure (e.g., Caltrac), but without an explicit presentation of the underlying algorithm used in making this estimate, it was difficult to extract and focus on the relative contribution of movement. Other instruments (e.g., the CSA, an early generation version of the ActiGraph accelerometer) provided activity count outputs, but as mentioned above these were not truly meaningful without some sort of reference frame to aid in their interpretation. Since public health guidelines have been expressed historically in terms of time spent in MVPA [65, 66], physical activity epidemiologists and other researchers have been very interested in studying the segments of daily time spent in various intensities of physical activity (including MVPA) and, more recently, sedentary behaviour.

Capitalizing on the time-stamped nature of the CSA's data capture, Patty Freedson's team [59] was the first to establish activity counts/min cut-points to categorize time spent in MVPA. This was a laboratory-based study linking their data to METs determined by indirect calorimetry during a range of incremental treadmill speeds from walking to running. Although their particular choice of cut-point has been criticized and other cut-points have since been established for categorizing MVPA and the full range of physical activity and sedentary behaviors, the number "1952" activity counts/min is a well-known cut-point in the epidemiology of objectively monitored physical activity. Cut-points for moderate and vigorous intensity physical activity have also been produced to aid the interpretation of data obtained using the Actical accelerometer [67].

3.6.4 Peak Effort Indicators

The StepWatch Activity Monitor provides a direct output that the manufacturer calls a "peak activity index" [68]. Essentially, this is an average of the steps accumulated in the most active minutes of the monitored day. Andrew Gardner (Fig. 3.5) and his collegaues [69] used the StepWatch and reported patterns of ambulatory activity, including peak activity index, in individuals with and without intermittent claudication. Inspired by this objective monitor's direct output, Tudor-Locke et al. [46] applied the same concept to minute-by-minute ActiGraph data, producing derived outputs generally called "peak cadence indicators." A peak 1-minute cadence represents the steps/minute value for the most active minute of the monitored day [46]. A peak 30-minute cadence is the mean steps/minute for the most active (but not necessarily consecutive) 30 minutes of the monitored day [46] and the peak 60-minute cadence is similarly based on the most active 60 minutes of the monitored day [47]. Peak cadence indicators are shaped not only by maximum effort, but also by the relative persistence of this free-living behavior pattern in a day and thus have been described as indices of "best natural effort."

Fig. 3.5 Andrew Gardner has spearheaded using the StepWatch Activity Monitor patient populations with limited mobility

3.6.5 Time Indicators

3.6.5.1 Time-Stamped Step Accumulation Patterns

Step accumulation patterns have been grouped into incremental cadence bands, including 0 (non-movement during wearing time), 1–19 (incidental movement), 20–39 (sporadic movement), 40–59 (purposeful steps), 60–79 (slow walking), 80–99 (medium walking), 100–119 (brisk walking), and 120+ steps/min (all faster locomotion) [49]. Based on free-living cadences detected by the ActiGraph 7164 accelerometer in 3744 adults ≥20 years of age in the 2005–2006 National Health and Nutrition Examination Survey (NHANES) [70], it appears that U.S. adults spend ≅4.8 hours/day in non-movement during device wearing time, ≅8.7 hours at 1–59 steps/minute, ≅16 minutes/day at cadences of 60–79 steps/minute, ≅8 minutes at 80–99 steps/minute, ≅5 minutes at 100–119 steps/minute, and ≅2 minutes at 120+ steps/minute. John Schuna (Fig. 3.6) and his associates [71] reported changes in step accumulation patterns as a result of sedentary workers participating in a workplace shared treadmill-desk intervention. Using ActiGraph accelerometers, he convincingly demonstrated that the primary apparent change as a result of this intervention was an increase in the time spent at 40–99 steps/minute (i.e., light physical activity).

3.6.5.2 Time-Stamped Activity Count Accumulation Patterns

Efforts to replicate the original ActiGraph accelerometer calibration study conducted by Freedson et al. [59] quickly revealed problems with measurement standardization and a number of alternative cut-points for interpreting the time spent in moderate (and higher) intensity activity appeared in the literature. The 2020 activity count/min cut point used in the descriptive presentation of the 2003–2004 NHANES adult ActiGraph accelerometer data was based on a weighted average of the cut-points established previously from four studies [64]. More recently, Loprinzi et al. [72] catalogued at least 12 unique adult-specific cut-points for moderate intensity. Further, the prevalence of adults meeting public guidelines

Fig. 3.6 John Schuna interpreted time-stamped step accumulation patterns to evaluate the effects of a workplace shared treadmill desk intervention

ranged from 4.7 to 97.5 %, depending on which cut-point was applied! Disagreement between cut-points is also apparent for the Actical accelerometer [67, 73]. Unfortunately, the use of differing cut-points affects conclusions about the time spent in moderate (and higher) intensities of physical activity and it undermines the ability to establish clear dose/response relationships with health indicators (see also Chap. 10). From a descriptive epidemiological point of view, prevalence estimates for participation in MVPA vary widely when differing cut-points are applied to the same data [74].

Cut-points have also been established for interpreting the time spent in sedentary behaviour and the full range of physical activity intensities when using the ActiGraph accelerometer (a sampling is presented in Table 3.3). Unfortunately the same problems are apparent, and the field is divisive in terms of the preferred cut-points for most derived outputs. The inability to establish valid and consistent cut-points for the ActiGraph accelerometer has frustrated researchers, and it has spurred interest in getting away from activity counts and moving towards more promising methods of presenting the data. Regardless of this thorny issue, cut-points continue to be used in epidemiological research, but the onus is on the researcher to select and justify their choices carefully.

3.6.6 *Event Counts*

Healy et al. [52] first presented the possibility of counting sedentary breaks, or the number of daily transitions between sedentary behavior and physical activity of any intensity. Building on the original sedentary behaviour cut-point suggested by Matthews et al. [50], Healy (Fig. 3.7) counted the number of daily occurrences where the ActiGraph activity counts rose from <100 activity counts in 1 minute to ≥ 100 activity counts in the subsequent minute [52].

Fig. 3.7 Genevieve Healy introduced the concept of counting daily breaks in sedentary time

3.6.7 *Energy Expenditure*

Perhaps the "Holy Grail" of objective monitoring is the pursuit of an illusively accurate estimate of energy expenditure due to physical activity. Underlying this pursuit is likely the desire to present physical activity in similar units of energy output to those used in the assessment of dietary intake or energy input.

The energy derived (expressed in kiloJoules) from ingesting a donut will be the same for every individual regardless of their sex, age, or body mass. In contrast, energy expenditure is shaped by sex, age, body mass, and the thermic effect of food in addition to physical activity. Within an individual (intra-individual), energy expenditure increases with physical activity intensity, for example, when a person is walking and running at increasing faster speeds and cadences on a treadmill [58]. Also, within an individual, day-to-day fluctuations in energy expenditure must logically be due to differences in physical activity, because sex, age, and body mass do not waiver. However, between individuals (or inter-individually, as is most relevant to epidemiology), body mass is the greatest predictor of energy expenditure [67]. To be clear, the amount of physical activity required to burn the kiloJoules that come from a given donut will vary widely between individuals, and will be related primarily to their body mass.

3.6.7.1 Laboratory-Based Estimates of Energy Expenditure

The predictive ability of accelerometer-based outputs has been assessed primarily under controlled conditions, using indirect calorimetry (the criterion standard of measurement) during treadmill walking/jogging and/or the performance of simulated activities of free-living (e.g., stair climbing, folding clothes, watching TV, etc.). Early attempts considered both body mass and activity counts/min when predicting metabolic cost (i.e., energy expenditure), using a single regression line [44]. Scott Crouter (Fig. 3.8) improved estimates by considering two regression

Fig. 3.8 Scott Crouter developed a two-regression line approach to estimating energy expenditure from accelerometer output

lines applied discriminately, based on discerned differences in the coefficient of variation of activity counts/min indicative of different types of activity [75, 76]. John Staudenmayer et al. [77] proposed using artificial neural networks (i.e., statistically-based learning systems) to improve the prediction of laboratory-based measures of energy expenditure. These methods of estimating energy expenditure have been challenging to translate successfully to estimates of energy expenditure made outside of the laboratory. However, Kate Lyden and colleagues [78] recently were able to demonstrate improved free-living energy expenditure estimates with two novel machine-learning methods relative to a neural network previously calibrated in the laboratory and direct observation.

3.6.7.2 Free-Living Energy Expenditure

The Caltrac, an early accelerometer-based device (whose direct output was framed in terms of kilocalories) required preprogramming by entering an individual's sex, age, body mass and height into its memory. Unfortunately, attempts to "back out" the proprietary algorithm to get at a more fundamental indicator of movement behaviour were unsuccessful [79] and moreover the device is no longer produced. Many of the current consumer-focused wearable technologies provide the user with a direct display of energy expenditure. However, the process by which these estimates are produced is typically not disclosed and few have been subjected to rigorous validity assessments under free-living conditions.

Johannsen et al. [32] pre-programmed the SenseWear Pro3 armband with each research participant's sex, age, height and body mass for a criterion validity study, and they reported good agreement (68–71 % of variability explained) between the armband and doubly labeled water estimates of energy expenditure. Manufacturers of the armband claim that their device considers detected movement as well as various heat-related sensors to estimate energy expenditures. However, the exact algorithm underlying this claim remains proprietary, and is therefore difficult to evaluate. Tudor-Locke et al. [79] explored the potential for developing a transparent algorithm for predicting free-living energy expenditure from ActiGraph accelerometer ambulatory outputs. Relative to the criterion standard of doubly labeled water, the strongest model included body mass, steps/day and the time spent in incremental cadence bands; it explained 79 % (in men) and 65 % (in women) of the predicted energy expenditure. The published algorithm has yet to be cross-validated.

3.6.8 Distance

Some objective monitors provide estimates of distance traveled (typically in miles). Unless based on a standard one-size-fits-all assumption, the manufacturer typically requires the user to input various pieces of information into the objective monitor's on-board memory, including stride length during normal walking (if an estimate of distance is required) and/or sex and age. An on-board calculator then executes a proprietary algorithm to produce the required output based on the directly counted steps or some other combination of movement-related data. As argued previously [54], this process of manipulating step counts to obtain the desired output is based on a number of assumptions and thus introduces additional layers of possible error. For example, shorter individuals with shorter stride lengths will appear less physically active than taller individuals with longer stride lengths who accumulate the same number of steps/day if an estimate of distance is misinterpreted. Further, multiplying a step count by stride length measured during normal walking assumes that the user walks like a robot much of the day with little variation in stride length whether walking for exercise, shopping, chores, engaging in personal care, or otherwise "milling about"! Although most researchers are well aware of the fallacy of estimating distance from detected movement, distance remains a popular data output offered by manufacturers of commercial devices.

3.6.9 Sedentary Behaviour

With the growing interest in the potential deleterious effects of prolonged and excessive time spent in sedentary behaviour, there became a need to process accelerometer data beyond providing only time-based estimates in MVPA. Chuck Matthews (Fig. 3.9) [50] was the first investigator to present the descriptive epidemiology of sedentary behaviour in the U.S., using the 2003–2004 NHANES ActiGraph accelerometer data. He reported that participants in NHANES spent 54.9 % of their monitored time, or 7.7 hours/day, in sedentary behaviours. Having no previously established cut-point for sedentary behaviour, Matthews defined sedentary behaviour as the time when <100 activity counts/min were detected. Although there continues to be debate about the appropriateness of different cut-points for various intensities of physical activity and most notably for the time spent in MVPA [80], Matthews' cut-point for defining sedentary behaviour continues to be broadly used. Further, although activity counts are notably different

Fig. 3.9 Chuck Matthews led the field in objective monitoring of time spent in sedentary behaviour

Fig. 3.10 Tiago Barreira refined a publicly available algorithm for distinguishing children's nocturnal sleep from day-time physical activity and sedentary patterns

between different accelerometers, this cut-point has also been validated for use with data collected using the Actical monitor [81].

3.6.10 Sleep-Related Outputs

Although sleep actigraphy is also based on accelerometer-detected movement and non-movement patterns, it is only recently that we have witnessed the blurring of edges between objective sleep and physical activity monitoring, with expanded 24-hour data collection protocols [82]. Tiago Barreira (Fig. 3.10) and co-workers [83] have published a validated and transparent (accessible and editable) algorithm to interpret 24-hour ActiGraph data and to distinguish children's nocturnal sleep from day-time physical activity and sedentary patterns. Its appropriateness in other populations has yet to be determined.

3.6.11 Raw Data Signals

A workshop was convened in 2012 to set best practices and future directions for objective monitoring. Freedson et al. [80] recommended to this meeting that researchers move away from cut-point based analyses, given the evidential chaos compromising researchers' ability to achieve a consensus. Interest was growing in developing more complex statistical models that could consider additional features of the raw data accelerometer signal, such as signal distributions and frequency spectra, and to identify activity mode or type in addition to duration and intensity. Although no definitive process has been established, statistical approaches that continue to be explored include artificial neural networks and multivariate adaptive regression splines and trees [84]. Logistically, the storage of such potentially vast population data bases for epidemiological purposes was initially challenging [85], but recent technological advances are reducing this concern.

3.7 Conclusions

A wide range of outputs are available from objective physical activity monitors. Simple pedometers may provide on-board digital displays of accumulated steps, distance, and/or energy expenditure. Some accelerometer-based devices also provide on-board digital displays and/or may provide outputs electronically following direct or wireless data downloads. Some of these accelerometer-based data are raw and require additional dedicated processing and treatment prior to interpretation; others are more readily available, using manufacturers' proprietary software. The science is complicated by the fact that, between different objective monitors, similarly named outputs (e.g., step, time in MVPA, sedentary behavior, etc.) are measured differently. The underlying mechanical and/or electronic configuration of the various objective monitors shape their validity, reliability, sensitivity and specificity and thus the precision of these various outputs. There is no clear universal consensus on how to define most outputs. Simply put, a step counted by one objective monitor is not necessarily the same as a step counted by another one. And the same goes for all other outputs; this is perhaps especially true for the time spent in MVPA. The consideration of additional physiological data in energy estimates complicates, but does not necessarily clarify outputs. Technical storage of immense sets of raw data signals may help future research, but currently they challenge the logistical and financial resources supporting large scale epidemiological studies.

In the pursuit of increasingly complex measurement and technical analysis, there is a potential to overlook reasonable outputs and analyses that may be more readily intuitive, public health-oriented, and therefore directly and immediately applicable to many epidemiological and public health applications. In particular, there is merit in being able to "translate" technological and analytical advances into more accessible terms, especially if public health goals are to be realized. A common metric that could be collected and interpreted across objective instruments, including those used in the laboratory and in the real world, would be invaluable. The ability to confidently interpret common cut-points across instruments would be ideal. Although the trend has been for increasing complex approaches to treating and interpreting data obtained from objective monitors, it may be time to re-think in terms of creating consensus for the measurement and analysis of simple step and cadence counting as more direct and comparable estimates of ambulatory behavior. Setting industry standards for measurement specification (i.e., like Japan) maybe a way to cut through the current chaos of technical and statistical variability.

References

1. Bassett Jr DR, Ainsworth BE, Leggett SR, et al. Accuracy of five electronic pedometers for measuring distance walked. Med Sci Sports Exerc. 1996;28(8):1071–7.
2. Melanson Jr EL, Freedson PS. Validity of the Computer Science and Applications, Inc. (CSA) activity monitor. Med Sci Sports Exerc. 1995;27(6):934–40.
3. Plasqui G, Westerterp KR. Physical activity assessment with accelerometers: an evaluation against doubly labeled water. Obesity. 2007;15(10):2371–9.
4. Plasqui G, Bonomi AG, Westerterp KR. Daily physical activity assessment with accelerometers: new insights and validation studies. Obes Rev. 2013;14(6):451–62.
5. Masse LC, Fulton JE, Watson KL, et al. Influence of body composition on physical activity validation studies using doubly labeled water. J Appl Physiol. 2004;96:1357–64.
6. Tudor-Locke C, Williams JE, Reis JP, et al. Utility of pedometers for assessing physical activity: construct validity. Sports Med. 2004;34(5):281–91.
7. Schulz LO, Schoeller DA. A compilation of total daily energy expenditures and body weights in healthy adults. Am J Clin Nutr. 1994;60:676–81.
8. Tudor-Locke C, Brashear MM, Johnson WD, Katzmarzyk PT. Accelerometer profiles of physical activity and inactivity in normal weight, overweight, and obese U.S. men and women. Int J Behav Nutr Phys Act. 2010;7:60.
9. Lamonte MJ, Ainsworth BE. Quantifying energy expenditure and physical activity in the context of dose response. Med Sci Sports Exerc. 2001;33:S370–8. discussion S419–320.
10. McClain JJ, Hart TL, Getz RS, et al. Convergent validity of 3 low cost motion sensors with the ActiGraph accelerometer. J Phys Act Health. 2010;7(5):662–70.
11. Tudor-Locke C, Williams JE, Reis JP, et al. Utility of pedometers for assessing physical activity: convergent validity. Sports Med. 2002;32(12):795–808.
12. Santos-Lozano A, Marin PJ, Torres-Luque G, et al. Technical variability of the GT3X accelerometer. Med Eng Phys. 2012;34(6):787–90.
13. Esliger DW, Rowlands AV, Hurst TL, et al. Validation of the GENEA accelerometer. Med Sci Sports Exerc. 2011;43(6):1085–93.
14. Metcalf BS, Curnow JS, Evans C, et al. Technical reliability of the CSA activity monitor: The EarlyBird Study. Med Sci Sports Exerc. 2002;34(9):1533–7.
15. Esliger DW, Tremblay MS. Technical reliability assessment of three accelerometer models in a mechanical setup. Med Sci Sports Exerc. 2006;38(12):2173–81.
16. Esliger DW, Probert A, Gorber SC, et al. Validity of the Actical accelerometer step-count function. Med Sci Sports Exerc. 2007;39(7):1200–4.
17. Takacs J, Pollock CL, Guenther JR, et al. Validation of the Fitbit One activity monitor device during treadmill walking. J Sci Med Sport. 2014;17(5):496–500.
18. Holbrook EA, Barreira TV, Kang M. Validity and reliability of Omron pedometers for prescribed and self-paced walking. Med Sci Sports Exerc. 2009;41(3):670–4.
19. McClain JJ, Sisson SB, Tudor-Locke C. Actigraph accelerometer interinstrument reliability during free-living in adults. Med Sci Sports Exerc. 2007;39(9):1509–14.
20. Jago R, Watson K, Baranowski T, et al. Pedometer reliability, validity and daily activity targets among 10- to 15-year-old boys. J Sports Sci. 2006;24(3):241–51.
21. Felton GM, Tudor-Locke C, Burkett L. Reliability of pedometer-determined free-living physical activity data in college women. Res Q Exerc Sport. 2006;77(3):304–8.
22. Newton Jr RL, Hongmei HM, Dubbert PM, et al. Pedometer determined physical activity tracks in African American adults: the Jackson Heart Study. Int J Behav Nutr Phys Act. 2012;9:44.
23. Ried-Larsen M, Brond JC, Brage S, et al. Mechanical and free living comparisons of four generations of the Actigraph activity monitor. Int J Behav Nutr Phys Act. 2012;9:113.
24. Cain KL, Conway TL, Adams MA, et al. Comparison of older and newer generations of ActiGraph accelerometers with the normal filter and the low frequency extension. Int J Behav Nutr Phys Act. 2013;10:51.
25. Le Masurier GC, Tudor-Locke C. Comparison of pedometer and accelerometer accuracy under controlled conditions. Med Sci Sports Exerc. 2003;35(5):867–71.

26. Tudor-Locke C, Johnson WD, Katzmarzyk PT. Accelerometer-determined steps per day in US adults. Med Sci Sports Exerc. 2009;41(7):1384–91.
27. Barreira TV, Tudor-Locke C, Champagne CM, et al. Comparison of GT3X accelerometer and YAMAX pedometer and steps/day in a free-living sample of overweight and obese adults. J Phys Act Health. 2013;10(2):263–70.
28. Crouter SE, Schneider PL, Karabulut M, et al. Validity of 10 electronic pedometers for measuring steps, distance, and energy cost. Med Sci Sports Exerc. 2003;35(8):1455–60.
29. Fortune E, Lugade V, Morrow M, et al. Validity of using tri-axial accelerometers to measure human movement—Part II: step counts at a wide range of gait velocities. Med Eng Phys. 2014;36(6):659–69.
30. Lee JM, Kim Y, Welk GJ. Validity of consumer-based physical activity monitors. Med Sci Sports Exerc. 2014;46(9):1840–8.
31. Karabulut M, Crouter SE, Bassett Jr DR. Comparison of two waist-mounted and two ankle-mounted electronic pedometers. Eur J Appl Physiol. 2005;95(4):335–43.
32. Johannsen DL, Calabro MA, Stewart J, et al. Accuracy of armband monitors for measuring daily energy expenditure in healthy adults. Med Sci Sports Exerc. 2010;42(11):2134–40.
33. Santos-Lozano A, Santin-Medeiros F, Cardon G, et al. Actigraph GT3X: validation and determination of physical activity intensity cut points. Int J Sports Med. 2013;34(11):975–82.
34. Rowlands AV, Rennie K, Kozarski R, et al. Children's physical activity assessed with wrist- and hip-worn accelerometers. Med Sci Sports Exerc. 2014;46(12):2308–16.
35. Ryan CG, Grant PM, Tigbe WW, et al. The validity and reliability of a novel activity monitor as a measure of walking. Br J Sports Med. 2006;40(9):779–84.
36. McClain JJ, Craig CL, Sisson SB, Tudor-Locke C. Comparison of Lifecorder EX and ActiGraph accelerometers under free-living conditions. Appl Physiol Nutr Metab. 2007;32:753–61.
37. Tudor-Locke C, Lutes L. Why do pedometers work? A reflection upon the factors related to successfully increasing physical activity. Sports Med. 2009;39(12):981–93.
38. Abel M. The effect of pedometer tilt angle on pedometer accuracy. Int J Fit. 2008;4(1):51–7.
39. Heesch KC, Dinger MK, McClary KR, et al. Experiences of women in a minimal contact pedometer-based intervention: a qualitative study. Women Health. 2005;41(2):97–116.
40. Crouter SE, Schneider PL, Bassett Jr DR. Spring-levered versus piezo-electric pedometer accuracy in overweight and obese adults. Med Sci Sports Exerc. 2005;37(10):1673–9.
41. John D, Freedson P. ActiGraph and Actical physical activity monitors: a peek under the hood. Med Sci Sports Exerc. 2012;44(1 Suppl 1):S86–9.
42. Troiano RP, Berrigan D, Dodd KW, et al. Physical activity in the United States measured by accelerometer. Med Sci Sports Exerc. 2008;40:181–8.
43. Colley RC, Garriguet D, Janssen K, et al. Physical activity of Canadian adults: accelerometer results from the 2007 to Outputs available from objective monitors. 2009 Canadian Health Measures Survey. Health Rep. 2007;2011(22):1–8.
44. Miller R, Brown W, Tudor-Locke C. But what about swimming and cycling? How to 'count' non-ambulatory activity when using pedometers to assess physical activity. J Phys Act Health. 2006;3(3):257–66.
45. Tudor-Locke C, Rowe DA. Using cadence to study free-living ambulatory behavior. Sports Med. 2012;42(5):381–98.
46. Tudor-Locke C, Brashear MM, Katzmarzyk PT, et al. Peak stepping cadence in free-living adults: 2005–2006 NHANES. J Phys Act Health. 2012;9:1125–9.
47. Barreira TV, Katzmarzyk PT, Johnson WD, et al. Cadence patterns and peak cadence in U.S. children and adolescents: NHANES 2005–2006. Med Sci Sports Exerc. 2012;44 (9):1721–7.
48. Tudor-Locke C, Camhi SM, Leonardi C, et al. Patterns of adults stepping cadence in the 2005–2006 NHANES. Prev Med. 2011;53:178–81.
49. Schuna Jr JM, Johnson WD, Tudor-Locke C. Adult self-reported and objectively monitored physical activity and sedentary behavior: NHANES 2005–2006. Int J Behav Nutr Phys Act. 2013;10(1):126.

50. Matthews CE, Chen KY, Freedson PS, et al. Amount of time spent in sedentary behaviors in the United States, 2003–2004. Am J Epidemiol. 2008;167(7):875–81.
51. Hagstromer M, Troiano RP, Sjostrom M, et al. Levels and patterns of objectively assessed physical activity—a comparison between Sweden and the United States. Am J Epidemiol. 2010;171(10):1055–964.
52. Healy GN, Dunstan DW, Salmon J, et al. Breaks in sedentary time: beneficial associations with metabolic risk. Diabetes Care. 2008;31(4):661–6.
53. Tudor-Locke C, McClain JJ, Hart TL, et al. Expected values for pedometer-determined physical activity in youth. Res Q Exerc Sport. 2009;80(2):164–74.
54. Tudor-Locke C, Myers AM. Methodological considerations for researchers and practitioners using pedometers to measure physical (ambulatory) activity. Res Q Exerc Sport. 2001;72 (1):1–12.
55. Hatano Y. Use of the pedometer for promoting daily walking exercise. J Int Comm Health Phys Educ Recreat. 1993;29:4–8.
56. Chen KY, Bassett Jr DR. The technology of accelerometry-based activity monitors: current and future. Med Sci Sports Exerc. 2005;37(11 Suppl):S490–500.
57. Bassett DR, Troiano RP, McClain JJ, et al. Accelerometer-based physical activity: total volume per day and standardized measures. Med Sci Sports Exerc. 2015;47(4):833–8.
58. Tudor-Locke C, Craig CL, Brown WJ, et al. How many steps/day are enough? For adults. Int J Behav Nutr Phys Act. 2011;8:79.
59. Freedson PS, Melanson E, Sirard J. Calibration of the Computer Science and Applications Inc. accelerometer. Med Sci Sports Exerc. 1998;30(5):777–81.
60. Tudor-Locke C, Sisson SB, Collova T, et al. Pedometer-determined step count guidelines for classifying walking intensity in a young ostensibly healthy population. Can J Appl Physiol. 2005;30(6):666–76.
61. Marshall SJ, Levy SS, Tudor-Locke CE, et al. Translating physical activity recommendations into a pedometer-based step goal: 3000 steps in 30 minutes. Am J Prev Med. 2009;36 (5):410–5.
62. Rowe DA, Welk GJ, Heil DP, et al. Stride rate recommendations for moderate intensity walking. Med Sci Sports Exerc. 2011;43(2):312–8.
63. Beets MW, Agiovlasitis S, Fahs CA, et al. Adjusting step count recommendations for anthropometric variations in leg length. J Sci Med Sport. 2010;13(5):509–12.
64. Abel M, Hannon J, Mullineaux D, et al. Determination of step rate thresholds corresponding to physical activity classifications in adults. J Phys Act Health. 2011;8(1):45–51.
65. U.S. Department of Health and Human Services. Physical activity and health: a report of the Surgeon General. Atlanta, GA: U.S. Department of Health and Human Services, Centers for Disease Control and Prevention, National Center for Chronic Disease Prevention and Health Promotion; 1996.
66. U.S. Department of Health and Human Services. 2008 Physical activity guidelines for Americans: be active, healthy, and happy! Washington, DC: ODPHP Publication No. U0036; 2008
67. Hooker SP, Feeney A, Hutto B, et al. Validation of the Actical activity monitor in middle-aged and older adults. J Phys Act Health. 2011;8(3):372–81.
68. Mudge S, Stott NS. Test–retest reliability of the StepWatch Activity Monitor outputs in individuals with chronic stroke. Clin Rehabil. 2008;22(10–11):871–7.
69. Gardner AW, Montgomery PS, Scott KJ, et al. Patterns of ambulatory activity in subjects with and without intermittent claudication. J Vasc Surg. 2007;46(6):1208–14.
70. Tudor-Locke C, Camhi SM, Leonardi C, et al. Patterns of adults stepping cadence in the 2005–2006 NHANES. Prev Med. 2011;53(3):178–81.
71. Schuna Jr JM, Swift DL, Hendrick CA, et al. Evaluation of a workplace treadmill-desk intervention: a randomized controlled trial. J Occup Environ Med. 2014;56(12):1266–76.
72. Loprinzi PD, Lee H, Cardinal BJ, et al. The relationship of Actigraph accelerometer cut-points for estimating physical activity with selected health outcomes: results from NHANES 2003–06. Res Q Exerc Sport. 2012;83(3):422–30.

73. Colley RC, Tremblay MS. Moderate and vigorous physical activity intensity cut-points for the Actical accelerometer. J Sports Sci. 2011;29(8):783–9.
74. Pedisic Z, Bauman A. Accelerometer-based measures in physical activity surveillance: current practices and issues. Br J Sports Med. 2015;49(4):219–23.
75. Crouter SE, Kuffel E, Haas JD, et al. Refined two-regression model for the ActiGraph accelerometer. Med Sci Sports Exerc. 2010;42(5):1029–37.
76. Crouter SE, Clowers KG, Bassett Jr DR. A novel method for using accelerometer data to predict energy expenditure. J Appl Physiol. 2006;100(4):1324–31.
77. Staudenmayer J, Pober D, Crouter S, et al. An artificial neural network to estimate physical activity energy expenditure and identify physical activity type from an accelerometer. J Appl Physiol. 2009;107(4):1300–7.
78. Lyden K, Keadle SK, Staudenmayer J, et al. A method to estimate free-living active and sedentary behavior from an accelerometer. Med Sci Sports Exerc. 2014;46:386–97.
79. Tudor-Locke C, Martin CK, Brashear MM, et al. Predicting doubly labeled water energy expenditure from ambulatory activity. Appl Physiol Nutr Metab. 2012;37(6):1091–100. doi:10.1139/h2012-097.
80. Freedson P, Bowles HR, Troiano R, et al. Assessment of physical activity using wearable monitors: recommendations for monitor calibration and use in the field. Med Sci Sports Exerc. 2012;44(1 Suppl 1):S1–4.
81. Wong SL, Colley R, Connor Gorber S, et al. Actical accelerometer sedentary activity thresholds for adults. J Phys Act Health. 2011;8(4):587–91.
82. Galland B, Meredith-Jones K, Terrill P, et al. Challenges and emerging technologies within the field of pediatric actigraphy. Front Psychiatry. 2014;5:99.
83. Barreira TV, Schuna Jr JM, Mire EF, et al. Identifying children's nocturnal sleep using 24-h waist accelerometry. Med Sci Sports Exerc. 2015;47(5):937–43.
84. Staudenmayer J, Zhu W, Catellier DJ. Statistical considerations in the analysis of accelerometry-based activity monitor data. Med Sci Sports Exerc. 2012;44(1 Suppl 1):S61–7.
85. Lee IM, Shiroma EJ. Using accelerometers to measure physical activity in large-scale epidemiological studies: issues and challenges. Br J Sports Med. 2014;48(3):197–201.

Chapter 4
Protocols for Data Collection, Management and Treatment

Catrine Tudor-Locke

Abstract Epidemiologists and other researchers must plan study protocols that incorporate objective monitoring of physical activity and/or sedentary behaviours by systematically considering all of the complex logistics associated with data collection, management and treatment. With regard to data collection, instrument choice is a foremost consideration, largely shaped by the researcher's questions and available resources. Instrument-specific features may provide greater analytical capacity, but can also greatly complicate planning and must be accommodated. Data collection decisions also include the duration of monitoring required to establish stable estimates of behaviour, without overburdening participants. Data management concerns include systematic processes for quality control, data cleaning, data organization and storage. Data treatment includes decision rules that further shape the accumulated information, including computation of derived variables (as catalogued in Chap. 3) in anticipation of subsequent data analyses.

4.1 Introduction

Objective monitoring of physical activity and sedentary behaviour is a multi-stage process that includes planning and implementing protocols for data collection, management and treatment. Protocols represent the system of rules that epidemiologists and other researchers follow to ensure data quality and ultimately confidence in the study findings. Typically, a manual of operations is developed recording in detail each stage of the protocol during planning of the study. By way of example, an operations manual for the International Study of Childhood Obesity Lifestyle and Environment (ISCOLE) has been published as a supplementary electronic file [1]. Although there are some recognized standard approaches at

C. Tudor-Locke (✉)
Department of Kinesiology, University of Massachusetts Amherst, Amherst, MA, USA
e-mail: ctudorlocke@umass.edu

R.J. Shephard, C. Tudor-Locke (eds.), *The Objective Monitoring of Physical Activity: Contributions of Accelerometry to Epidemiology, Exercise Science and Rehabilitation*, Springer Series on Epidemiology and Public Health,
DOI 10.1007/978-3-319-29577-0_4

each stage of data collection, management, and treatment, unique research questions ultimately shape the protocols that are used. A fine balance between standardization and creativity must be tolerated if scientific discovery is to be encouraged [2]. This chapter provides a starting point to understand the many complex decisions that must be made during protocol planning and implementation.

4.2 Data Collection Protocols

4.2.1 Choice of Instrument

There are many commercially available objective monitors from which to choose. Of course, the research question driving the study should be the most important consideration underlying the choice of objective monitor. A relative ranking of ambulatory behaviours may only require use of a simple pedometer. In contrast, a need to identify patterns of time accumulated in sporadic bouts of moderate-to-vigorous physical activity (MVPA) throughout the day will require a more powerful accelerometer. And precisely discriminating the time spent sitting may require a combination of accelerometer and inclinometer that is sensitive to changes in an individual's posture.

Additional instrument features and increased capabilities such as an extended memory capacity and flexibility in manipulating data aggregation intervals (i.e., epoch lengths) are associated with higher costs, and they also increase the time and technical expertise necessary to process data. Therefore, the choice of instrument must also be shaped by available personnel, time, and financial resources [3]. Study personnel will require time to train and practice all stages of data collection, management, and treatment with the proposed equipment before any research participants are recruited. Budgets must consider these personnel salaries, as well as planning appropriately for the number of instruments (including replacements) that are needed to complete the study in a timely manner, together with ancillary equipment such as any requisite interfaces, batteries or charging hubs, connector cables, and software, and separately ordered attachment straps or belts. Provision must also be made for appropriate technical support, and service agreements covering any anticipated maintenance or repair fees [4].

Other considerations shaping the choice of objective monitor include comfort, safety, the availability of evidence concerning reliability and validity (collected specifically in the planned study's target population), the overall burden on participants and researchers, and the relative need to compare findings with other previously published studies. Laura Rogers (Fig. 4.1) published a simple decision tree to guide researchers in the selection of an objective monitor appropriate to their unique study design [3]. Ultimately, the choice about which objective monitor to use is critical to the success of a research study and it should be informed by careful and systematic deliberation. It is advisable to compare candidate objective monitors

Fig. 4.1 Laura Rogers has focused on the benefits of objectively measured exercise in cancer survivors

using a study-customized matrix representing all the desired factors, and to augment this with expert consultation when possible [1].

4.2.1.1 Attachment Site

Objective monitors have been tested on almost every possible body attachment site. Traditionally, researchers have attached accelerometers on the trunk at the waist/hip, near the body's centre of mass. The science providing equations to interpret accelerometer output collected at other body attachment sites such as the wrist, is new and currently unverified [5]. Multi-sensor accelerometers collect movement data from a number of different body parts simultaneously. For example, the Mini-Sun IDEEA captures information via a waist-worn microprocessor wired to 5 different body sensors; these are attached to the bottom of the feet, the thighs, and the chest, identifying and recording 32 different postures and gaits [6]. Gait parameters that are typically assessed in the laboratory under contrived conditions could be captured reliably under "real-world" conditions, using this technology [7]. Despite the powerful information capture system, the device could be considered cumbersome for most free-living studies and unfortunately is no longer commercially available.

The ActivPal device is adhered directly to the anterior surface of the thigh, using specially prepared adhesive strips. Its placement is intended to be sensitive to ambulation, while also discriminating between sitting/lying and upright postures. When the internal inclinometer detects the adoption of a horizontal plane, then a period of sitting/lying is identified. Unfortunately, the ActivPal cannot discriminate between sitting and lying [6]. When the internal inclinometer detects adoption of a vertical plane, then an upright posture is identified. The internal accelerometer detects movement and the combined accelerometer and inclinometer signals are used to separate standing from ambulation when a person is upright. Although the

ActivPal has been used successfully in studies of free-living behaviour [8], researchers have noted that the thigh attachment may irritate the skin and is poorly tolerated by some wearers [9]. Some researchers have reported supplementing the manufacturer's adhesive attachment system with externally applied bandages [10].

The two most popular accelerometer attachment sites are currently the waist and the wrist. Waist attachments are achieved by using either a manufacturer's clip or an accessory belt/pouch system. Built-in clips are popular, because most clothing has a waist band, and small clipped devices can be easily covered with clothing if so preferred by the wearer. The requirement for an elasticized accessory belt/pouch system has made the ActiGraph accelerometer less acceptable for some study participants, as it occasionally interferes with fashion preferences [11].

Wrist actigraphy is popular with sleep researchers [12]. A number of commercially available wrist-worn personal fitness tracking devices have also emerged recently, and some of these offer sleep-related outputs. The most recent roll-out of accelerometer monitoring in the U.S. National Health and Nutrition Examination Survey (NHANES) has shifted to a wrist-worn ActiGraph [13], based primarily on a desire to reduce a threat to validity resulting from concerns over optimizing wear-time compliance [14]. Original data collection protocols required that study participants only wear accelerometers during waking hours, removing them upon retiring for bed. Large between-subject differences in collected data suggested non-compliance with wearing regimens, largely attributed to premature removal at the end of the day [15]. It was believed that the wrist attachment might be more convenient and/comfortable for research participants [16]. The assumption was made that this would translate to increased wear time compliance. However, with the shift to a waist attachment site, NHANES also simultaneously increased their wear time regimen from waking hours to 24 hours [17]. Since both methods have changed simultaneously, it is difficult to attribute any evidence of increased compliance to just the change in attachment site. Differences in wear time compliance could be just as easily attributed to a simple change in wearing time instructions, increasing from waking hours only to 24-hour continuous wear. Aoyagi et al. [18] have collected 24-hour waist-worn pedometer/accelerometer data in a sample of older adults for $>$12 years!

As a challenge to this assumption, Tudor-Locke et al. [1] compared waking hours wear-time derived from 24-hour waist-worn ActiGraph accelerometer data collected in 10 year old children participating in the U.S. site of the ISCOLE study to waist-worn accelerometer data collected using a waking hours regimen in similarly aged children participating in the 2003–2006 U.S. NHANES. Since both protocols were based on waist-worn accelerometry, the fundamental difference between these two protocols was the difference in wearing time instructions. Wear-time averaged 1357.0 ± 4.2 minutes per 24-hour day in ISCOLE. Since the NHANES protocol requested nighttime removal of the accelerometer, comparable data were not available. When expressed just as wear time during waking hours, ISCOLE children accumulated 884.4 ± 2.2 minutes/day and U.S. NHANES children accumulated 822.6 ± 4.3 minutes/day, a significant difference of 61.8 minutes/day in favor of the 24-hour regimen.

There is evidence indicating that waist-worn devices produce similar sleep outputs to wrist-worn devices [19], and are superior in terms of recording the time spent in MVPA [13], energy expenditure [13, 20], sedentary time [13] and steps/day [21]. For example, Tudor-Locke et al. [21] recently demonstrated that when evaluated against the criterion standard of observed steps taken on a treadmill, a waist-worn ActiGraph accelerometer performed better than the same brand of accelerometer attached to the wrist. Further, they demonstrated that the wrist-worn accelerometer *under-counted* steps on the treadmill and *over-counted* steps in "real life" when compared to the output of the waist-worn accelerometer. The take-home message from this study was that accelerometers worn at the wrist do not indicate the same number of steps/day as those worn at the waist, so using one to track or translate messages based on data obtained from the other is fallible. Therefore, researchers need to keep body attachment sites in mind when planning their data collection protocols.

4.2.1.2 Metric Choice

In the early stages of protocol planning, investigators must clearly identify and define which specific metrics of physical activity and/or sedentary behaviour they wish to collect with their chosen objective monitor. Once again, the choice is shaped primarily by the research question, but *a priori* decisions based on sound justifications will streamline data management and treatment decisions later on in the process. The range of potential outputs obtained from objective monitors is catalogued and discussed thoroughly in Chap. 3.

4.2.1.3 Epoch Choice

An epoch is the term typically used to describe an accelerometer's time-stamped data sampling interval. A 60-second epoch indicates that accelerometer data are recorded every minute. A 15-second epoch samples every 15 seconds. In some devices, sampling intervals are fixed and pre-set by the manufacturer, but others allow manipulation by the researcher upon initialization.

Data processed over longer epochs may obscure details of interest to the researcher [22]. Shorter epochs are better able to capture the uniquely intermittent and sporadic physical activity patterns of children, for example [23]. The choice of accelerometer cut-point may also drive epoch choice, since different cut-points have been set using different epochs [24].

Once selected, it is impossible to re-process the data to reflect shorter epochs. However, it is possible to pre-select a brief epoch (e.g., 1 second), and subsequently to re-integrate the data over longer time segments as may be needed during data treatment. As noted above, some accelerometers do not offer the capacity to edit the epoch length. And some now offer the researcher ultimate flexibility by providing access to the raw accelerometer signal, thus rendering epoch selection moot.

However, it is still not clear what is to be done with such raw accelerometer data. As I-Min Lee [25] has noted: "*the technological capability for data capture using accelerometers has outpaced the current knowledge on how to reduce and process these data*." Of course, data storage requirements are greatly affected by this decision so it is not easily deferred. Lee et al. [25] estimated that storage requirements for the ActiGraph raw data (with no clear analysis plan) for the Women's Health Study will be ~80 terabytes.

4.2.2 Monitoring Frame (How Many and What Type of Days?)

The monitoring frame is the length of time (typically in days) that objectively monitored data are collected to provide a confident estimate of habitual behaviour. Monitoring frames in the scientific literature have ranged from 1 day [26] to several weeks [27] and up to a year [28], and there is still no consensus on the most appropriate monitoring frame for any specific objectively monitored output (Fig. 4.2). The concept of "how many days are enough?" is complicated by the research question, the population under study, the choice of objectively monitored output, the day-to-day variability of the behaviour, the timing of monitoring (e.g., effects of day of the week and season), the choice of statistic for determining and expressing reliability, what, exactly should be considered the criterion standard for an estimate of "habitual" behaviour, and research participants' tolerances of extended monitoring.

A single day of monitoring may be sufficient to address population surveillance needs for a nationally representative estimate of average steps/day. Here the question may be "how many people are enough?" rather than "how many days are enough?" This would provide an index of national (or group level) behaviour, useful for surveillance purposes for example. This is a different purpose than trying

Fig. 4.2 Stewart Trost was one of the first to tackle the "how many days" question in the objective monitoring of childrens' physical activity

to estimate an individual's habitual behaviour related to a health outcome or for intervention purposes. Trends in pedometer-determined steps/day from 1995–2007 have been reported for 6502–9833 Japanese adults sampled each year, based on a single-day protocol implemented by the Japanese Ministry of Health, Labour and Welfare [26]. These data have been used to demonstrate a decreasing trend in steps/day from a peak around 1998–2000. At the other extreme, a much longer-term monitoring frame may be necessary if the research question is focused on the effects of meterological factors (precipitation, temperature, etc.) on activity patterns [29].

Day-to-day variability in behaviour will impact the number of days required to establish a stable estimate. A number of different statistics have been used to quantify and explain this variability. For example, a coefficient of variation (CV) can be calculated as [(SD/mean) × 100] and used to describe the day-to-day variability of objectively monitored outputs. A study of free-living nursing students indicated that the CV calculated over the course of a monitored week averaged 35–36 % for pedometer-determined steps/day [30]. The CV for 4 consecutive days of pedometer monitoring in 6–12 year old children was 23–24 % [31]. Intraclass correlations (ICC) are estimates of the consistency of repeated measurements, for example, days of monitoring. In general, an ICC of 0.70 is considered minimally acceptable and anything over 0.80 provides little additional meaningful value [32]. Fewer days may be required to establish stable estimates of habitual behaviour in notoriously sedentary populations. For example, Rowe et al. [33] reported that between-day variability was high (ICC = 0.90 for Yamax pedometer-determined steps and 0.87 for ActiGraph-determined steps) and 2 days of monitoring were sufficient to obtain a stable estimate of older adults' steps/day. Stewart Trost (Fig. 4.2) and his colleagues [34] used the ICC to determine the minimal number of days of monitoring required to achieve an ICC of 0.80 in children and adolescents relative to the criterion of 1 week. Specifically, Trost et al. demonstrated that children had less day-to-day variability in time spent in MVPA than adolescents. The difference in variability in this specific output indicated that relatively fewer days (i.e., 4–5 days) of accelerometer monitoring were required to estimate children's time in MVPA compared to adolescents who required up to 8–9 days of monitoring.

We may also question not only how many days of monitoring are required, but what types of days. Objectively monitored physical activity outputs are known to vary according to day of the week. Brooke et al. [35] conducted a meta-analyses of accelerometer studies, comparing school-children's mean activity counts/minute and time in MVPA and concluded that, overall, children were more active on weekdays than on weekend days. Any apparent patterns by which weekend days and weekdays differ seem to vary between populations [36] and are likely shaped by other little understood factors like personal schedules, local customs, involvement in sports, sex, age, etc. Seasonal effects are also recognized in objective monitoring [37]. Silva et al. [38] demonstrated that Portuguese boys accumulated less accelerometer-determined time in MVPA (and more time in sedentary behaviours) during the winter months than in the summer. Comparatively, girls were

Fig. 4.3 Cathy Chan studied the effects of weather on steps/day in Prince Edward Island

much more stable in their behaviour across the year. Precipitation reduces step-determined physical activity [29]. Pacific children's accumulated time in MVPA is associated with parent reports of rain-free days [39]. Extreme temperatures (both very cold and very hot) also negatively impact markers of objectively monitored physical activity [29]. Cathy Chan (Fig. 4.3) and her associates [40] calculated that pedometer-determined steps/day decreased by almost 4 % for every 20 cm of accumulated snow on the ground in Prince Edward Island, Canada. Although these effects can be attributed in part to meteorological events, they may also be shaped by cultural customs and schedules. For example, Tudor-Locke et al. [41] demonstrated relatively low steps/day on both winter holidays and summer vacation days (as indicated by days not at work).

The number of monitoring days required have also been shaped by the applied criterion of "habitual behaviour." As indicated above, Trost et al. [34] determined that 4–5 days were predictive of the mean values obtained in a week of monitoring. Togo et al. [42] defined habitual physical activity based on a year of monitoring and determined that at least 25 and 8 consecutive days (for men and women respectively) were required to reach an ICC of 0.80.

As described above, protocol planning requires considerable attention to the optimal monitoring frame for any given study. The best plans will also consider research participants' tolerances for extended monitoring. We know that the longer the monitoring regimen, the greater the extent of non-compliance [43]. The duration of monitoring is thus a delicate balance. The best strategy is to choose, rationalize, and implement the best possible monitoring frame up front, after considering all possible factors, while retaining a separate (but related) plan for what will be minimally acceptable once data are finally prepared for analysis (described in more detail below).

The answer to the deceptively simple query of "how many days are enough" is complicated. In the end, investigators are best advised to consider carefully their specific research question, their study population, and the study setting (e.g., school, workplace, free-living, etc.) in the context of the most relevant scientific literature. Where no prior information is available, a 7-day monitoring frame is sensible for most populations given the widespread understanding and acceptance of a week as a human-scaled unit of time with demonstrated periodicity [44]. For more particular questions, prudence dictates that collection of pilot data is warranted to firmly establish an appropriate monitoring frame.

4.2.2.1 Reactivity

As previously introduced, reactivity is an apparent change in behaviour that can be attributed to a research participant's awareness of being monitored. It is difficult to hide the fact that a participant is wearing an objective monitor, and therefore it is impossible to completely rule out the potential threat of reactivity. In theory, its effect would be to inflate estimates of true behaviour, so a test of this threat is to examine data for such evidence. There is evidence both supporting [45] and refuting [46] reactivity with objective monitoring. It may be that reactivity is more of a concern in smaller samples (where more susceptible individual's behaviour may have a greater statistical impact) and that the threat is minimized in large population type surveys. The Canadian Physical Activity Levels Among Youth (CANPLAY) study, initiated by Cora Craig (Fig. 4.4), is a large on-going pedometer-based surveillance program focused on children and adolescent's step-defined physical activity [47]. Early on in data collection, CANPLAY researchers reported no evidence of reactivity, defined as a significant mean different in steps/day between the first day of data collection and subsequent days monitored during the same week-long time frame.

If reactivity is considered a real concern, minimizing its potential threat to validity starts with a well-devised protocol. It is possible to choose an objective monitor that does not display any data to the wearer. It is also possible to seal the objective monitor to prevent access to any digital data display [48]. Some researchers have extended their monitoring frame to include a planned familiarization period of one or more days to de-sensitize research participants [49]. The

Fig. 4.4 Cora Craig initiated CANPLAY, the largest children's pedometer surveillance program in the world

intent is that these data would not be considered in any subsequent analyses. Alternatively these data can be used to test for reactivity in the data set [47].

4.2.3 Enhancing Compliance

Objective monitoring requires that research participants comply optimally with measurement protocols that require hours, days, and sometimes weeks or even years [50] of continuous monitoring. Missing data can result from reduced wear-time, due to equipment loss or damage, forgetfulness, fatigue, apathy or attrition. Since reduced wear-time and shortened monitoring periods leading to data loss threaten validity, it is important that investigators carefully consider ways to enhance compliance to their measurement protocols and proactively prevent missing data.

Compliance enhancing strategies include plans for participant incentives, regular contact (face-to-face, telephone, electronic media, and/or mail, etc.), continuous data checking in as real time as possible for accuracy and completeness, immediate follow-up contact in the case of questionable or missing data, and/or plans for a second round of monitoring if the first round is inadequate. For example, ISCOLE data collection sites had the option to ask children to wear the accelerometer a second week if during the first week they did not accumulate sufficient wear-time to provide valid data (as determined by on-site data checks), or if the accelerometer had malfunctioned in some way [1].

Other tactics include planning for larger sample sizes and/or longer monitoring frames to maximize the number of research participants who yield minimally acceptable amounts of data. As noted above, this is a balancing act, since we know that more days of monitoring are also associated with increased amounts of data loss [43]. Success is ultimately dependent on a well thought-out and executed protocol. Vigilance at all stages of data collection is needed to ensure data quality proactively.

4.2.4 Collection and Retrieval Logistics

Data collection logistics begin with determining the number of objective monitors needed to complete the planned study, considering losses/damage, staffing, time, and expense.

Time allowances must include the periods required when preparing equipment for data collection. For example, quality control procedures may dictate that some threshold of validity is required before any specific instrument is used in the field. Pedometers have frequently been checked before deployment, using for example simple short distance walking tests [47]. Although accelerometer manufacturers frequently claim that the calibration of their devices is not necessary, a best practice for accelerometry may be to assess each unit's coefficient of variation (CV) while strapped to some type of mechanical shaking device. Good quality instruments demonstrate <3 % error [51].

Batteries will need to be checked and/or replaced for all electronic equipment and some accelerometers require time for charging and possibly initialization. Initialization is a process whereby some objective monitors require input data (e.g., the participant's sex, age, body mass, etc.) to function correctly; there may also be a need to set data collection start and stop dates and times, to select optional measurement features, and to de-select default features that may not be desired and may take up valuable memory storage and/or battery life.

Decisions must also be made about how to distribute objective monitors to research participants, provide instructions to participants on correctly attaching and wearing the devices, and retrieve the instruments when the study is completed. The 2003–2006 NHANES distributed accelerometers and instructed research participants in one-on-one clinical visits and then retrieved them through the postal system [43]. In ISCOLE [1], where Mark Tremblay was the Canadian site's principal investigator (Fig. 4.5), accelerometers were distributed (and retrieved) in school classrooms. Schoolchildren participants were instructed how to wear accelerometers in class and a printed instruction sheet was sent home. At some sites, research staff made regular appearances at the schools to check visually for correct accelerometer attachment [1]. In CANPLAY [47], parents initially contacted by random digit dialing were mailed data collection kits that included instructions on how children were to wear pedometers and record their data on provided logs before mailing them back to the study centre. This was a phenomenonally successful surveillance and on-going system; CANPLAY reported pedometer data collected from 21,271 children and 12,956 adolescents between 2005 and 2011 [52].

Fig. 4.5 Mark Tremblay was Principal Investigator at the Canadian ISCOLE data collection site

4.2.4.1 Data Upload and Transfer

Immediately upon retrieval of an instrument, it is important to check data for completeness and any irregularities. This may be as simple as a quick scan of a paper record, scrolling through an objective monitor's on-board memory, and/or downloading and electronically checking data for completeness evaluated against selected wear-time thresholds. For example, in ISCOLE, study staff retrieved ActiGraph accelerometers from children in the school and then immediately uploaded the data onto laptops prepared with the manufacturer's software to determine whether adequate data were obtained during the designated monitoring time-frame [1]. Adequate data were defined as at least 4 days of data, including 1 weekend day with greater than 10 hours/day of wear time. Based on these immediately processed results, researchers then had the option to ask participants to wear an accelerometer for a second week.

The data uploading process produces a data file that is unique to each brand of objective monitor. For example, depending on settings chosen during initialization, the uploaded ActiGraph file will include data representing three axes of orientation, step counts, lux (an indicator of ambient light), and an inclinometer reading; all of this information is processed according to the researcher-selected epoch and data filter. Paper records of self-recorded data (e.g., step counts collected using a device without electronic transfer capabilities, removal of the device for any purpose, contextual information if desired, etc.) must be double-entered into an electronic data base to reduce data transfer errors. Commercial software products (e.g., Fitabase) are becoming available that facilitate the combining of data obtained from multiple FitBit personal tracking devices, thereby facilitating remote data transfer. Whatever process is implemented, data then need to be backed up and systematically transferred to a secure storage site in preparation for data management.

4.2.4.2 Checklists

Administrative checklists are useful for tracking participants' progress through each stage of data collection, serving to prompt data collectors with protocol reminders, and capturing important logistical details that can be later mined for various purposes. Checklists should be developed for each study, considering each unique aspect of the protocol. At a minimum, checklists should include identification numbers associated with individual research participants, identification numbers assigned to individual objective monitors, distribution and retrieval dates, records of any compliance checks, data quality control checks, and any type of progress notes that may later prove useful to help explain any anomalies that are noted. If used in this manner, administrative checklists can provide valuable paradata, supporting the integrity of data collection.

4.3 Data Management

Data management includes systematic processes for combining individual data files into larger sample/population-level data sets prior to analyses. It also includes running simple descriptive data queries to identify missing and unusual values quickly and to verify minimal wear time characteristics (if available). The magnitude of data management is related to the complexity of the objective monitor's measurement characteristics and the sheer volume of data that is collected.

4.3.1 Automated Quality Control Checks

Automated quality control checks include running data queries to verify that the protocol was implemented as intended. If the protocol required 7-days of monitoring, the data are checked to determine the number of participants who met this criterion, the number who did not, and the number who met some other, *a priori* determined, minimal number of monitored days (e.g., 4 days). Initialization errors are often uncovered during these initial quality control checks. The administrative checklist described above can be useful in resolving any noted discrepancies and other clerical errors during this process. For example, if the checklist indicates that the accelerometer was distributed on a specific date and the accelerometer data suggest that data collection unexpectedly started before or after that date, steps can be taken to maximize best quality data from this particular accelerometer at this stage.

4.3.2 Data Organization and Storage

Data organization is a necessary step in preparing the data for further cleaning and analyses. It is at this time that data files are typically created using the preferred statistical software package (e.g., SAS, STATA, SPSS, etc.) by linkage to some common variable (for example, a participant identification number); variables are renamed for consistency and ease of interpretation, and otherwise systematically organized, backed-up, and stored. Several distinct data sets may be created if accelerometer data need to be processed using different epochs and/or filters. One data organizational step that might be undertaken at this time is to separate or otherwise distinguish accelerometer data collected during any familiarization period from the actual intended monitoring time-frame. Similarly, all data collected beyond the intended monitoring period, inconsistent with the documented retrieval date, should be removed.

4.3.3 Data Cleaning

4.3.3.1 Missing Data

Missing data are common to objective monitoring protocols with long monitoring time-frames, even when preventive compliance strategies have been implemented Missing data are usually due to non-wear, inadequate amounts of wear, and/or equipment malfunction. Deleting participants from a data set is not usually a desirable strategy; however, an *a priori* decision to consider only adequate data (e.g., those with a minimum amount of wear-time over a minimum number of monitored days) may facilitate this decision during data cleaning. The choice may include a decision to tolerate short amounts (e.g., 1–2 h) of non-wear during the monitored day [53]. Alternatively, there are strategies for replacing missing data by individual imputation, both for pedometry [54] and for accelerometry [55]; these improve the statistical characteristics of the data, although it is not yet clear whether the effort of such data manipulations actually alters conclusions about behaviour [56].

4.3.3.2 Identifying and Addressing Outliers

At this stage, descriptive data queries should be run to generate time-based graphic plots displaying estimates of central tendency and distribution (particularly range values) for each data output, in order to identify and remove any unusual values or potential outliers. A common tactic is to eliminate the most extreme data points (i.e., <1st percentile and >99th percentile) at either end of the data continuum. Comparable published data should also be consulted, in order to assess what is plausible for the target population. David Rowe (Fig. 4.6) and colleagues [56]

Fig. 4.6 David Rowe is credited for developing "Rowe's rules" to identifyi outliers in children's daily pedometer data

suggested that for children's pedometer data, recordings of <1000 steps or $>30{,}000$ steps on any single day could be used to identify potential outliers. Evidence of outliers should trigger immediate verification, for example, by consulting the associated participant checklist. The analyst can then decide whether to delete the record or, less drastically (and as executed in the CANPLAY [45]) just truncate the data to the appropriate minimum or maximum value. Questionable records in the 2003–2006 NHANES accelerometer data sets were flagged if they contained >10 minutes with (1) 32,767 activity counts/minute (maximum saturation value for the ActiGraph model used at the time), (2) 0 steps and >250 activity counts/minute, and (3) >200 steps/minute [57].

4.4 Decision Rules for Data Treatment

Decision rules guide analysis of objectively monitored data. Louise Masse (Fig. 4.7) and co-workers [14] elegantly presented the effects that different types and combinations of decision rules have on nearly every outcome variable of interest. Data analysts must consider and implement each decision carefully in order to rationalize their choices and defend their findings.

4.4.1 Wear Time

Wear time is an indicator of monitoring compliance that should be addressed early on in data collection by clearly instructing research participants and implementing protocol features to enhance compliance to either a waking hours and minimal wear time monitoring protocol (e.g., at least 10 hours/day) or an extended wear time (e.g., 24 hours) monitoring protocol. Even with these data collection plans in place, however, it will be necessary to implement data rules regarding acceptable amounts of wear time during the data treatment phase of complex accelerometer-based data.

Fig. 4.7 Louise Masse advanced understanding about how data treatment decision rules affect accelerometer data

The U.S. National Cancer Institute published SAS syntax for investigators treating the NHANES accelerometer data (http://riskfactor.cancer.gov/tools/nhanes_pam), including separating wear time from non-wear time, as identified using their waking hours protocol. Researchers have applied the editable syntax broadly. However, a widespread convention seems to require at least 10 hours of wear-time in a 24 hour period in order to define a valid day of data capture [2]. This is despite the fact that an increase of wear-time beyond 10 hours provides an even better estimate of daily physical activity [58]. Although reduced wear-time can affect all estimates of time spent in the various intensities of physical activity negatively, the most profound effect is on estimates of sedentary time [14], perhaps because of premature removal of accelerometers in the evening before retiring to bed, but still engaging in primarily sedentary behaviours [15]. All this being said, Schmidt et al. [59] demonstrated that manipulating pedometer data to adjust for differences in wear-time did not alter conclusions about how step-defined physical activity related to cardiovascular risk factors.

The implementation of 24-hour protocols reduce the impact of non-wear time on physical activity and sedentary behaviour estimates, while also opening up the opportunity to study the nocturnal sleeping time [60]. However, treating 24-hour data to identify wear-time is complicated by the fact that the analyst must first identify nocturnal sleep patterns before categorizing the remaining data in terms of wear-time vs. non-wear time. Fortunately, Barreira et al. [61] have made their refined algorithm publically available for other researchers to apply to their own 24-hour waist-worn accelerometer data. This algorithm identifies sleep onset and offset from distinct movement and non-movement patterns, and it then organizes the period between these two clock times into sleep and wake episodes, before applying more common definitions of non-wear time to the remaining data collected over the full 24-hour period.

4.4.2 Data Transformation

Following the identification of the non-wear time, accelerometer data signals can be scored and used to derive a number of variables, as catalogued in Chap. 3. Many objective monitors actually begin the data transformation process through on-board processing [53], summarizing the raw data signal into some index value, for example an activity count, a step, or a minute of MVPA. Some monitors may also provide initial data processing when the monitored data are uploaded to the manufacturer's proprietary software. There are both advantages (e.g., time and simplicity through automation) and disadvantages (e.g., "black box" measurements using unknown formulae) to these manufacturer-determined data transformation processes. Although there is little opportunity for researchers to reverse on-board processing, they can further transform outputs using available calibration algorithms to get them to a preferred state. For example, algorithms that have been used to transform activity count-based data into estimates of energy expenditure have been based on

Fig. 4.8 John Staudenmayer is a leader in developing statistical considerations for the analysis of accelerometry-based data

linear regression [62] and 2-regression models [63]. Step data have been processed in a similar manner, to derive cadence outputs expressed as steps/minute [64]. With the movement to make raw accelerometer data more widely available, more complex approaches to data transformation have included artificial neural networks [63], Hidden Markov models [65], etc. Data transformation choices are driven to a large extent by the research questions, the chosen objective monitor, the nature of the data signal acquired, and the analytical skills and experience of the research team [66]. John Staudenmayer (Fig. 4.8) is currently one leader in developing statistical considerations for the analysis of accelerometry-based data.

4.5 Conclusions

Investigators must plan and implement data collection, management and treatment of objective monitored data carefully if their study is to be successful and to move science forward. Data handling is a complex process, with many decisions impacting on other decisions. Decisions in data collection, management and treatment are further complicated by the fact that the technology supporting objective monitoring does not stand still. The best strategy is to keep up with the burgeoning relevant scientific literature in order to be able to justify decisions made at each stage of the process.

References

1. Tudor-Locke C, Barreira TV, Schuna JM, et al. Improving wear time compliance with a 24-hour waist-worn accelerometer protocol in the International Study of Childhood Obesity, Lifestyle and the Environment (ISCOLE). Int J Behv Nutr Phys Act. 2015;12:11.

2. Tudor-Locke C, Camhi SM, Troiano RP. A catalog of rules, variables, and definitions applied to accelerometer data in the National Health and Nutrition Examination Survey, 2003–2006. Prev Chronic Dis. 2012;9:E113.
3. Rogers LQ. Objective monitoring of physical activity after a cancer diagnosis: challenges and opportunities for enhancing cancer control. Phys Ther Rev. 2010;15(3):224–37.
4. Trost SG, McIver KL, Pate RR. Conducting accelerometer-based activity assessments in field-based research. Med Sci Sports Exerc. 2005;37(11 Suppl):S531–43.
5. Crouter SE, Flynn JI, Bassett DR. Estimating physical activity in youth using a wrist accelerometer. Med Sci Sports Exerc. 2015;47(5):944–51.
6. Hart TL, McClain JJ, Tudor-Locke C. Controlled and free-living evaluation of objective measures of sedentary and active behaviors. J Phys Act Health. 2011;8(6):848–57.
7. Gorelick ML, Bizzini M, Maffiuletti NA, et al. Test–retest reliability of the IDEEA system in the quantification of step parameters during walking and stair climbing. Clin Physiol Funct Imaging. 2009;29(4):271–6.
8. Kozey-Keadle S, Libertine A, Staudenmayer J, et al. The feasibility of reducing and measuring sedentary time among overweight, non-exercising office workers. J Obes. 2012;2012:282303.
9. Stanton R, Guertler D, Duncan MJ, et al. Validation of a pouch-mounted activPAL3 accelerometer. Gait Posture. 2014;40(4):688–93.
10. Harrington DM, Dowd KP, Tudor-Locke C, et al. A steps/minute value for moderate intensity physical activity in adolescent females. Pediatr Exerc Sci. 2012;24:399–408.
11. Audrey S, Bell S, Hughes R, et al. Adolescent perspectives on wearing accelerometers to measure physical activity in population-based trials. Eur J Public Health. 2013;23(3):475–80.
12. Meltzer LJ, Montgomery-Downs HE, Insana SP, et al. Use of actigraphy for assessment in pediatric sleep research. Sleep Med Rev. 2012;16(5):463–75.
13. Rosenberger ME, Haskell WL, Albinali F, et al. Estimating activity and sedentary behavior from an accelerometer on the hip or wrist. Med Sci Sports Exerc. 2013;45(5):964–75.
14. Masse LC, Fuemmeler BF, Anderson CB, et al. Accelerometer data reduction: a comparison of four reduction algorithms on select outcome variables. Med Sci Sports Exerc. 2005;37 (11 Suppl):S544–54.
15. Tudor-Locke C, Johnson WD, Katzmarzyk PT. U.S. population profile of time-stamped accelerometer outputs: impact of wear time. J Phys Act Health. 2011;8:693–8.
16. Choi L, Ward SC, Schnelle JF, et al. Assessment of wear/nonwear time classification algorithms for triaxial accelerometer. Med Sci Sports Exerc. 2012;44(10):2009–16.
17. Troiano RP, McClain JJ, Brychta RJ, et al. Evolution of accelerometer methods for physical activity research. Br J Sports Med. 2014;48(13):1019–23.
18. Aoyagi Y, Shephard RJ. Sex differences in relationship between habitual physical activity and health in the elderly: practical implications for epidemiologists based on pedometer/accelerometer data from the Nakanojo Study. Arch Gerontol Geriatr. 2013;56:327–38.
19. Kinder JR, Lee KA, Thompson H, et al. Validation of a hip-worn accelerometer in measuring sleep time in children. J Pediatr Nurs. 2012;27(2):127–33.
20. Swartz AM, Strath SJ, Bassett DR, et al. Estimation of energy expenditure using CSA accelerometers at hip and wrist sites. Med Sci Sports Exerc. 2000;32(9 Suppl):S450–6.
21. Tudor-Locke C, Barreira TV, Schuna Jr JM. Comparison of step outputs for waist and wrist accelerometer attachment sites. Med Sci Sports Exerc. 2015;47(4):839–42.
22. Nilsson A, Ekelund U, Yngve A, et al. Assessing physical activity among children with accelerometers using different time sampling intervals and placements. Pediatr Exerc Sci. 2002;14(1):87–96.
23. McClain JJ, Abraham TL, Brusseau TA, et al. Epoch length and accelerometer outputs in children: comparison to direct observation. Med Sci Sports Exerc. 2008;40(12):2080–7.
24. Evenson KR, Catellier DJ, Gill K, et al. Calibration of two objective measures of physical activity for children. J Sports Sci. 2008;26(14):1557–65.
25. Lee IM, Shiroma EJ. Using accelerometers to measure physical activity in large-scale epidemiological studies: issues and challenges. Br J Sports Med. 2014;48(3):197–201.

26. Inoue S, Ohya Y, Tudor-Locke C, et al. Time trends for step-determined physical activity among Japanese adults. Med Sci Sports Exerc. 2011;43(10):1913–9.
27. Matthews CE, Ainsworth BE, Thompson RW, et al. Sources of variance in daily physical activity levels as measured by an accelerometer. Med Sci Sports Exerc. 2002;34(8):1376–81.
28. Yoshiuchi K, Nakahara R, Kumano H, et al. Yearlong physical activity and depressive symptoms in older Japanese adults: cross-sectional data from the Nakanojo study. Am J Geriatr Psychiatry. 2006;14(7):621–4.
29. Togo F, Watanabe E, Park H, et al. Meteorology and the physical activity of the elderly: the Nakanojo Study. Int J Biometeorol. 2005;50(2):83–9.
30. Felton GM, Tudor-Locke C, Burkett L. Reliability of pedometer-determined free-living physical activity data in college women. Res Q Exerc Sport. 2006;77(3):304–8.
31. Vincent SD, Pangrazi RP. An examination of the activity patterns of elementary school children. Pediatr Exerc Sci. 2002;14:432–41.
32. Nunnally JC, Bernstein IH. Psychometric theory. New York: McGraw-Hill; 1994.
33. Rowe DA, Kemble CD, Robinson TS, et al. Daily walking in older adults: day-to-day variability and criterion-referenced validity of total daily step counts. J Phys Act Health. 2007;4(4):434–46.
34. Trost SG, Pate RR, Freedson PS, et al. Using objective physical activity measures with youth: how many days of monitoring are needed? Med Sci Sports Exerc. 2000;32(2):426–31.
35. Brooke HL, Corder K, Atkin AJ, et al. A systematic literature review with meta-analyses of within- and between-day differences in objectively measured physical activity in school-aged children. Sports Med. 2014;44(10):1427–38.
36. Tudor-Locke C, McClain JJ, Hart TL, et al. Expected values for pedometer-determined physical activity in youth. Res Q Exerc Sport. 2009;80(2):164–74.
37. Dasgupta K, Joseph L, Pilote L, et al. Daily steps are low year-round and dip lower in fall/winter: findings from a longitudinal diabetes cohort. Cardiovasc Diabetol. 2010;9:81.
38. Silva P, Santos R, Welk G, et al. Seasonal differences in physical activity and sedentary patterns: the relevance of the PA context. J Sports Sci Med. 2011;10(1):66–72.
39. Oliver M, Schluter PJ, Schofield GM, et al. Factors related to accelerometer-derived physical activity in Pacific children aged 6 years. Asia Pac J Public Health. 2011;23(1):44–56.
40. Chan CB, Ryan DA, Tudor-Locke C. Relationship between objective measures of physical activity and weather: a longitudinal study. Int J Behav Nutr Phys Act. 2006;3:21.
41. Tudor-Locke C, Bassett DR, Swartz AM, et al. A preliminary study of one year of pedometer self-monitoring. Behav Med. 2004;28(3):158–62.
42. Togo F, Watanabe E, Park H, et al. How many days of pedometer use predict the annual activity of the elderly reliably? Med Sci Sports Exerc. 2008;40(6):1058–64.
43. Troiano RP, Berrigan D, Dodd KW, et al. Physical activity in the United States measured by accelerometer. Med Sci Sports Exerc. 2008;40(1):181–8.
44. Rowlands AV, Gomersall SR, Tudor-Locke C, et al. Introducing novel approaches for examining the variability of individuals' physical activity. J Sports Sci. 2014;21:1–10.
45. Clemes SA, Deans NK. Presence and duration of reactivity to pedometers in adults. Med Sci Sports Exerc. 2012;44(6):1097–101.
46. Matevey C, Rogers LQ, Dawson B, et al. Lack of reactivity during pedometer self-monitoring in adults. Meas Phys Educ Exerc Sci. 2006;10(1):1–11.
47. Craig CL, Tudor-Locke C, Cragg S, et al. Process and treatment of pedometer data collection for youth: the CANPLAY study. Med Sci Sports Exerc. 2010;42(3):430–5.
48. Raustorp A, Ekroth Y. Eight-year secular trends of pedometer-determined physical activity in young Swedish adolescents. J Phys Act Health. 2010;7(3):369–74.
49. Dossegger A, Ruch N, Jimmy G, et al. Reactivity to accelerometer measurement of children and adolescents. Med Sci Sports Exerc. 2014;46(6):1140–6.
50. Yasunaga A, Togo F, Watanabe E, et al. Yearlong physical activity and health-related quality of life in older Japanese adults: the Nakanojo Study. J Phys Act Health. 2006;14(3):288–301.
51. Ward DS, Evenson KR, Vaughn A, et al. Accelerometer use in physical activity: best practices and research recommendations. Med Sci Sports Exerc. 2005;37(11 Suppl):S582–8.

52. Craig CL, Cameron C, Tudor-Locke C. CANPLAY pedometer normative reference data for 21,271 children and 12,956 adolescents. Med Sci Sports Exerc. 2013;45(1):123–9.
53. Heil DP, Brage S, Rothney MP. Modeling physical activity outcomes from wearable monitors. Med Sci Sports Exerc. 2012;44(1 Suppl 1):S50–60.
54. Kang M, Rowe DA, Barreira TV, et al. Individual information-centered approach for handling physical activity missing data. Res Q Exerc Sport. 2009;80(2):131–7.
55. Catellier DJ, Hannan PJ, Murray DM, et al. Imputation of missing data when measuring physical activity by accelerometry. Med Sci Sports Exerc. 2005;37(11 Suppl):S555–62.
56. Rowe DA, Mahar MT, Raedeke TD, et al. Measuring physical activity in children with pedometers: reliability, reactivity, and replacement of missing data. Pediatr Exerc Sci. 2004;16:343–54.
57. Schuna Jr JM, Johnson WD, Tudor-Locke C. Adult self-reported and objectively monitored physical activity and sedentary behavior: NHANES 2005–2006. Int J Behav Nutr Phys Act. 2013;10(1):126.
58. Herrmann SD, Barreira TV, Kang M, et al. How many hours are enough? Accelerometer wear time may provide bias in daily activity estimates. J Phys Act Health. 2012;10(5):742–9.
59. Schmidt MD, Blizzard CL, Venn AJ, et al. Practical considerations when using pedometers to assess physical activity in population studies: lessons from the Burnie Take Heart Study. Res Q Exerc Sport. 2007;78(3):162–70.
60. Tudor-Locke C, Barreira TV, Schuna JM, et al. Fully automated waist-worn accelerometer algorithm for detecting children's sleep period time separate from 24-hour physical activity or sedentary behaviors. Appl Physiol Nutr Metab. 2014;39(1):53–7.
61. Barreira TV, Schuna Jr JM, Mire EF, et al. Identifying children's nocturnal sleep using 24-hour waist accelerometry. Med Sci Sports Exerc. 2015;47(5):937–43.
62. Freedson PS, Melanson E, Sirard J. Calibration of the Computer Science and Applications, Inc. accelerometer. Med Sci Sports Exerc. 1998;30(5):777–81.
63. Crouter SE, Clowers KG, Bassett DR. A novel method for using accelerometer data to predict energy expenditure. J Appl Physiol. 2006;100(4):1324–31.
64. Tudor-Locke C, Sisson SB, Collova T, et al. Pedometer-determined step count guidelines for classifying walking intensity in a young ostensibly healthy population. Can J Appl Physiol. 2005;30(6):666–76.
65. Staudenmayer J, Pober D, Crouter S, et al. An artificial neural network to estimate physical activity energy expenditure and identify physical activity type from an accelerometer. J Appl Physiol. 2009;107(4):1300–7.
66. Pober DM, Staudenmayer J, Raphael C, et al. Development of novel techniques to classify physical activity mode using accelerometers. Med Sci Sports Exerc. 2006;38(9):1626–34.

Chapter 5
Resources for Data Interpretation and Reporting

Catrine Tudor-Locke

Abstract Investigators need to be aware of the growing resources to aid the interpretation and reporting of objectively monitored physical activity and sedentary time. Effective interpretation requires that the researcher considers deeply his or her own data in order to derive its meaning relative to the study design, and also in the context of similar data (in terms of population, instrumentation, metrics, etc.) collected by others, as well as suggesting indices for data classification. This chapter covers expected values and/or normative data currently available for various objectively monitored metrics, including suggested indices, and expected incremental changes with longitudinal tracking or intervention. Available standards, checklists, and flowcharts to support clear, complete, and transparent reporting are also described.

5.1 Introduction

In order to report objectively monitored data effectively, and to interpret (i.e., derive meaning from it), investigators need access to resources that include expected values and/or normative data (including variance estimates) and indices for classification purposes.

5.2 Expected Values

Expected values are comparative estimates of central tendency and variability for targeted outcome variables. Expected values can be distilled from single relevant studies that most closely reflect the target population of interest, but as the literature

C. Tudor-Locke (✉)
Department of Kinesiology, University of Massachusetts Amherst, Amherst, MA, USA
e-mail: ctudorlocke@umass.edu

R.J. Shephard, C. Tudor-Locke (eds.), *The Objective Monitoring of Physical Activity: Contributions of Accelerometry to Epidemiology, Exercise Science and Rehabilitation*, Springer Series on Epidemiology and Public Health,
DOI 10.1007/978-3-319-29577-0_5

base grows, they are more commonly derived from systematic reviews. Such compilations are useful not only for guiding data interpretation and reporting, but also for research and programme planning purposes. A comparison of similarly collected data to expected values can increase confidence in data when it demonstrates consistency, and if inconsistencies must be reconciled, it also helps to illuminate unusual or unexpected findings, thus pushing the envelope of understanding.

The body of research compiling steps/day values is more developed than for any other objectively monitored metric at this time, perhaps because the need for alternatives was neither readily apparent nor urgent even 20 years ago.

5.2.1 Systematic Reviews

The first compilation of expected values for steps/day brought together the then novel and fractured science, reporting free-living pedometer-determined data. Tudor-Locke and Anita Myers (Fig. 5.1) [1] identified 32 unique scientific studies, published between 1980 and 2000, and they noted that we can expect 8–10 years old children to take 12,000–16,000 steps/day (a lower total for girls than for boys); relatively healthy adults aged 20–50 years old take 7000–13,000 steps/day (lower scores for women than for men); healthy older adults take 6000–8500 steps/day; and individuals living with disabilities and chronic illness take no more than 3500–5500 steps/day. Estimates of variance for each of the compiled studies were tabulated (and can therefore be sourced as needed from the cited original publications), but were not summarized in any manner. The authors acknowledged that these preliminary expected values would need to be modified and refined, as the growing interest in the use of objective monitors, and specifically pedometers was just gaining momentum. Many more studies are now published; these shape expected values for steps/day, and a number of reviews have subsequently compiled expected values for specific population groups.

Fig. 5.1 Anita Myers is a Canadian leader in programme evaluation; she helped develop a methodological framework for pedometry, including the first compilation of expected values

Fig. 5.2 Richard Bohannon is a physical therapy researcher who has studied activity measurement issues

5.2.1.1 Adults

Richard Bohannon [2] (Fig. 5.2) compiled results from 42 studies of adults aged <65 years, published between 1983 and 2004. He used a meta-analytic approach to report that expected values for adults ranged widely, from 5400 to 18,000 steps/day (the maximum value being observed in a sample of Amish men with an average age of 34 years) [3]. Excluding the Amish sample, the overall mean value was 9448 steps/day, with a 95 % confidence interval of 8889–9996 steps/day. A later review that considered additional studies expanded the range for expected values for adults to 4000–18,000 steps/day [4].

5.2.1.2 Children and Adolescents

The first compilation [1] of expected values for pedometer-determined steps/day in children identified only a single study published after 1980; it focused on a free-living population. In that study Rowlands et al. [5] reported that 8–10 year old children took an average of between 12,000 (girls) and 16,000 (boys) steps/day. Tudor-Locke et al. [6] subsequently summarized 31 additional research studies, published between 1999 and 2007. Considering both weekdays and weekend days, boys averaged 12,000–16,000 steps/day and girls 10,000–16,000 steps/day. In the same review, expected values for physical education classes, recess, lunch breaks, and after school activities were published. In summary, boys accumulated 42–49 % and girls accumulated 41–47 % of their total daily steps during school hours. Physical education accounted for 9–24 % of daily steps in boys and 11–17 % in girls. Recess accounted for 8–11 % and lunch-time for 15–16 %; after school activities represented 47–56 % of daily steps for boys and 47–59 % for girls. Sex and age associated patterns of physical activity were represented graphically. The most notable patterns were that, with few exceptions, boys took more steps/day than girls between ages 6–19 years; that for both sexes the youngest age groups tended to

Fig. 5.3 Michael Beets published a 13 country review of children's pedometer-determined daily step counts

take fewer steps/day than those immediately older; and, that both boys and girls experienced a decline in steps/day throughout adolescence, but that the slope of the decline was much steeper for boys (on account of their comparatively higher peak values). In 2010 Michael Beets (Fig. 5.3) and associates [7] published a 13 country review, covering 43 reports on children's pedometer-determined steps/day, published up to 2009. They were able to confirm further the clear patterning of data by age, sex, and day of the week, and noted additional patterns characteristic of geographic regions.

5.2.1.3 Older Adults

Based on four studies identified in the original compilation of expected values [1], it was estimated that ostensibly healthy adults aged >50 years accumulated between 6000–8500 steps/day. These expectations were updated in a review of 28 studies published between 2001 and 2008 [8]; a broader range, from 2000 to 9000 steps/day, was proposed, and likely reflects the true variability of physical activity behaviors among older adults. Men generally took more steps/day than similarly aged women; steps/day decreased with advancing age, and BMI-defined normal weight individuals took more steps/day than overweight/obese older adults.

5.2.1.4 Special Populations

The first compilation [1] of expected values of pedometer-determined steps/day for special populations located 12 studies; together, these represented individuals with heart and vascular diseases, chronic obstructive pulmonary disease, diabetes mellitus, joint replacement, and other disabilities that included blindness, physical handicaps, and mental retardation. Overall, these studies indicated that adults living with chronic illnesses and disabilities accumulate 3500–5500 steps/day. An updated review covered studies published between 2000–2008; 60 new unique studies were located, providing sufficient data to calculate median steps/day for heart and vascular diseases, chronic obstructive lung disease, diabetes mellitus, renal dialysis, breast cancer survivors, neuromuscular diseases, arthritis, joint replacement, fibromyalgia, disability (including mental retardation/intellectual difficulties), and other special populations. Additionally, expected values were presented separately for waist-worn monitors and the StepWatch; the latter is attached to the ankle and is known to be more sensitive to low force accelerations (and therefore to provide higher estimates of daily step counts) [9]. For waist-worn monitors, the lowest reported median values were for disabled older adults (1214 steps/day), followed by people living with COPD (2237 steps/day). The highest values were reported for individuals with Type 1 diabetes mellitus (8008 steps/day), mental retardation/intellectual disability (7787 steps/day), and HIV (7545 steps/day).

As the literature base has continued to grow, there are now more focused systematic reviews compiling expected values for specific medical conditions. English et al. [10] located 22 studies (15 of which used the StepWatch monitor) measuring free-living physical activity in individuals surviving stroke. Expected values ranged from 1400 to 7500 steps/day in studies using the StepWatch. Naal and Impellizzeri [11] compiled steps/day values from 25 studies of objectively monitored patients undergoing total joint arthroplasty that were published prior to 2008. They also were able to separate pedometer-based data from that obtained using accelerometers (primarily the StepWatch). They provided weighted mean expected values for the total literature base (6721 steps/day, 95 % confidence interval 5744–7698), for pedometers (5023 steps/day, 95 % confidence interval 4305–5741) and for accelerometers (11,250 steps/day, 95 % confidence interval 10,290–12,210).

5.3 Representative National Normative Data

Normative data are a special type of expected values, based upon specific surveillance efforts. Whereas expected values are frequently means estimated by amalgamating the results of a varied and fractured body of unique research studies, each with their distinctive protocols and instrumentation, normative data are produced from a single large and representative population-based study that has implemented a standardized objective monitoring protocol and has used a single type, brand and model of monitoring device. Presentations of normative data for objectively monitored physical activity and sedentary time metrics increase our sense of their meaning by defining estimates of central tendency and detailed distribution characteristics. Although there are examples of city, community, state and other geographically based representative studies, the focus here will be primarily on representative national normative data.

5.3.1 Steps/Day

Normative data collected on U.S. adults indicated that they average between 5100 and 6800 steps/day. This conclusion is based on three published surveillance-type studies. The 2005–2006 NHANES accelerometer data were processed to facilitate data interpretation on a pedometer-based scale [12]. The conclusion was that U.S. adults ≥20 years of age took approximately 6500 steps/day (5800 for women and 7400 for women), not too different from another state-based pedometer-determined surveillance effort conducted in Colorado, which reported a mean of approximately 6800 steps/day (6600 for women and 7000 for men) [13]. Another pedometer-based U.S. surveillance study estimated that U.S. adults took 5100 steps/day (4900 for women and 5300 for men) [14].

Other representative population studies around the world indicate that Japanese people (≥15 years of age) average approximately 6200 (women) and 7300 (men) steps/day [15], Norwegians aged 20–85 years take approximately 8100 (women) and 8000 (men) steps/day [16], Danish adults aged 18–75 years take approximately 8000 (women) and 8500 (men) steps/day [17] Canadians aged 20–79 years of age take approximately 8400 (women) and 9500 (men) steps/day [18], Belgians aged 25–75 years [19] take approximately 9200 (women) and 9900 (men) steps/day, Western Australians aged 18+ years take approximately 9200 (women) and 10,000 (men) steps/day [20], and Swiss adults aged 25–74 years take approximately 8900 (women) and 10,400 (men) steps/day [21].

There are fewer representative population studies documenting children's and adolescents' steps/day. Tudor-Locke et al. [22] reported accelerometer-determined normative data (adjusted to a pedometer-based scale) for U.S. children (6–11 years) and adolescents (12–19 years). Female children averaged 8600–10,300 steps/day and boys averaged 9600–11,700 steps/day. Adolescents, girls averaged 5700–7300 steps/day and boys averaged 7800–10,200 steps/day. Rachel Colley and colleagues (Fig. 5.4) [23] reported accelerometer-determined normative data for 6–19 year olds collected from the 2007–2009 Canadian Health Measures Survey (CHMS). Overall, girls averaged 10,300 steps/day and boys averaged 12,100 steps/day. Craig et al. [24] reported Canadian pedometer-determined normative data collected during the 2005–2011 CANPLAY surveillance study of 21,271 children and 12,956 adolescents. Girls averaged 9200–11,600 steps/day and boys averaged 10,100–13,000 steps/day.

Fig. 5.4 Rachel Colley reported normative accelerometer-determined physical activity data for Canadians

Pushing beyond the simple presentation of mean values, Craig et al. [24] published the most comprehensive set of normative data for children and adolescents currently available, based on the 2005–2011 CANPLAY data. They published sex-specific percentile values, organizing the full data distribution into 5 percentile increments for each single year between 5 and 19 of age. Publication of such detailed normative data is similar to what has been done to track and communicate children's BMI-related growth charts [25]. Craig et al. [24] also provided more practically organized categories of quintile-defined normative data for combined age groups, labelling these in ascending order as: lowest, lower than average, average, higher than average, and highest levels of physical activity. Taking this Canadian team's lead, Barreira et al. [26] have since published similarly comprehensive 5 percentile increments for accelerometer-determined steps/day (including values adjusted to a pedometer-based scaling) for U.S. children and adolescents, and Tudor-Locke et al. [27] have published the older adult data, both based on the 2005–2006 NHANES surveillance public use data set. Such detailed normative data provide important resources to aid future research interpretation and to support planning and policy evaluations.

5.3.2 Total Activity Counts/Day

Detailed age- and sex-specific normative data for total activity counts/day has only been recently published. Dana Wolff-Hughes (Fig. 5.5) and her associates [28] re-purposed the 2003–2006 NHANES accelerometer data to develop age- and sex-specific percentiles, based on 6093 U.S. adults aged ≥20 years. Men in the 50th percentile accumulated 288,240 total activity counts/day; the corresponding value for women was 235,741 total activity counts/day. In a separate publication, Wolff-Hughes et al. [29] also presented detailed normative data culled from the 2003–2006 NHANES accelerometer data collected on U.S. youth aged 6–19 years of age. Boys in the 50th percentile accumulated 234,322 total activity counts/day; the corresponding value for girls and young women was 234,322 total activity

Fig. 5.5 Dana Wolff-Hughes published detailed normative data for total activity counts

counts/day. No other normative data for this accelerometer-based metric have yet been published.

5.3.3 MVPA Time

Troiano et al. [30] was the first to publish accelerometer-determined the duration of moderate-to-vigorous physical activity (MVPA) as observed in a representative national survey. Based on data collected from the 2003–2004 NHANES, he and his colleagues reported that the mean MVPA in U.S. males ranged from 8.7 minutes/ day (in those aged ≥70 years) to 95.4 minutes/day in 6–11 year olds. The corresponding mean figures for U.S. females were 5.4 and 75.2 minutes/day.

Maria Hagstromer (Fig. 5.6) et al. [31] provided normative accelerometer data for Swedish adults aged 18–75 years and compared them directly with age matched and similarly processed 2003–2004 NHANES data. Overall, 18–39 year old Swedish men averaged 41 minutes/day of MVPA and similarly aged U.S. men averaged 42 minutes/day. MVPA for 40–59 year old men averaged 35 (in Sweden) and 31 (in the U.S.) min/day and for those aged 60–75 years the corresponding values were 29 and 15 minutes/day. MVPA for 18–39 year old Swedish women averaged 36 minutes/day compared to 23 minutes/day in similarly aged U.S. women. The country-specific values for 40–59 years old women were 32 vs. 18 minutes, and 23 vs. 10 minutes for 60–75 year olds.

Hansen et al. [16] also reported normative data for daily MVPA in Norwegian adults. Men aged 20–64 years of age averaged 36.5 minutes/day in MVPA and similarly aged women averaged 34.3 minutes/day. Men and women aged 65–85 years of age averaged 30.2 and 25.6 minutes/day in MVPA, respectively.

The studies of Troiano et al. [30], Hagstromer et al. [31], and Hansen et al. [16] all used the ActiGraph accelerometer (the 7164 model in the first two studies, and the GT1M model in the last study) to collect representative national data; they form

Fig. 5.6 Maria Hagstromer published detailed normative data for Swedish adults, comparing her results with those obtained in the U.S.

the bulk of normative accelerometer-based data currently available. Uniquely, Colley et al. [18] reported normative data collected using the ActiCal accelerometer in the Canadian Health Measures Survey (CHMS). MVPA averaged 33 (men) and 24 (women) min/day in 20–39 year old Canadians, 26 and 21 in 40–59 year olds, and 17 and 12 in 60–79 year olds. In a separate publication, Colley et al. [23] also reported normative data for Canadian youth aged 6–19 years. MVPA averaged 69 (boys) and 58 (girls) minutes/day in 6–10 year old students, 59 and 47 in 11–14 year olds, and 53 and 39 in 15–19 year olds. The only sources of representative national normative values for accelerometer-variables beyond steps/day for youth are the 2003–2006 U.S. NHANES data [30] and the 2007–2009 Canadian CHMS data [23].

5.3.4 Mean Daily Activity Counts/Minute

Troiano et al. [30] used the 2003–2004 NHANES accelerometer data to calculate sex- and age-group specific normative data for mean daily activity counts/minute; these investigators divided total activity counts for a valid day by the number of minutes of wear in that day across all valid days. For males, mean daily activity counts/minute ranged from 188.9 in those aged ≥70 years to 646.5 in 6–11 year olds. For females, corresponding mean values ranged from 169.8 to 567.6 activity counts/min. This study provides the only normative data for this specific metric that includes young people as well as adults.

Hagstromer et al. [31] also published normative values for mean daily activity counts/minute for Swedish adults aged 18–75 years old. The means for men were 409 (18–39 years of age), 368 (40–59 years of age), and 321 (60–75 years of age). Corresponding values for Swedish women were 389, 366, and 304. They also organized the U.S. 2003–2004 NHANES data into similar age groupings. Mean daily activity counts/minute for U.S. men in the designated age groups were 436, 365, and 242 and for women they were 329, 294, and 321.

Hansen et al. [16] reported normative sex-specific mean daily activity counts/minute for Norwegian adults aged 20–64 and 65–85 years of age. Mean values for men in these two age groupings were 349 and 305 activity counts/min. Corresponding values for women were 345 and 287 activity counts/min.

5.3.5 Sedentary Time

Matthews et al. [32] used the 2003–2004 U.S. NHANES accelerometer data to report the first representative national objectively monitored normative data for sedentary time. Adults aged 20–85 years of age averaged 7.3–9.3 hours/day of sedentary time, defined as <100 activity counts/minute; this amounted to 50–58 % of the monitored day. Adults aged 70–85 years of age were the most sedentary group.

Fig. 5.7 Bjorge Herman Hansen published normative data on sedentary time for Norwegian adults

Hagstromer et al. [31] provided normative data for Swedish adults aged 18–75 years of age; this was collected in 2001–2002, and was expressed in minutes/day (rather than hours/day). It was compared directly with similarly aged U.S. adults drawn from the 2003–2004 NHANES accelerometer data set. Daily time accumulated in sedentary behaviors averaged 497 (Sweden) vs. 444 (U.S.) minutes/day for 18–39 year old men, 505 vs. 484 minutes/day for 40–59 years old men, and 505 vs. 539 minutes/day for 60–75 year old men. Corresponding values for women were 486 vs. 462, 485 vs. 483, and 485 vs. 501 minutes/day.

Bjorge Herman Hansen (Fig. 5.7) et al. [16] published normative data for sedentary time in Norwegian adults. Overall, 62 % of monitored time was spent being sedentary. Grouped by age, 20–64 year olds averaged 555 (men) and 530 (women) min/day and 65–85 year olds averaged 567 and 545 minutes/day.

As noted above, the U.S., Swedish, and Norwegian studies all used the ActiGraph accelerometer. The Canadian normative data. collected on 20–79 year olds in 2007–2009, were based on the ActiCal accelerometer. Overall, approximately 68–69 % of monitored time in the Canadian adults was classified as sedentary (575 minutes for men and 585 minutes for women).

Objectively monitored and representative national normative data for sedentary time for young people are currently only available from the U.S. (using the ActiGraph) [32] and Canada (using the ActiCal) [23]. Total daily sedentary time for U.S. youth averaged 6.1–8.1 hours/day (6.0–7.9 hours/day for boys and 6.1–8.1 hours/day for girls) [32]. Canadian youth accumulated an average of 8.6 hours (507 minutes/day for boys and 524 minutes/day for girls) of sedentary time [23].

5.3.6 Mean Daily Steps/Minute

Schuna et al. [33] reported representative national normative data for mean daily steps/minute (adjusted to a pedometer scaling), originally defined as the total accelerometer-determined steps accumulated over 1440 minutes, divided by accelerometer wear time [34]. They found that U.S. adults ≥20 years of age accumulated 7.7 mean daily steps/minute (results not reported by sex). Matthiessen et al. [17] reported a similar metric collected using a pedometer among Danish adults aged 18–75 years. Danes accumulated 9.4 mean daily steps/minute (9.5 for men and 9.3 for women).

5.3.7 Peak Cadence Indicators

Cadence-related representative national normative data are currently only available from the 2005–2006 U.S. NHANES data set. Tudor-Locke et al. [34] reported that U.S. adults accumulate ≅4.8 hours/day of zero cadence during wearing time, ≅8.7 hours between 1 and 59 steps/minute, ≅16 minutes/day at cadences of 60–79 steps/minute, ≅8 minutes at 80–99 steps/minute, ≅5 minutes at 100–119 steps/minute, and ≅2 minutes at 120+ steps/minute. They also observed that U.S. adults accumulated ≅30 minutes/day at cadences of 60+ steps/minute. Barreira et al. [35] published similarly categorized cadence-related normative data for children and adolescents. On average, U.S. children and adolescents spent ≅4 hours/day at zero cadence during wearing time, ≅8.9 hours/day between 1 and 59 steps/minute, ≅22 minutes/day at cadences of 60–79 steps/minute, ≅13 minutes/day at 80–99 steps/minute, ≅9 minutes/day at 100–119 steps/minute, and ≅3 minutes/day at cadences ≥120 steps/minute.

Normative data for peak cadence indicators are also available based on these NHANES data [36]. U.S. adults aged ≥20 years old average a peak 30-minute cadence of 71.1 (men: 73.7, women: 69.6) steps/minute and a peak 1-minute cadence of 100.7 (men: 100.9, women: 100.5) steps/minute [36]. Barreira et al. [37] provided peak 1-minute, 30-minute, and 60-minute cadence values for age groupings of 8–11, 12–15, and 16–18 years of age pulled from the U.S. NHANES data set. They reported that peak cadence deceased across ascending age groups from 123 to 111 for peak 1-minute cadence, from 87 to 82 steps/minute for peak 30-minute cadence, and from 74 to 68 steps/minute for peak 60-minute cadence. Barreira et al. [35] have published sex- and age-group specific quintile-defined categories and also more detailed percentiles values for peak 60-minute cadence for children and adolescents.

5.4 Indices

An index is an indicator, signal or measure of something and is frequently used in epidemiology to classify individuals within a population or to screen individuals for clinical applications. Indices based on objectively monitored metrics may be set relative to selected health outcomes, and other behaviors, and/or selected points (either relatively high or relatively low) along the distribution of data.

5.4.1 Graduated Step Index

Tudor-Locke and Bassett [38] introduced the concept of a graduated step index as a means of organizing and classifying echelons of free-living steps/day. They anchored the index at 5000 (a sedentary lifestyle index) and 10,000 (an index of high physical activity) steps/day. They used 2500 step increments to define each echelon and then applied appropriate qualitative descriptors. Specifically, <5000 was indicative of a sedentary lifestyle, 5000–7499 was "low active", 7500–9999 was "somewhat active," 10,000–12,499 was active, and ≥12,500 was "highly active." This graduated step index was confirmed in 2008 [39] and was expanded in 2009 [12] to provide additional separation in the lowest echelon as <2500 steps/day ("basal activity") and 2500–4999 steps/day ("limited activity"). Based on accumulating evidence that 7000–8000 steps/day was an appropriate translation of minimal public health guidelines [4], Tudor-Locke et al. [40] proposed that the ≥7500 steps/day echelon be re-classified as "physically active."

The U.S. 2005–2006 accelerometer data have been adjusted to a pedometer-based scale and analyzed to provide prevalence estimates for the activity of adults, based on the original graduated step index. Specifically, Tudor-Locke et al. [41] reported that 14.1 % of men took <2500 steps/day, 20.6 % took 2500–4999 steps/day, 24.2 % took 5000–7999 steps/day, 19.3 % took 7500–9999 steps/day, 19.3 % took 10,000–12,499 steps/day, and 10.8 % took ≥12,500 steps/day. The corresponding values for women were 14.1, 20.6, 24.2, 19.3, 13.2, and 10.8 %.

A number of studies have been compiled that report the percentages of their samples achieving various combinations of these steps/day thresholds [4]. Notably, representative national accelerometer-based studies (other than the U.S. study, indicated above) report that 22.7 % of Norwegian adults [16] and ≥35 % of Canadian adults [18] accumulated ≥10,000 steps/day. Based on pedometer surveillance, 23.3 % of Japanese men and 16 % of women accumulated ≥10,000 steps/day [15]. In Denmark, 31.7 % of men and 30.5 % of women accumulated ≥10,000 steps/day [17].

A youth-specific graduated step index has been proposed. Building on previous work linking children's BMI and steps/day [42], Tudor-Locke et al. [39], suggested sex-specific indices. For boys they were: (1) <10,000; (2) 10,000–12,499; (3) 12,500–14,999; (4) 15,000–17,499; and (5) ≥17,500 steps/day. The corresponding values for girls were: (1) <7000; (2) 7000–9499; (3) 9500–11,999; (4) 12,000–14,499; and (5) ≥14,500 steps/day. For both sexes, these levels are congruent with the "sedentary," "low active," "somewhat active," "active," and "highly active" labels originally used to define the adult graduated step index. Prevalence estimates have been presented graphically [40] and serve as another resource that can be accessed as needed to guide data interpretation.

5.4.2 Steps/Day Translation of Public Health Guidelines

Public health guidelines have been written and promoted by many different governmental and non-governmental agencies around the world. Typically, these have been written in terms of frequency, intensity and time dimensions, and they are based on a scientific foundation that has been shaped largely by self-reported behaviours. They are used for programme/policy planning and evaluation, as well as for clinical and personal applications. The push to endorse a steps/day translation of public health guidelines was in part a reaction to the growing needs of researchers and the public, who were increasingly using objective monitors both for research and for clinical applications including personal health tracking. Responding to this need, in 2010 the Public Health Agency of Canada (PHAC) commissioned a set of literature reviews designed to identify optimal steps/day target translations of public health guidelines in adults [4], children/adolescents [43], and older adults/special populations [44]. Researchers from around the world participated in the process to review the science, identify gaps and opportunities, and ultimately to achieve a consensus. Based on studies using objective monitors to quantify both steps/day and time in MVPA, 7000–8000 was the public health translation for approximately 30 minutes/day of MVPA in ostensibly healthy adults (aged 20–50 years) [4]. The range was a little broader for healthy older adults (i.e., 7000–10,000 steps/day) and somewhat reduced in special populations living with disability or chronic illness (i.e., 6500–8500 steps/day) [44]. In children (aged 6–11 years), public health guidelines of 60 minutes/day of MVPA translated to 11,000–12,000 steps/day in girls and 13,000–15,000 steps/day for boys [43]. For adolescents (both boys and girls aged 12–19 years), this translated to 10,700–11,700 steps/day [43].

At this time, no governmental agency has adopted any steps/day translation of public health guidelines. However, in 2011, the American College of Sports Medicine wrote "Increasing pedometer step counts by ≥2000 steps/day to reach a daily step count ≥7000 steps/day is beneficial" in an updated Position Stand on the quantity and quality of exercise needed for fitness and health, prepared by Carol Ewing Garber and her colleagues (Fig. 5.8).

5.4.3 Steps/Day Associated with Obesity Indicators

Steps/day indices have been evaluated in relation to obesity indicators including BMI and body fat-based definitions of obesity and these have been previously reviewed [43]. In 2004, Tudor-Locke et al. [42] combined three international (Australia, Sweden, and the U.S.) data sets focused on 6–12 year old children and used a contrasting groups method to identify 12,000 steps/day for girls and 15,000 steps/day for boys as cut-points associated with the separation between normal and overweight/obesity as defined by BMI. Since that time, a number of other children's

Fig. 5.8 Carol Ewing Garber was the lead author on the 2011 ACSM Position Stand on the quantity and quality of exercise needed for fitness and health

studies have been conducted on different populations, using different obesity indicators (e.g., body fat percent), and using different analytical approaches [43]. Overall, the studies of children that have used a contrasting groups technique have produced higher cut-points (e.g., 15,000–16,000 steps/day) than those produced using receiver operating curve (ROC) analyses (e.g., 10,000–13,500 steps/day). However, as has been previously suggested [43], differences may result from the different samples of children studied and the fact that factors other than physical activity shape obesity.

Steps/day indices related to BMI-defined overweight/obesity have also been computed from an amalgamated adult data set (n = 3127, age 18–94 years) formed with the data of collaborating researchers from Australia, Canada, France, Sweden, and the U.S. Best estimated cut-points for normal vs. overweight/obese have ranged from 11,000 to 12,000 steps/day for men and 8000–12,000 steps/day for women (consistently higher for younger age groups in both sexes.) No similar analyses have been conducted in adults at this time, but it may be that the same factors underlying the inability to identify a consistent cut-point in children's research will also emerge with additional analyses of adult data.

5.4.4 Steps/Day and Other Health Outcomes

Ewald et al. [45] have reported detailed dose/response values relating steps/day to a number of health markers derived from a representative community sample of 2458 Australian older adults (aged 55–85 years). They noted linear consistent relationships between steps/day and markers of inflammation in both men and women.

Similar linear relationships were apparent for BMI and high-density lipoprotein in women. For other health markers (e.g., waist-to-hip ratio, fasting blood glucose, depression, and quality of life), the benefit was mostly apparent in the lower half of the steps/day distribution. They tabulated steps/day values matching 60 and 80 % of the range of each health marker as potentially useful indices related to each of the health markers that were studied. Overall, they supported ≥8000 steps/day as an optimal level of activity associated with many health benefits.

Most remarkably, however, is the extensive tracking of Japanese older adults since 2000 as part of the Nakanojo Study that examines year-long objectively monitored metrics in relation to various aspects of health [46]. This study has used a specially adapted uniaxial pedometer/accelerometer (modified Kenz Lifecorder) to collect on a continuous basis steps/day and daily time accumulated ≥3 METs (equivalent to MVPA) for 24 hours/day and over 12 years [46] The multiple studies emerging from this effort were recently updated in tabular form [46]. Dose-response estimates have been continually clarified and refined [47]. Overall, objectively monitored physical activity thresholds associated with better mental and psychosocial health in older adults are 4000–5000 steps/day and/or 5–7.5 minutes/day accumulated ≥3 METs; 7000–8000 steps/day and/or 15–20 minutes/day accumulated ≥3 METs for markers of aortic arteriosclerosis, osteoporosis, sarcopenia, and poor physical fitness,; and 8000–10,000 steps/day and/or 20–30 minutes/day accumulated ≥3 METs for components of the metabolic syndrome, especially hypertension and hyperglycemia.

5.4.5 *Sedentary Lifestyle Index*

The concept of a step-defined sedentary lifestyle index was first proposed to quantify and objectively classify those likely at most risk for hypokinetic diseases [48]. Tudor-Locke et al. [48] cut a sample's steps/day distribution into quartiles and reported that those in the lowest quartile (taking <5000 steps/day) were more likely to be overweight/obese than individuals in the highest quartile. Tudor-Locke and Bassett [4] went on to use this threshold to anchor the lower end of their graduated step index. A recent review [40] expounded upon its use as a sedentary lifestyle index, and included a suggestion to use <7000 steps/day as a sedentary lifestyle index for youth.

5.4.6 *Adherence to Public Health Recommendations for MVPA*

Troiano et al. [30] were the first to report normative data for accelerometer-determined adherence to physical activity recommendations, defined as the proportion of the 2003–2004 NHANES accelerometer sample that accumulated 30 minutes of MVPA (for adults) or 60 minutes (for youth) on most days of the week. The data treatment for adults considered only MVPA that was accumulated in "modified 10-minute bouts" (10 consecutive minutes above the relevant activity counts/minute cut-points, allowing for 1–2 minutes below the cut-point) on 5 or more of the monitored days of the week. Applying this very literal and stringent interpretation, the researchers concluded that <5 % of U.S. adults adhered to public health recommendations. Unrestricted in MVPA accumulation pattern and employing child-specific accelerometer cut-points, the analysis indicated that 48.9 % of boys and 34.7 % of girls adhered to youth-specific public health recommendations.

Hansen et al. [16] used a similar data treatment plan to provide normative data for accelerometer-determined adherence to public health recommendations in Norwegian adults. Overall, 20.4 % of the study sample met the accelerometer-based definition.

As mentioned above, Colley et al. [18] analyzed Canadian normative data from the CHMS; they used the ActiCal accelerometer, in contrast to the ActiGraph used in the U.S. and Norwegian studies. They also applied a number of different approaches to classifying adherence to public health recommendations, including determining the proportion accumulating 150 minutes/week of MVPA in modified 10-minute bouts (defined similarly to the NHANES analyses) and patterns of accumulation. On average, approximately 15 % of Canadian adults achieved 150 minutes/week of MVPA and approximately 5 % managed this by accumulating ≥30 minutes/day on ≥6 days/week. In a separate publication, Colley et al. [23], reported that approximately 9 % of boys and 4 % of girls achieved public health recommendations to accumulate 60 minutes/day of MVPA on ≥6 days/week.

It is important to note here than there are known and dramatic differences in population prevalence of meeting public health physical activity guidelines based on accelerometry vs. self-reported behaviour [49]. Physical activity guidelines have been formed from decades of epidemiological studies based primarily on measured health outcomes linked to self-reported behaviours. Troiano et al. [50] have recently suggested that evaluating adherence to physical activity guidelines as derived from accelerometry is actually inappropriate, because the two measurement approaches (accelerometry vs. self-reported behaviour) differ greatly both in metric and conceptualization. As more objectively monitored data is collected relative to health outcomes, it may eventually come to pass that consistent and rigorous objectively monitored metrics will inform health-related guidelines and be used to track adherence. A good example of this potential is apparent from the collective research produced from the Nakanojo Study as described above [46].

5.5 Expected Change

Investigators need resources to interpret meaning from observed changes in objectively monitored data, for example, with age/time and intervention. Sources for this information include point cross-sectional studies of age, panel cross-sectional surveillance of trends, longitudinal studies, and systematic reviews of intervention studies.

5.5.1 *Cross-Sectional Studies of Age*

As noted above, two reviews [6, 7] have compiled steps/day values of children and adolescents by age. Both have confirmed a generalized pattern, whereby the youngest (5–6 year olds) accumulate relatively fewer steps/day than those immediately older, up to a peak around 9–10 years of age; then there is a steady decline into and through adolescence. As a large and representative national study, the CANPLAY data have displayed this type of pattern among Canadian youth [51]. In contrast to these established and robust age-associated patterns, the U.S. NHANES representative national data [22] displayed an earlier peak, at 6 years of age (that is, the peak of activity was shifted to a younger age in the U.S. data), followed by a steady decline throughout the rest of childhood and into adolescence.

5.5.2 *Surveillance of Trends*

There are now two shining examples of the surveillance of time trends available for objectively monitored physical activity, captured as steps/day for Japanese adults [15] and for Danish adults [17]. The Japanese data [15] were originally collected as part of the Japan Health and Nutrition Survey, administered annually by the Ministry of Health, Labour, and Welfare of Japan between 1995 and 2005. Reported time trends indicate a reduction of 529 steps/day in men and 857 steps/day in women from peak values observed in 1998–2000 to 2007. Over the same period, the percentage of individuals accumulating ≥10,000 steps/day decreased by approximately 5 % (both in men and in women) and the percentage accumulating ≤4000 steps/day increased by 4.8 % in men and by 8.2 % in women. The Danish data [17] are from 2007 to 2008 and from 2011 to 2012. They reported a borderline significant decline ($p = 0.077$) from 9788 to 8341 steps/day over this time period. The overall decrease in step-defined physical activity was driven mostly by an apparent decrease of activity in women. Specifically, the odds of a woman (controlling for age, level of education, and seasonality of assessment) accumulating ≥10,000 steps/day was 45.4 % lower in 2011–2012 than in 2007–2008. There was no difference in the proportion of Danes taking <5000 steps/day over time.

Gortmaker et al. [52] reported changes in objectively monitored mean activity counts/minute and minutes/day in MVPA for children and adolescents 6–19 years of age participating in the 2003–2004 and the 2005–2006 NHANES accelerometer survey. For 6–11 year olds there was a statistically significant increase of 31.6 mean activity counts/minute between the two survey cycles. No change was apparent for the 12–19 year olds. No changes were apparent in MVPA time between the two cycles for any age group.

5.5.3 Longitudinal Studies

Longitudinal studies track the same participants over time. Although once very scarce, a growing number of longitudinal studies of objectively monitored physical activity metrics are now emerging. Unfortunately, they have not yet been pulled together in a systematic review. Investigators should take the time to evaluate the literature for relevant longitudinal studies to guide the interpretation and reporting of their own data. What follows is only a brief sampling of the studies published to date.

Anders Raustorp (Fig. 5.9) et al. [53] reported the first longitudinal study of objectively monitored physical activity. They reported steps/day for 93 adolescents (46 girls) who were first assessed in 2000 at 12–14 years of age and subsequently in 2003 when they were 15–18 years of age. They averaged 14,612 steps/day in 2000 (boys: 15,736 vs. girls: 13,512) and 12,065 steps/day in 2003 (boys: 11,205 vs. girls: 12,944).

Telford et al. [54] reported longitudinal patterns of objectively monitored physical activity in children monitored annually for 5 years from the time when they were approximately 8 years of age. Each year pedometers were used and in the final 2 years of monitoring they also included an accelerometer. There was no clear 5-year trend in steps/day from 8 to 12 years of age, but there was a single-year trend toward lowered MVPA and light intensity physical activity, with a corresponding increase in sedentary time between 11–12 years of age.

The first known longitudinal pedometer-based study in adults was a tracking study that monitored Western Australian participants before and after they

Fig. 5.9 Anders Raustorp conducted the first known longitudinal study of pedometer-determined physical activity in adolescents

re-located their place of residence [55]. Overall there was a small and non-significant decrease of 81 steps/day over the year. The researchers demonstrated that, despite re-location, this Australian sample held their relative rank position as defined by steps/day to a moderate to moderate-high extent over the course of the study.

Yamamoto et al. [56] reported the 8-year tracking of pedometer-determined physical activity among older Japanese adults. Steps/day were measured (for 1 week in January, April, July, and October during each year of assessment) initially when the participants were 72 years of age and then again after 2 (n = 177 and 76 women), 5 (n = 183 and 73 women), and 8 years (n = 145 and 54 women). Participants remained relatively stable in their rank order over the 8-year tracking period (correlation coefficient >0.60 at every follow-up point relative to baseline).

In 2004 Tudor-Locke et al. [55] reported the results of continuous year long pedometer tracking in 23 adults (mean age 38 ± 9.9 years). Participants recorded 8197 person-days of step data (of a possible 8395 person-days, or 98 %). Mean values for each month were displayed graphically and showed consistent patterns by season, day of the week, workday vs. non-workday, and sport/exercise day vs. no sport/exercise. A single exception to the seasonal pattern was an apparent dip in mean steps/day by approximately 600–700 in July relative to June and August; this could be attributed to a greater number of logged non-work days during July. Similarly, year-long month-to-month patterns in steps/day and time accumulated at ≥3 METs have been presented graphically on a subsample of older adult participants aged 56–83 years engaged in longitudinal tracking in the Nakanojo Study [57].

5.5.4 Intervention Studies

A number of systematic reviews of change in steps/day [58–60] have been conducted and are useful resources to guide interpretation and reporting. Together, they indicate that an increase of 2000–2500 steps/day can be expected in adults with a pedometer-based intervention. Kang et al. [60] computed the effect-size for such an intervention to be 0.68 (moderately positive). Lubans et al. [61] conducted a systematic review of studies using pedometers to promote physical activity among youth. Although they were unable to present a simple summarized incremental value of expected change, expressed as steps/day appropriate for all ages, their effort still represents an important resource to consider when interpreting change in step-determined physical activity as a result of an intervention in young people. Tudor-Locke et al. [44] compiled the results of interventions focused on older adult samples ranging in age from 55 to 95 years. They reported that the expected increase with an intervention was approximately 800 steps/day with an effect size of 0.26 (generally considered a small effect).

5.6 Reporting

Investigators have a scientific and ethical obligation to make their study findings publicly available. This includes a duty to report their findings clearly, completely, and transparently, so as to allow their peers an opportunity to evaluate the science and/or replicate the findings.

5.6.1 *Reporting Study Protocols*

There was a time when print journal page/word restrictions conflicted with an author's obligations to report detailed study methods, let alone other extensive issues associated with the objective monitoring of physical activity and sedentary time. This has changed with the advent of electronic and open access journals, which allow more detailed presentations. There are also journals that now specialize in publishing study protocols without any report of findings. For example, *BMC Public Health* is an open access journal that published a study protocol for accelerometry assessment that was implemented in five countries (Belgium, Greece, Hungary, Switzerland and the Netherlands) for the European ENERGY-project (focused on developing obesity interventions in schoolchildren). In this publication, Yildirim et al. [62] describe the decisions for each stage of data collection and the processing of accelerometer-determined physical activity and sedentary time collection.

Some journals now also provide linkages to supplementary electronic files that are not restricted in size, allowing the publication of additional study details as desired. For example, the entire accelerometer Manual of Operations for the International Study of Childhood Obesity, Lifestyle, and Environment (ISCOLE) was published as a supplementary file linked to an original research article focused on wear-time findings resulting from implementation of a 24-hour protocol with >7000 9–11 year old children from 12 different study sites around the world [60]. Author teams are also choosing to supplement their reporting by adding hyperlinks to analysis syntax that is held on web-pages maintained by their own institutions. For example, the SAS syntax used to analyze the 2003–2006 NHANES accelerometer data, as reported in Troiano et al. [30], are available at http://riskfactor.cancer.gov/tools/nhanes_pam. This publically available resource has been used, adapted, and extended by countless others [63]. In fact it forms the basis for the syntax used to extract sleep from 24-hour accelerometry in ISCOLE and is also publically available at http://www.pbrc.edu/pdf/PBRCSleepEpisodeTimeMacroCode.pdf, as reported in Barreira et al. [64]. We are now also witnessing the publication of catalogues of decision rules and treatment decisions applied to the analysis and reporting of objectively monitored data as practiced by users of the same data set [63]. Despite these advances, we are not yet at a point where we can standardize reporting of objectively monitored data by requiring publication of a minimal data set in terms of variable names and definitions. It may be folly to pursue overly rigid standardization, however, as research is an innately creative process. As Tudor-Locke et al. [63] have acknowledged, "unique research questions may require equally unique analytical approaches; some inconsistency in approaches must be tolerated if scientific discovery is to be encouraged."

5.6.2 Checklists and Flowcharts

Incomplete or inadequate reporting of research methods and results impairs interpretation and evaluation of the strengths and weakness of a scientific study.

All scientific journals publish author guidelines to assist investigators in describing their study design and findings. In addition, the International Committee of Medical Journal Editors (ICMJE) has published general standards for crafting scientific articles: the "Uniform Requirements for Manuscripts Submitted to Biomedical Journals: Writing and Editing for Biomedical Publication," also known at the "Uniform Requirements." The Uniform Requirements are maintained electronically at http://www.icmje.org/recommendations/ and are freely downloadable. By design, these are general guidelines, and they lack sufficient detail to ensure that all relevant information is reported in a standardized manner.

To further standardize and simplify the process, epidemiologists and other researchers have been collaborating with journal editors to provide checklists to guide reporting. STROBE (Strengthening the Reporting of OBservational studies in Epidemiology) is one such collaborative. It maintains an internet presence at http://www.strobe-statement.org, offering links to a number of downloadable checklists that cover several study designs and topic areas directly relevant to epidemiological observational studies. Additional checklists and guidance are provided by other international collaboratives, include the CONSORT (Consolidated Standards of Reporting Trials) guidelines for randomized trials [65], QUORUM (Quality of Reporting of Meta-analyses) [66], and MOOSE (Meta-analysis Of Observational Studies in Epidemiology) [67].

Specific to objective monitoring of physical activity and sedentary time, in 2005 Ward et al. [68] called for the creation of accelerometer reporting standards on wearing time, monitoring frames and decision rules as part of best practice recommendations. Freedson et al. [69] further pointed out that since many objective monitoring devices are continuously being updated and modified, it is essential to report monitor models and versions of firmware and any digital filtering processes that are used. They [69] and others [70] have called for researchers to report practical lessons learned from using objective monitors in their studies, including sharing financial and human time costs related to implementation. In 2012, Matthews et al. [71] offered a checklist and flowchart as tools to aid investigators reporting objectively monitored physical activity and sedentary time data. Checklist items included reporting: (1) a rationale for instrument selection along with information on its basic validity and reliability, (2) detailed data collection protocols, including site of attachment, (3) an *a priori* determined monitoring frame and its rationale, (4) wearing time estimations and/or any adjustments, (5) quality control checks and compliance criteria used in determining valid data capture, and (6) decision rules guiding data treatment and analyses. The suggested flowchart tabulated recruitment efforts and subsequent participant flow resulting in valid data capture, while also providing an accounting of any excluded or otherwise lost data (and the reasons for these losses).

Recently, Tudor-Locke et al. [72] presented a model based on the ISCOLE study for reporting accelerometer paradata (process-related data produced from survey administration). Accelerometer paradata includes tallies of consenting and eligible participants, the number of accelerometers distributed, a listing of inadequate data (e.g., due to loss or malfunction, insufficient wear-time), and averages for waking wear-time, valid days of data, participants with valid data ($\geq$4 valid days of data, including 1 weekend day), and minutes with implausibly high values ($\geq$20,000 activity counts/min). They also presented an evolved version of the original flow chart of Matthews et al. [71], organizing the study data flow into distinct enrollment, data collection, and data processing phases cross-tabulated with trackable data derived from accelerometers, participants/data files, and reasons for data loss at each stage. Such detailed reporting of accelerometer paradata would be useful in standardizing communication and in facilitating comparisons between studies.

5.7 Conclusions

Deriving meaning from objectively monitored physical activity and sedentary data is a cognitive process that can be strengthened by carefully considering features of the study design, the instrument used, the population studied, and other contextual factors. There are growing resources to aid investigators with the reporting and interpretation of their own objectively monitored physical activity and sedentary time data. The current movement towards more detailed disclosure and a sharing of resources is a welcome advance, facilitated in part by the evolution of electronic publishing options. Researchers must continue to extend the knowledge base supporting objective monitoring as they publish, compile, and update expected values, normative data, and useful indices for data classification. The development of more detailed reporting guidelines would also be welcome, allowing a more standardized sharing of study details. On the other hand, this should not be considered prescriptive, nor used to evaluate and restrain the innately creative nature of science too rigidly.

References

1. Tudor-Locke C, Myers AM. Methodological considerations for researchers and practitioners using pedometers to measure physical (ambulatory) activity. Res Q Exerc Sport. 2001;72 (1):1–12.
2. Bohannon RW. Number of pedometer-assessed steps taken per day by adults: a descriptive meta-analysis. Phys Ther. 2007;87(12):1642–50.
3. Bassett Jr DR, Schneider PL, Huntington GE. Physical activity in an old order Amish community. Med Sci Sports Exerc. 2004;36(1):79–85.
4. Tudor-Locke C, Craig CL, Brown WJ, et al. How many steps/day are enough? For adults. Int J Behav Nutr Phys Act. 2011;8:79.
5. Rowlands AV, Eston RG, Ingledew DK. Relationship between activity levels, aerobic fitness, and body fat in 8- to 10-yr-old children. J Appl Physiol. 1999;86(4):1428–35.
6. Tudor-Locke C, McClain JJ, Hart TL, et al. Expected values for pedometer-determined physical activity in youth. Res Q Exerc Sport. 2009;80(2):164–74.
7. Beets MW, Bornstein D, Beighle A, et al. Pedometer-measured physical activity patterns of youth: a 13-country review. Am J Prev Med. 2010;38(2):208–16.
8. Tudor-Locke C, Hart TL, Washington TL. Expected values for pedometer-determined physical activity in older populations. Int J Behav Nutr Phys Act. 2009;6:59.
9. Karabulut M, Crouter SE, Bassett Jr DR. Comparison of two waist-mounted and two ankle-mounted electronic pedometers. Eur J Appl Physiol. 2005;95(4):335–43.
10. English C, Manns PJ, Tucak C, et al. Physical activity and sedentary behaviors in people with stroke living in the community: a systematic review. Phys Ther. 2014;94(2):185–96.
11. Naal FD, Impellizzeri FM. How active are patients undergoing total joint arthroplasty?: a systematic review. Clin Orthop Relat Res. 2010;468(7):1891–904.
12. Tudor-Locke C, Johnson WD, Katzmarzyk PT. Accelerometer-determined steps per day in US adults. Med Sci Sports Exerc. 2009;41(7):1384–91.
13. Wyatt HR, Peters JC, Reed GW, et al. A Colorado statewide survey of walking and its relation to excessive weight. Med Sci Sports Exerc. 2005;37(5):724–30.
14. Bassett Jr DR, Wyatt HR, Thompson H, et al. Pedometer-measured physical activity and health behaviors in United States adults. Med Sci Sports Exerc. 2010;42(10):1819–25.
15. Inoue S, Ohya Y, Tudor-Locke C, et al. Time trends for step-determined physical activity among Japanese adults. Med Sci Sports Exerc. 2011;43(10):1913–9.
16. Hansen BH, Kolle E, Dyrstad SM, et al. Accelerometer-determined physical activity in adults and older people. Med Sci Sports Exerc. 2012;44(2):266–72.
17. Matthiessen J, Andersen EW, Raustorp A, et al. Reduction in pedometer-determined physical activity in the adult Danish population from 2007 to 2012. Scand J Public Health. 2015;43 (5):525–33.
18. Colley RC, Garriguet D, Janssen K, et al. Physical activity of Canadian adults: accelerometer results from the 2007 to 2009 Canadian Health Measures Survey. Health Rep. 2011;22(1):1–8.
19. De Cocker K, Cardon G, De Bourdeaudhuij I. Pedometer-determined physical activity and its comparison with the International Physical Activity Questionnaire in a sample of Belgian adults. Res Q Exerc Sport. 2007;78(5):429–37.
20. McCormack G, Giles-Corti B, Milligan R. Demographic and individual correlates of achieving 10,000 steps/day: use of pedometers in a population-based study. Health Promot J Austr. 2006;17(1):43–7.
21. Sequeira MM, Rickenbach M, Wietlisbach V, et al. Physical activity assessment using a pedometer and its comparison with a questionnaire in a large population survey. Am J Epidemiol. 1995;142(9):989–99.
22. Tudor-Locke C, Johnson WD, Katzmarzyk PT. Accelerometer-determined steps/day in U.S. children and youth. Med Sci Sports Exerc. 2010;42(12):2244–50.

23. Colley RC, Garriguet D, Janssen I, et al. Physical activity of Canadian children and youth: accelerometer results from the 2007–2009 Canadian Health Measures Survey. Health Rep. 2011;22(1):1–9.
24. Craig CL, Cameron C, Tudor-Locke C. CANPLAY pedometer normative reference data for 21,271 children and 12,956 adolescents. Med Sci Sports Exerc. 2013;45(1):123–9.
25. Kuczmarski RJ, Ogden CL, Guo SS, et al. 2000 CDC growth charts for the United States: methods and development. Washington, DC: National Center for Health Statistics; 2002. Contract No.: 246.
26. Barreira TV, Schuna Jr JM, Mire EF, et al. Normative steps/day and peak cadence values for United States children and adolescents: National Health and Nutrition Examination Survey 2005–2006. J Pediatr. 2015;166(1):139–43. e3.
27. Tudor-Locke C, Schuna Jr JM, Barreira TV, et al. Normative steps/day values for older adults: NHANES 2005–2006. J Gerontol A Biol Sci Med Sci. 2013;68(11):1426–32.
28. Wolff DL, Fitzhugh EC, Bassett DR, et al. Waist-worn actigraphy: population-referenced percentiles for total activity counts in U.S. adults. J Phys Act Health. 2015;12(4):447–53.
29. Wolff-Hughes DL, Bassett DR, Fitzhugh EC. Population-referenced percentiles for waist-worn accelerometer-derived total activity counts in U.S. youth: 2003–2006 NHANES. PLoS One. 2014;9(12):e115915.
30. Troiano RP, Berrigan D, Dodd KW, et al. Physical activity in the United States measured by accelerometer. Med Sci Sports Exerc. 2008;40(1):181–8.
31. Hagstromer M, Troiano RP, Sjostrom M, et al. Levels and patterns of objectively assessed physical activity—a comparison between Sweden and the United States. Am J Epidemiol. 2010;171(10):1055–64.
32. Matthews CE, Chen KY, Freedson PS, et al. Amount of time spent in sedentary behaviors in the United States, 2003–2004. Am J Epidemiol. 2008;167(7):875–81.
33. Schuna Jr JM, Johnson WD, Tudor-Locke C. Adult self-reported and objectively monitored physical activity and sedentary behavior: NHANES 2005–2006. Int J Behav Nutr Phys Act. 2013;10(1):126.
34. Tudor-Locke C, Camhi SM, Leonardi C, et al. Patterns of adults stepping cadence in the 2005–2006 NHANES. Prev Med. 2011;53:178–81.
35. Barreira TV, Katzmarzyk PT, Johnson WD, et al. Cadence patterns and peak cadence in U.S. children and adolescents: NHANES 2005–2006. Med Sci Sports Exerc. 2012;44 (9):1721–7.
36. Tudor-Locke C, Brashear MM, Katzmarzyk PT, et al. Peak stepping cadence in free-living adults: 2005–2006 NHANES. J Phys Act Health. 2012;27(9):1125–9.
37. Barreira TV, Katzmarzyk PT, Johnson WD, et al. Walking cadence and cardiovascular risk in children and adolescents. Am J Prev Med. 2013;45(6):e27–34.
38. Tudor-Locke C, Bassett Jr DR. How many steps/day are enough? Preliminary pedometer indices for public health. Sports Med. 2004;34(1):1–8.
39. Tudor-Locke C, Hatano Y, Pangrazi RP, et al. Revisiting "how many steps are enough?". Med Sci Sports Exerc. 2008;40(7 Suppl):S537–43.
40. Tudor-Locke C, Craig CL, Thyfault JP, et al. A step-defined sedentary lifestyle index: <5000 steps/day. Appl Physiol Nutr Metab. 2013;38:100–14.
41. Tudor-Locke C, Johnson WD, Katzmarzyk PT. Relationship between accelerometer-determined steps/day and other accelerometer outputs in U.S. adults. J Phys Act Health. 2011;8:410–9.
42. Tudor-Locke C, Pangrazi RP, Corbin CB, et al. BMI-referenced standards for recommended pedometer-determined steps/day in children. Prev Med. 2004;38(6):857–64.
43. Tudor-Locke C, Craig CL, Beets MW, et al. How many steps/day are enough? For children and adolescents. Int J Behav Nutr Phys Act. 2011;8:78.
44. Tudor-Locke C, Craig CL, Aoyagi Y, et al. How many steps/day are enough? For older adults and special populations. Int J Behav Nutr Phys Act. 2011;8:80.

45. Ewald B, Attia J, McElduff P. How many steps are enough? Dose-response curves for pedometer steps and multiple health markers in a community-based sample of older Australians. J Phys Act Health. 2014;11(3):509–18.
46. Aoyagi Y, Shephard RJ. Sex differences in relationship between habitual physical activity and health in the elderly: practical implications for epidemiologists based on pedometer/accelerometer data from the Nakanojo study. Arch Gerontol Geriatr. 2013;56:327–38.
47. Aoyagi Y, Shephard RJ. Steps per day: the road to senior health? Sports Med. 2009;39 (6):423–38.
48. Tudor-Locke C, Ainsworth BE, Whitt MC, et al. The relationship between pedometer-determined ambulatory activity and body composition variables. Int J Obes Relat Metab Disord. 2001;25(11):1571–8.
49. Tucker JM, Welk GJ, Beyler NK. Physical activity in U.S. adults compliance with the physical activity guidelines for Americans. Am J Prev Med. 2011;40(4):454–61.
50. Troiano RP, McClain JJ, Brychta RJ, et al. Evolution of accelerometer methods for physical activity research. Br J Sports Med. 2014;48(13):1019–23.
51. Craig CL, Cameron C, Griffiths JM, et al. Descriptive epidemiology of youth pedometer-determined physical activity: CANPLAY. Med Sci Sports Exerc. 2010;42(9):1639–43.
52. Gortmaker SL, Lee R, Cradock AL, et al. Disparities in youth ohysical activity in the United States: 2003–2006. Med Sci Sports Exerc. 2011;44(5):888–93.
53. Raustorp A, Mattsson E, Svensson K, Stahle A. Physical activity, body composition and physical self-esteem: a 3-year follow-up study among adolescents in Sweden. Scand J Med Sci Sports. 2006;16(4):258–66.
54. Telford RM, Telford RD, Cunningham R, et al. Longitudinal patterns of physical activity in children aged 8 to 12 years: the LOOK study. Int J Behav Nutr Phys Act. 2013;10:81.
55. Tudor-Locke C, Giles-Corti B, Knuiman M, et al. Tracking of pedometer-determined physical activity in adults who relocate: results from RESIDE. Int J Behav Nutr Phys Act. 2008;5:39.
56. Yamamoto N, Shimada M, Nakagawa N, et al. Tracking of pedometer-determined physical activity in healthy elderly Japanese people. J Phys Act Health. 2015;12(10):1421–9.
57. Yasunaga A, Togo F, Watanabe E, et al. Sex, age, season, and habitual physical activity of older Japanese: the Nakanojo study. J Aging Phys Act. 2008;16(1):3–13.
58. Bravata DM, Smith-Spangler C, Sundaram V, et al. Using pedometers to increase physical activity and improve health: a systematic review. JAMA. 2007;298(19):2296–304.
59. Richardson CR, Newton TL, Abraham JJ, et al. A meta-analysis of pedometer-based walking interventions and weight loss. Ann Fam Med. 2008;6(1):69–77.
60. Kang M, Marshall SJ, Barreira TV, et al. Effect of pedometer-based physical activity interventions: a meta-analysis. Res Q Exerc Sport. 2009;80(3):648–55.
61. Lubans DR, Morgan PJ, Tudor-Locke C. A systematic review of studies using pedometers to promote physical activity among youth. Prev Med. 2009;48(4):307–15.
62. Yildirim M, Verloigne M, de Bourdeaudhuij I, et al. Study protocol of physical activity and sedentary behaviour measurement among schoolchildren by accelerometry—cross-sectional survey as part of the ENERGY-project. BMC Public Health. 2011;11:182.
63. Barreira TV, Schuna Jr JM, Mire EF, et al. Identifying children's nocturnal sleep using 24-hour waist accelerometry. Med Sci Sports Exerc. 2015;47(5):937–43.
64. Tudor-Locke C, Camhi SM, Troiano RP. A catalog of rules, variables, and definitions applied to accelerometer data in the National Health and Nutrition Examination Survey, 2003–2006. Prev Chronic Dis. 2012;9:E113.
65. Schulz KF, Altman DG, Moher D. CONSORT 2010 statement: updated guidelines for reporting parallel group randomised trials. BMJ. 2010;340:c332.
66. Moher D, Cook DJ, Eastwood S, et al. Improving the quality of reports of meta-analyses of randomised controlled trials: the QUOROM statement. Quality of reporting of meta-analyses. Lancet. 1999;354(9193):1896–900.

67. Stroup DF, Berlin JA, Morton SC, et al. Meta-analysis of observational studies in epidemiology: a proposal for reporting. Meta-analysis Of Observational Studies in Epidemiology (MOOSE) group. JAMA. 2000;283(15):2008–12.
68. Ward DS, Evenson KR, Vaughn A, et al. Accelerometer use in physical activity: best practices and research recommendations. Med Sci Sports Exerc. 2005;37(11 Suppl):S582–8.
69. Freedson P, Bowles HR, Troiano R, et al. Assessment of physical activity using wearable monitors: recommendations for monitor calibration and use in the field. Med Sci Sports Exerc. 2012;44(1 Suppl 1):S1–4.
70. Intille SS, Lester J, Sallis JF, et al. New horizons in sensor development. Med Sci Sports Exerc. 2012;44(1 Suppl 1):S24–31.
71. Matthews CE, Hagstromer M, Pober DM, et al. Best practices for using physical activity monitors in population-based research. Med Sci Sports Exerc. 2012;44(1 Suppl 1):S68–76.
72. Tudor-Locke C, Mire EF, Dentro KN, Barreira TV, Schuna Jr JM, Zhao P, Tremblay MS, Standage M, Sarmiento OL, Onywera V, Olds T, Matsudo V, Maia J, Maher C, Lambert EV, Kurpad A, Kuriyan R, Hu G, Fogelholm M, Chaput JP, Church TS, Katzmarzyk PT, ISCOLE Research Group. A model for presenting accelerometer paradata in large studies: ISCOLE. Int J Behav Nutr Phys Act. 2015;12:52. doi:10.1186/s12966-015-0213-5.

Chapter 6
New Information on Population Activity Patterns Revealed by Objective Monitoring

Richard Larouche, Jean-Philippe Chaput, and Mark S. Tremblay

Abstract Objective monitors allow a variety of new and established physical activity variables to be evaluated. Time-stamped devices that can measure movement intensity with high resolution permit the determination and assessment of a myriad of movement measurements. Whereas traditional self-report metrics are limited by the burdens of recall and response, new objective measures allow for multiple indices of intensity, duration, frequency, amount, pattern and sequence of movement activities to be calculated and used in epidemiological analyses. These new metrics enable an examination of all movement behaviours (physical activity, sedentary behaviours and sleep), together with their interactions and their combined relationships with health indicators. Although monitors can still be enhanced, and more research is required, the new generation of metrics already provides opportunity to advance the understanding of movement behavior epidemiology significantly and to validate previous work carried out with less robust or complete measures.

R. Larouche (✉)
Healthy Active Living and Obesity Research Group, CHEO Research Institute, Ottawa, ON, Canada
e-mail: rlarouche@cheo.on.ca

J.-P. Chaput
Healthy Active Living and Obesity Research Group, CHEO Research Institute, Ottawa, ON, Canada

Faculty of Medicine, University of Ottawa, Ottawa, ON, Canada

M.S. Tremblay
Healthy Active Living and Obesity Research Group, CHEO Research Institute, Ottawa, ON, Canada

Department of Pediatrics, University of Ottawa, Ottawa, ON, Canada

R.J. Shephard, C. Tudor-Locke (eds.), *The Objective Monitoring of Physical Activity: Contributions of Accelerometry to Epidemiology, Exercise Science and Rehabilitation*, Springer Series on Epidemiology and Public Health,
DOI 10.1007/978-3-319-29577-0_6

6.1 Introduction

Contemporary objective monitoring devices provide new opportunities to measure and monitor physical activity. Time-stamped devices that can determine the intensity of movement with high resolution allow for a myriad of measurements to be made and assessed. Whereas traditional self-report metrics have been limited by difficulties in recall and the burden of detailed responses, new objective evaluations allow investigators to calculate multiple indices of the intensity, duration, frequency, amount, pattern and sequence of physical activity that can be used in epidemiological analyses. This chapter explores the new opportunities afforded by the range of movement (and non-movement) metrics that can now be generated by objective monitors of movement behaviour. Specifically, this chapter approaches physical activity epidemiology from the perspective of an entire day, and it underlines the importance and indeed the necessity of integrating all movement behaviours (i.e., physical activity, sedentary behaviour and sleep times, sequences and patterns) when performing epidemiological analyses and when making physical activity recommendations. The inherent limitations of current objective monitoring devices are also noted, and the likely directions of future research are also indicated.

6.2 Physical Activity

6.2.1 Surveillance of Physical Activity: Self-Reports Versus Objective Measurements

Given the lack of availability of practical and inexpensive measures of physical activity (PA), until recently the majority of population-based studies have relied on self-reports including questionnaires, logs, recalls and diaries [1]. Such measurements have generally shown a satisfactory reliability for the measurement of leisure-time PA, and their low cost and high feasibility for use on large samples has contributed to their extensive use over the last 50 years [1, 2]. However, self-reports and proxy-reports of PA also have important limitations, which have been extensively documented [2–5].

Of particular relevance to the surveillance of PA, such measures tend to overestimate individuals' PA levels [3, 6]. For instance, Basterfield and colleagues compared measures of moderate- to vigorous-intensity PA (MVPA) from the Health Survey for England parental questionnaire and Actigraph accelerometer data in a sample of 6–7 year olds. The respective estimates of MVPA were 146 and 24 minutes/day, a difference of 608 %.

Similarly, in the context of population-based estimates of PA prevalence, the use of self-reports may lead to substantial overestimates. A comparison of nationally-representative data from the Canadian Community Health Survey (using self-reported PA) and the Canadian Health Measures Survey (which used Actical accelerometers) illustrates this point. Specifically, based on self-reports, 55.2 % of adults who participated in the 2013 Canadian Community Health Survey were classified as "moderately active" (i.e., equivalent to walking at least 30 minutes a day or taking an hour-long exercise class at least three times a week) [7]. In contrast, accelerometry data from the 2012–2013 Canadian Health Measures Survey revealed that only about one in five adults met the current PA guidelines [8]. The discrepancy between the two surveys was even greater for 12–19 year olds. For instance, only 3 % of girls were sufficiently active when PA was measured by accelerometry [9], whereas 65 % were classified as "moderately active" when assessed by self-report [7]!

In addition, self-reports may be affected by a changing awareness of what is perceived as MVPA over time. For example, the recognition of activities such as walking and gardening as sources of MVPA may explain in part the apparent increase in leisure-time PA reported in Canada [10] and the United States [11] in recent decades. Given this limitation, caution is needed when interpreting self-reports of a changing prevalence of PA, especially knowing that other domains of PA (e.g., occupational, transportation and domestic activity) are not as well documented in most surveillance studies [10]. In contrast, objective measures should not be affected by changing awareness or perceptions of the definition of MVPA.

Finally, while individuals may recall their vigorous PA rather precisely, self-reported measures are notoriously limited when assessing light intensity PA [2]. The contribution of accelerometers in addressing this issue is discussed in Sect. 6.2.3. As discussed in Sects. 6.2.4 and 6.2.5, accelerometers are also useful when examining the fractionalization of PA (i.e., the accumulation of PA in short bouts throughout the day), and in recording unstructured forms of PA (e.g., playing, climbing stairs, etc.) which can be very difficult for an individual to recall accurately.

6.2.2 Objective Monitoring of Physical Activity in the Population

Technological advances have made research grade pedometers and accelerometers more affordable over the last 10–20 years. As a result, these tools have been used in a growing number of population-based studies. Noteworthy examples include the National Health and Nutrition Examination Survey (NHANES) in the United States [12], the Canadian Physical Activity Level Among Youth (CANPLAY) study [13], the Canadian Health Measures Survey [14, 15], and the Health Survey for England [16] (see Table 6.1 for details and additional examples). These various studies provide consistent evidence that physical inactivity is a major public health problem in children and adults alike.

Table 6.1 Overview of selected large scale population-based studies using objective measures of physical activity and sedentary behaviour

Study, year and references	Population represented and sample size	Measure of physical activity	Key findings
National Health and Nutrition Examination Survey (NHANES)—Annual survey, includes accelerometry data in 2003–2006 and 2011–2014 surveys [12, 17–19] (Fig. 6.1)	Americans aged 3–79 (but only 6–79 in 2003–2006) ~5000 participants per year (valid accelerometry data available from 68 % of participants in 2003–2004 survey)	Hip-worn ActiGraph 7164 in 2003–2006; Wrist-worn ActiGraph GT3X+ accelerometer in 2011–2014.	In the 2003–2004 survey, 42 % of 6–11 year olds met the PA guideline, compared to 6–8 % of adolescents and <5 % of adults. Participants spent on average 7.7 hours/day sedentary. SED increased from childhood to adolescence, decreased in young adulthood and increased afterwards. Prevalence estimates from NHANES 2003–2006 vary widely depending on which MVPA cut-points are used in analyses.
Canadian Health Measures Survey (CHMS)—Biennial survey launched in 2007 [14, 15, 20]	Canadians aged 3–79 (but only 6–79 in the 1st cycle). ~5500 participants per cycle (valid accelerometry data available from ~70–75 % of participants)	Hip-worn Actical accelerometer	Cycle 1 (2007–2009) of the CHMS indicated that children and youth spent 8.6 hours/day sedentary and only 9 % of boys and 4 % of girls (6–19 year olds) met the Canadian PA guidelines. Similarly, adults (20–79 year olds) spent 9.5 hours/day sedentary and only 15 % were sufficiently active. Subsequent cycles showed similar low prevalence. Among 3–4 year olds, 84 % met the PA guidelines (e.g., 180 minutes of PA of any intensity).

(continued)

Table 6.1 (continued)

Study, year and references	Population represented and sample size	Measure of physical activity	Key findings
Health Survey for England [16]	Britons aged 4+. 6214 participants were asked to wear an accelerometer. Only 43–49 % of participants provided valid accelerometry data	Hip-worn ActiGraph GT1M accelerometer	Among 4–15 year olds, 33 % of boys and 21 % of girls met the physical activity guidelines. PA was higher in younger children while SED increased from childhood to adolescence. Only 6 % of men and 4 % of women ≥16 year old were sufficiently active. PA was lower among older participants and those with higher BMI/waist circumference.
Canadian Physical Activity Level Among Youth (CANPLAY) study [13, 21]	Canadians aged 5–19. N = 34,227 for 2005–2011. >95 % of participants wore the pedometer for ≥ 5 days	Hip-worn Yamax SW-200 pedometer	Boys and girls took 12,259 and 10,906 steps/day, respectively. Daily steps were higher among boys than girls and declined with age. Weekday steps/day were generally higher than weekend day steps/day and varied by season. Normative reference data were developed.
European Youth Heart Study [22–24]	Regional samples of 9 and 15 year olds from Denmark, Estonia, Norway and Portugal. N = 2185. 75 % of participants provided valid accelerometer data	Hip-worn ActiGraph 7164	Boys were more active than girls at age 9 (784 vs. 649 counts/minute) and 15 (615 vs. 491 counts/minute). With respect to time engaged in MVPA, gender differences were apparent at age 9 (192 vs. 160 minutes/day) and 15 (99 vs. 73 minutes/day). At age 9, 97.4 % of boys and 97.6 % of girls achieved PA recommendations. At

(continued)

Table 6.1 (continued)

Study, year and references	Population represented and sample size	Measure of physical activity	Key findings
			age 15, 81.9 % of boys and 62.0 % girls met the guidelines. There was a strong dose-response relationship between PA and reduced cardiometabolic risk.
Healthy Lifestyle in Europe by Nutrition in Adolescence (HELENA) study [25]	Regional samples of 12–17 year olds in 10 European cities within 9 different countries (Austria, Belgium, France, Germany, Greece, Hungary, Italy, Portugal, Spain, Sweden). N = 2200. 88 % of participants had valid accelerometer data	ActiGraph GT1M accelerometer worn on the lower back	56.8 % of boys and 27.5 % of girls met the PA guidelines. Boys and girls accumulated respectively 9.0 and 9.1 hours/day of SED. Fitter youth had lower SED and higher counts/minute and MVPA. Youth from Central/Northern Europe were more active and less sedentary than their Southern Europe counterparts.
Avon Longitudinal Study of Parents and Children (ALSPAC) [26, 27]	Birth cohort study in the county of Avon (England). 5595 participants had valid accelerometry data at age 12. 84.5 % of participants had valid accelerometer data	Hip-worn ActiGraph 7164	Boys had higher values than girls for both minutes MVPA (25.4 vs. 15.8) and counts/minute (663 vs. 552). Odds ratio for obesity between top and bottom quintiles of MVPA was 0.03 (95 % CI = 0.01–0.13) in boys and 0.36 (95 % CI = 0.17–0.74) in girls. Prospective relationships between PA and lower adiposity were also reported.
Identification and Prevention of Dietary and Lifestyle Induced Health Effects in Children and Infants (IDEFICS) [28]	Regional samples of 2.0–10.9 year olds in 8 European countries (Belgium, Cyprus, Estonia, Germany, Hungary, Italy, Spain and Sweden). The	Hip-worn ActiGraph accelerometers (ActiTrainer or GT1M model)	The percentage of girls accumulating ≥60 minutes of MVPA per day ranged from 2.0 % in Cyprus to 14.7 % in Sweden. For boys it ranged

(continued)

Table 6.1 (continued)

Study, year and references	Population represented and sample size	Measure of physical activity	Key findings
	study included 18,745 children of whom 12,014 provided some data and 7684 had valid accelerometer data (41.0 % of the total sample).		from 9.5 % in Italy to 34.1 % in Belgium. This study provides sex and age-specific reference data on children's PA in Europe.

MVPA moderate-to-vigorous physical activity, *PA* physical activity, *SED* sedentary time

Fig. 6.1 Charles Matthews has undertaken extensive research on physical activity and cancer in adults and older adults, using a variety of objective movement metrics (see Table 6.1)

The following sections highlight new information that can be obtained through the objective monitoring of PA in the population.

6.2.3 Moving Beyond Leisure-Time Physical Activity

One of the most important advantages of using accelerometers rather than relying upon self-reports is the ability of objective monitors to measure light PA accurately. Recent studies have shown that displacing sedentary time by light PA is associated with substantial health benefits [29, 30]. Of particular interest, in a randomized controlled trial of overweight or obese middle-age adults, Dunstan et al. [29] observed that interrupting sedentary time with short bouts of light PA or MVPA every 20 minutes was associated with reduced postprandial glucose and insulin levels, suggesting that this may be a useful tactic in the prevention and management of type II diabetes mellitus. Moreover, using data from NHANES, Carson et al. [17] noted that light PA was associated with a lower diastolic blood pressure and higher high density lipoprotein cholesterol among youth. Objective data from multiple population-based studies indicate that children and adults accumulate several hours/day of light PA, although there is a large inter-individual variation [14, 15,

17, 18]. Although the rate of energy expenditure during light PA is low (1.50–2.99 metabolic equivalents), it nevertheless represents a substantial proportion of the total daily activity energy expenditure for most individuals [31]. At the population level, declining levels of incidental PA (including light PA), may be partly responsible for the increased levels of obesity, hypertension and type II diabetes seen over the last few decades [10, 31]. In support of the importance of a decline in light PA, Church and colleagues [32] observed that the proportion of American adults having active occupations has decreased markedly over the last 50 years.

Nevertheless, objective measures of physical activity have only been used relatively recently in the surveillance of PA. Thus, detailed "baseline" measures of PA before the transition to more sedentary occupations and the reduction of incidental PA are not readily available [31]. To overcome this limitation, researchers can compare PA levels between populations who do not have access (or refuse access) to "modern" technologies, including motorized modes of transportation. Two strands of studies have used this tactic. In the first strand, researchers compared objectively-measured PA between "modern" North American samples and samples of Amish and Old Order Mennonite communities [33, 34], both of whom reject modern equipment as a part of their beliefs. The second strand of studies is based on the PA transition in developing countries, as they experience a large shift in habitual PA from high-energy expenditure activities such as hunting and gathering, mining, forestry and farming to low-energy expenditure activities such as desk work and car travel [35]. Researchers have compared objectively-measured PA between children residing in urban Kenya and those living in rural Kenya. The latter were at a less advanced stage in the transition and had significantly more physical activity [36, 37]. Such cross-cultural comparisons can provide important insights into the sources of PA that have been "engineered" out of many Westerners' daily lives. However, this type of study has yet to be conducted at the population level.

Given the difficulty of recalling light PA accurately, earlier population-based studies of the relationships between light PA and health-related indicators should be repeated, using time-stamped devices such as accelerometers. In turn, such evidence could be used to develop new and more appropriate guidelines for the amount of time that children and adults should allocate to light PA.

6.2.4 Fractionalization of Physical Activity

Another major contribution of objective activity monitors is their ability to measure the intermittent bouts of incidental physical activity that occur in daily life [31, 38]. Although current public health guidelines recommend that adults accumulate MVPA in bouts of at least 10 minutes duration [39], this recommendation is predominantly based on previous studies that used subjective measures of PA. In their review of the effectiveness of interventions using single bouts vs. accumulated bouts of PA, Murphy et al. [40] concluded that bouts $\geq$10 minutes appeared to be as effective as longer bouts to improve health-related fitness indicators. However, they also noted that very few studies had examined the effectiveness of shorter bouts. An emerging body of evidence suggests that shorter bouts of PA may be as beneficial as longer bouts among children [41], youth [42], and adults [18, 43]. Of particular interest, the studies by Holman et al. [31] and Camhi et al. [18] used population data from NHANES. In a randomized controlled trial, Miyashita et al. [44] also observed that accumulating 30 minutes of brisk walking in a single bout or in 10 bouts of 3 minutes had similar effects on measures of postprandial lipemia and blood pressure.

The current body of evidence indicates that short bouts of PA carry important health benefits in both children and adults. However, unstructured PA accumulated in short bouts is very likely to be assessed inaccurately by self-report or proxy-report methods, thereby underscoring the usefulness of objective physical activity monitors. The opportunity to assess PA in short bouts is particularly important for studies of children, because they are most likely to accumulate their daily PA in short bouts, rather than in the longer bouts that would be more characteristic of deliberate exercise training [41, 45].

Time-stamped activity monitors also enable researchers to investigate whether the pattern in which PA is accumulated throughout the week influences health-related outcomes. The importance of examining the weekly cycle of PA was emphasized by Esliger and Tremblay [38]. Some individuals may engage in a daily 30 minute brisk walk, but others may accumulate several hours of MVPA on a single day and undertake little physical activity for the remainder of the week (i.e., “weekend warriors”). Examining whether there are differences in health-related outcomes between weekend warriors and individuals who have a more regular pattern of PA may help inform PA promotion initiatives. In a follow-up of the Harvard Alumni Study, Lee and colleagues [46] found that the risk of mortality among weekend warriors who had at least one known chronic disease risk factor (e.g., smoking, overweight, hypertension and hypercholesterolemia) did not differ from that of inactive participants. One potential explanation for these findings is that the acute benefits of a PA session on indicators such as blood pressure, cholesterol and glycemia wane after 48–72 hours [47]. This underscores the importance of regular PA, particularly for “at-risk” individuals.

In addition, knowing when PA occurs on weekdays and weekend days may help identify promising periods for the promotion of PA at the population level. For instance, the 2011 Active Healthy Kids Canada Report Card on Physical Activity for Children and Youth [48] put a strong emphasis on the importance of the “after-school” period. This is supported by consistent evidence showing that children who spend more time outside during the after-school period engage in substantially more overall physical activity, as measured by accelerometry [49–51]. Of particular interest to this stream of research, the combined use of accelerometers and GPS can broaden our understanding of both when and where individuals are active or sedentary [51, 52].

Although it is possible to collect detailed data on PA patterns using time-use surveys, this imposes a substantial burden upon participants, and it may lead to selection bias and to inaccurate reporting of activities.

6.2.5 Non-exercise Activity Thermogenesis

As emphasized above, unstructured or incidental PA is an important source of energy expenditure [31]. The concept of non-exercise activity thermogenesis (NEAT) was popularized by James Levine (Fig. 6.2) in a paper published in *Science* in 1999. Levine and colleagues [53] reported that NEAT was the key factor that explained individual variations in the increase of body mass following an overfeeding experiment. Specifically, 16 volunteers were overfed by 4.16 MJ (1000 kcal) per day for 8 consecutive weeks, and they were told to maintain their usual PA level. The investigators noted that the gain in body mass varied substantially from 1.4 to 7.2 kg, and the main explanation advanced for this variation was inter-individual differences in NEAT. These observations highlight the importance of NEAT as a source of energy expenditure when seeking to avoid weight gain. A subsequent study used a combination of accelerometers and inclinometers (the latter can distinguish between sitting and standing still); Levine and colleagues [54] observed that, on average, obese individuals sat 2 hours more per day than leaner individuals.

An attractive aspect of the NEAT concept is that it requires no specific location, no equipment and no teammates in order to increase personal energy expenditure [31]. Furthermore, it does not necessarily require allocation of time; for example, one can stand rather than sit during a business meeting. Given these advantages, it may be easier for many individuals to increase NEAT than to increase MVPA [55]. In contrast, individuals participating in prescribed PA interventions may compensate by reducing their NEAT, thereby limiting the effectiveness of the intervention [31, 56].

Fig. 6.2 James Levine popularized the term NEAT (Non-Exercise Activity Thermogenesis) while examining measures associated with total energy expenditure

Due to its very nature, NEAT cannot be measured accurately by self-report; therefore, objective measures are necessary. To date, NEAT studies have generally been conducted on small groups of volunteers (e.g., [53, 54, 56, 57]). Hence, there remains a need for investigating the contribution of NEAT to total energy expenditure at the population level.

6.3 Sedentary Time

Sedentary behavior is defined as any waking behaviour characterized by an energy expenditure $\leq$1.5 METs while a person is in a sitting or reclining posture [58]. Sedentary behavior is now being regarded as a distinctive form of behaviour, not merely the opposite of being physical activity [59]. Thus, it is suggested that the term "inactive" be used to describe those who are performing insufficient amounts of MVPA (i.e., not meeting specified physical activity guidelines), whereas the term sedentary behaviour be reserved for actual measurements of when a person is sitting or lying [58]. Although *sedentary time* can be adequately captured by accelerometers, *sedentary behaviours* (e.g., watching television, reading, eating, travelling in a car) are typically not well captured, so the context of sedentary time is often lacking, though some assumptions can be made (e.g., sedentary time when children are at school may generally be assumed to be sitting at a desk doing school work). Chapter 7 is devoted to the measurement of sedentary behaviours and sedentary time, so these issues are not discussed in detail here. Nevertheless, it is essential to include sedentary time as a key component of 24-hour movement behaviour. Not only is sedentary behaviour significantly related to health [59, 60], but it is a dominant behaviour in terms of its overall weighting over the 24-hour period; a typical behaviour pattern includes <5 % of MVPA and >40 % of sedentary behavior [61]. As shown by Esliger and Tremblay [38], there are numerous indices of sedentary time that accelerometers can capture (Chap. 7).

6.4 Sleep

Typically, sleep is not considered in texts about physical activity, but in the context of whole-day movement behaviours it is highly relevant. Objective monitoring devices allow sleep to be examined as a part of the epidemiology of integrated behaviours. Over the past decade, the assessment of sleep by epidemiologists has grown in popularity, because inadequate sleep patterns have been linked not only to unhealthy eating and physical activity behaviours, but also to chronic diseases including obesity and type 2 diabetes mellitus [62]. Epidemiologic studies generally rely on self-reported sleep patterns, but recent advances in technology have allowed scientists to monitor sleep characteristics by actigraphy (i.e., accelerometry for exercise science). An important advantage of using 24-hour accelerometry is

Fig. 6.3 Avi Sadeh has carried out extensive research on sleep disorders in infants and children

that all intensities of physical activity, sedentary time and sleep can be assessed simultaneously with a single monitor, thereby reducing the burden on participants in gathering information for the whole day as opposed to waking hours only. The 24-hour assessment of movement and non-movement behaviours is increasingly important, given the interactions that occur between sleep, sedentary time and physical activity [63]. For example, the duration of sleep can be an important confounding factor when examining associations between physical activity and/or sedentary behaviours and health outcomes. Not assessing sleep in exercise science is probably as problematic as not assessing diet if the outcome measure is obesity or type 2 diabetes, given the documented contribution of sleep (or lack thereof) to the development of chronic diseases [64].

Although polysomnography is the gold standard for research sleep assessment, it is used mainly for experimental studies and detecting sleep problems. This section focuses on the advantages and possibilities of using actigraphy in epidemiology under free-living conditions. Such measures cannot be obtained through self-report.

Actigraphs operate on the basis that motion infers wakefulness, and that inactivity infers sleep. As such, the conventional actigraph unit comprises an accelerometer-based motion sensor, a microprocessor and memory for data storage. During operation, such devices apply simple algorithms (typically, those of Sadeh et al. [65] (Fig. 6.3) or Cole et al. [66]) to summarize the overall intensity of measured accelerometry data across defined epochs (typically 15, 30, or 60 seconds) as "activity counts;" these are stored on the device [67]. The activity count data are then processed, and periods of wakefulness (high activity count) and sleep (low activity count) are inferred.

There are a number of key sleep and wake actigraph variables, including (but not limited to) sleep duration (24-hour, nocturnal, and diurnal), daytime naps, nighttime wakings (number, duration, and distribution), the longest sleep period overnight (signalling sleep continuity), sleep efficiency, sleep latency, and sleep onset and offset times [67, 68]. Some of the most commonly reported actigraphy variables are presented in Table 6.2. With the rapid growth in the use of actigraphy over the past years, there is an urgent need to standardize the scoring and reporting of actigraphy variables. There is a lack of consistency across studies in reporting on the methodological aspects of actigraphy use (e.g., devices, software, placement,

Table 6.2 Actigraphy variables commonly used in sleep research

Sleep onset	Clock time for first of a predetermined number of consecutive minutes of sleep following reported bedtime
Sleep offset	Clock time for last of a predetermined number of consecutive minutes of sleep prior to reported wake time
Midpoint	Clock time halfway between sleep onset and sleep offset
Sleep period	Duration between sleep onset and sleep offset
Total sleep time (TST)	Duration of sleep in sleep period
Sleep onset latency	Time between bedtime and sleep onset
Wake after sleep onset (WASO)	Number of minutes scored as wake during sleep period
Sleep efficiency	(TST $\div$ time in bed) $\times$ 100 (expressed in percent)
Night waking	Predetermined number of minutes of wake preceded and followed by predetermined number of minutes of sleep
Night waking frequency	Number of night wakings
Night waking duration	Sum of average of minutes scored as night waking
24 hour sleep duration	Amount of sleep in a 24 hour period

Adapted from Meltzer et al. [68]

algorithms, epoch length, scoring rules) and in the reporting of measured variables makes comparisons of results between studies difficult.

When compared with polysomnography, actigraphy shows reasonable validity and reliability in assessing sleep-wake patterns in normal individuals with a relatively good quality of sleep [69]. However, the validity of actigraphy in special populations (e.g., elderly people, or individuals with poor sleep quality) is more questionable [69]. The most problematic validity issue is the low specificity of actigraphy in detecting wakefulness within sleep periods. Despite this limitation, actigraphy is a cost-effective method to assess sleep and sleep disorders objectively. Extended monitoring (5 days or longer) reduces the measurement errors inherent in actigraphy and increases reliability [69].

Despite the advantages of actigraphy in assessing sleep patterns, self-reports still remain the most practical and cost-effective approach in epidemiologic sleep studies that involve large populations. However, recent evidence shows that sleep questions typically used in epidemiologic studies correspond poorly with objective measures of sleep as assessed using actigraphy; kappas range from −0.19 to 0.14 [70]. These findings have implications for the interpretation of studies that have used self-reports of sleep. Future studies looking at sleep as a risk factor for long-term health outcomes should be cognizant of the biases associated with self-reports, and they may benefit from the addition of a better tool (e.g., actigraphy).

6.5 Integrated Movement

Across a 24-hour day, a person distributes his or her time between sleep, sedentary and active behaviours. Increasing time in one behavior inevitably requires a decreased time in another, giving the fixed-time nature in which these behaviours occur. Thus, the health impacts of sleep, sedentary behaviour and physical activity depend on not only the behaviours themselves, but also on the behaviours that they displace [63]. Objective, time-stamped activity monitoring devices provide data that allow for analyses of all three behaviours. Recent isotemporal substitution analyses have explored associations of alternating allocations of time in one behaviour with another while holding the total time constant [71]. For example, Buman et al. [63] explored the effects of reallocating time to sleep, sedentary or active behaviours with cardiovascular disease risk biomarkers from the cross-sectional 2005–2006 US NHANES. Isotemporal substitution modeling indicated that, independent of potential confounders and time spent in other activities, beneficial associations with cardiovascular disease risk biomarkers were associated with the reallocation of 30 minutes/day of sedentary time with equal time to either sleep, light-intensity physical activity, or MVPA. Collectively, these findings provide additional evidence (using all movement and non-movement behaviours in the same analysis) that MVPA may be the most potent health-enhancing behaviour, with additional benefit conferred by the reallocation of sedentary time to light-intensity activities and extended sleep duration.

Cluster analysis is another technique now being used to examine multi-dimensional patterns of time use. Past research has focused mainly on examining the independent associations between movement behaviours (such as MVPA) and health indicators. However, recent health research suggests that combinations or patterns of behaviour may affect health in ways that cannot be explained by the effect of the individual behaviours acting alone. Cluster analysis is a pattern recognition technique that partitions objects or individuals to a single cluster, with the complete set of clusters containing all the objects or individuals. Cluster analysis has been widely used in other fields of research such as nutrition. For example, dietary patterning has successfully identified a Mediterranean pattern. This diet, although relatively high in fat, which when analyzed separately may be positively associated with cardiovascular disease, appears to be protective against cardiovascular disease when viewed in the cluster context [72]. In the field of exercise science, sedentary and physical activities were traditionally assumed to have an inverse relationship, where increased time in one detracted from time spent in the other [73]. However, recent research using adolescent time-use clusters shows that sedentary behaviour and physical activities can coexist, and that the risk of being overweight is not increased despite the high sedentary component of behaviour [74].

Compositional data analysis is another emerging statistical approach that will help scientists who are working with 24-hour accelerometer data (i.e., data that are part of a finite whole). Data quantifying total time spent in all mutually exclusive behaviours by nature violate the basic assumptions of linear regression; time spent in each behaviour (sleep, sedentary and active behaviour) is intrinsically collinear, correlated and finite. Thus, using linear regression with movement and non-movement behaviors coming from a single accelerometer worn over 24 hours can produce misleading results; scientists must stop looking at these behaviours in isolation from each other. Doing compositional analysis requires a change of perspective. We must abandon thinking of each behaviour as an independent variable. Instead, we need to view the behaviours relative to each other (i.e., ratios or log ratios) and we need to interpret the trade-offs in the balance among all behaviours over a finite time.

Historically, the impact of physical activity (especially MVPA) upon overall health has dominated discussions, but emerging evidence indicates that a broader, more inclusive and more integrated approach is needed to understand and address current public health crises. Existing global physical activity guidelines only focus on MVPA, and recently sedentary behavior, despite accumulating evidence that light-intensity physical activity such as walking can provide important health benefits [17, 63, 75]. Moreover, there is mounting support for the importance of adequate sleep and that these several behaviours moderate the health impact of each other [76–78]. Ignoring the other components of the movement continuum (i.e., sleep, sedentary time, light-intensity physical activity) while focusing interest exclusively upon MVPA (which usually accounts for <5 % of a 24 hour period) limits the potential to optimize the health benefits of movement behaviours, and to utilize the measurement potential of contemporary monitoring devices.

In order to address this limitation, Canadian experts are currently developing the world's first *Integrated 24 Hour Movement Behaviour Guidelines for Children and Youth* (aged 5–17 years) to help advance an integrated healthy active lifestyle agenda, an approach that has the potential to improve the overall health and well-being of children and youth significantly. The new guidelines will include all intensities of physical activity (light, moderate, vigorous), sedentary behavior, and sleep. The guideline development process [79] will subsequently be extended to other age groups, and it involves a large team of researchers, knowledge users, and international collaborators [61].

6.6 Limitations of Direct Activity Measurements

Despite their usefulness in providing objective measures of PA, direct activity measurements have some important limitations. Firstly, hip-worn devices such as pedometers and accelerometers markedly underestimate PA when the wearers are engaged in activities such as cycling, swimming and weight lifting [4, 80]. This may result in a substantial underestimation of total PA. Furthermore, because there is virtually no movement at the hip level, hip-worn accelerometers and pedometers cannot distinguish sitting from standing posture [30]. This limitation can be circumvented by using such devices in combination with inclinometers. Both of these devices have been integrated in the ActivPal monitor (PAL Technologies Ltd., Glasgow, UK) [81].

Another limitation of accelerometers and pedometers is that they do not provide information on the contexts in which PA, sedentary behaviour (SB) and sleep occur, unless they are combined with other devices (e.g., GPS) or self-reports. In the last couple of years, there has been a large increase in the number of studies that have used a combination of accelerometers, GPS and geographic information systems (GIS) to gain a richer understanding of the context in which different types of PA/SB occur [52]. Such an approach currently remains expensive and may be burdensome for both staff and participants. Nevertheless, future technological advancements may make it more feasible for population-based studies in the future.

Although accelerometers can provide valid and reliable estimates of the time spent sedentary or in light, moderate and vigorous intensities of physical activity, there remains no consensus on the cut-points that should be used to distinguish these categories of intensity [82, 83] (see also Chap. 5). The lack of consensus on which cut-points should be used can have a tremendous impact on estimates of PA prevalence. In their review, Cain and colleagues [82] observed that the cut-points used for MVPA (400–3600 counts per minute) and sedentary time (9–1259 counts per minute) varied widely. One study estimated that the prevalence of PA among US adults based on data from NHANES varied from 6.3 to 98.3 %, depending on which accelerometer cut-points were used [84]! Cain et al. [82] also observed marked inconsistencies across studies with respect to the definition of non-wear time, and the minimum number of hours/day and valid days of measurement that justified inclusion in statistical analyses. Such inconsistent methodology make comparisons between studies very difficult, and in many cases, inappropriate [82, 83].

Accelerometry can invalidate population PA levels because of missing data. For instance, Pedisic & Bauman [85] estimated that between 6 and 32 % (median = 17.6 %) of participants in population-based studies did not provide valid accelerometry data, and were therefore excluded from analyses. Results from NHANES [84–86] and the Health Survey for England [87] showed significant socio-demographic differences between participants who provided accelerometry data and those who did not. While reweighting the sample can reduce the magnitude of this selection bias, non-compliance to the accelerometry protocol raises questions about the generalizability of the data [83]. One tactic to address this limitation is to ask participants to wear the accelerometer throughout the day so that they do not forget to wear it. This approach was successfully used in the International Study of Childhood Obesity, Lifestyle, and the Environment [88, 89]. Furthermore, the use of the ActiGraph GT3X+ accelerometer (ActiGraph LLC, Pensacola, FL) provided objective measures of children's sleep patterns.

Finally, participants may change their habitual PA merely as a result of wearing PA monitors [80, 83]. Such Hawthorne effect (or measurement reactivity) may spuriously inflate estimates of PA prevalence and introduce measurement error when examining relationships with health-related outcomes. To date, studies examining the reactivity to PA monitoring provide conflicting evidence. It has been suggested that the use of longer monitoring periods and/or sealed pedometers help to reduce these effects [80].

6.7 Future Research Directions

In order to enhance comparability between studies and populations, there is a need for standardization of accelerometry protocols [82, 83]. Agreement on which accelerometer cut-points to use in which populations and for which movement intensities is particularly important. Furthermore, future studies should monitor activity throughout the day (24 h) as a strategy to: (1) increase device wear-time; (2) reduce the magnitude of non-wear time; (3) expand evidence on objectively-measured sleep quality and quantity, and (4) allow for the study of whole-day movement behaviours. Smartphone apps for actigraphy have become widely available over the last few years; the accelerometer is built into the mobile phone or an external device, but the validity and reliability of these devices for the monitoring of movement behaviours needs further evaluation. Finally, the combined use of accelerometers and GPS/GIS provides new opportunities to examine the context in which activities of different intensities occur, with the ultimate goal of informing the development of more effective tactics for the promotion of PA.

6.8 Conclusions

In conclusion, objective measuring devices provide many opportunities for improving the surveillance of PA, SB and sleep at the population level. Examination of a variety of intensities, durations, frequencies, volumes and patterns of movement is possible with data captured from currently available objective activity monitors. This rich array of new information provides an opportunity to develop integrated 24 hour movement guidelines, presenting optimal behaviour patterns to reduce the risk of non-communicable diseases (e.g., cancer, cardiovascular diseases and diabetes) and to enhance overall health. To improve the epidemiology of physical activity, it will be important to develop better harmonized cut-points for PA intensity, so that results can be more compared accurately between studies from different laboratories.

References

1. Haskell WL. Physical activity by self-report: a brief history and future issues. J Phys Act Health. 2012;9 Suppl 1:S5–10.

2. Shephard RJ. Limits to the measurement of physical activity by questionnaires. Br J Sport Med. 2003;37:197–206.
3. Adamo KB, Prince SA, Tricco AC, Connor-Gorber S, Tremblay MS. A comparison of indirect versus direct measures for assessing physical activity in the pediatric population: a systematic review. Int J Pediatr Obes. 2009;4(1):2–27.
4. Corder K, Ekelund U, Steele RM, Wareham NJ, Brage S. Assessment of physical activity in youth. J Appl Physiol. 2008;105:977–87.
5. Prince SA, Adamo KB, Hamel ME, Hardt J, Connor Gorber S, Tremblay MS. A comparison of direct versus self-report measures for assessing physical activity in adults: a systematic review. Int J Behav Nutr Phys Act. 2008;5:56.
6. Basterfield L, Adamson AJ, Parkinson KN, et al. Surveillance of physical activity in the UK is flawed: validation of the health survey for England Physical Activity Questionnaire. Arch Dis Child. 2008;93:1054–8.
7. Statistics Canada. Physical activity during leisure time, 2013. http://www.statcan.gc.ca/pub/82-625-x/2014001/article/14024-eng.htm. Accessed 2 Mar 2015.
8. Statistics Canada. Directly measured physical activity of adults, 2012 and 2013. http://www.statcan.gc.ca/pub/82-625-x/2015001/article/14135-eng.htm. Accessed 2 Mar 2015.
9. Statistics Canada. Directly measured physical activity of children and youth, 2012 and 2013. http://www.statcan.gc.ca/pub/82-625-x/2015001/article/14136-eng.htm. Accessed 2 Mar 2015.
10. Katzmarzyk PT, Tremblay MS. Limitations of Canada's physical activity data: implications for monitoring trends. Appl Physiol Nutr Metabol. 2007;32(Suppl 2E):S185–94.
11. Bouchard C, Blair SN, Haskell WL. History and current status of the study of physical activity and health. In: Bouchard C, Blair SN, Haskell WL, editors. Physical activity and health. 2nd ed. Champaign, IL: Human Kinetics; 2012.
12. Troiano RP, Berrigan D, Dodd KW, Mâsse LC, Tilert T, McDowell M. Physical activity in the United States measured by accelerometer. Med Sci Sports Exerc. 2008;40(1):181–8.
13. Craig CL, Cameron C, Griffiths JM, Tudor-Locke C. Process and treatment of pedometer data collection for youth: the Canadian physical activity levels among youth study. Med Sci Sports Exerc. 2010;42:1639–43.
14. Colley RC, Garriguet D, Janssen I, Craig CL, Clarke J, Tremblay MS. Physical activity of Canadian adults: accelerometer results from the Canadian Health Measures Survey. Health Rep. 2011;22(1):7–14.
15. Colley RC, Garriguet D, Janssen I, Craig CL, Clarke J, Tremblay MS. Physical activity of Canadian children and youth: accelerometer results from the Canadian Health Measures Survey. Health Rep. 2011;22(1):15–23.
16. Craig R, Mindell J, Hirani V. Health Survey for England 2008: physical activity and fitness. Summary of key findings. http://www.hscic.gov.uk/catalogue/PUB00430/heal-surv-phys-acti-fitn-eng-2008-rep-v1.pdf. Accessed 2 Mar 2015.
17. Carson V, Ridgers ND, Howard BJ, Winkler EAH, Healy GN, Owen N, Dunstan DW, Salmon J. Light-intensity physical activity and cardiometabolic biomarkers in US adolescents. PLoS ONE. 2013;8(8), e71417.
18. Camhi SM, Sisson SB, Johnson WD, Katzmarzyk PT, Tudor-Locke C. Accelerometer-determined moderate intensity lifestyle activity and cardiometabolic health. Prev Med. 2011;52:358–60.
19. Matthews CE, Chen KY, Freedson PS, Buchowsky MS, Beech BM, Pate RR, Troiano RP. Amount of time spent in sedentary behaviors in the United States, 2003–2004. Am J Epidemiol. 2008;167(7):875–81.
20. Colley RC, Garriguet D, Adamo KB, Carson V, Janssen I, Timmons BW, Tremblay MS. Physical activity and sedentary behavior during the early years in Canada: a cross-sectional study. Int J Behav Nutr Phys Act. 2013;10:54.
21. Craig CL, Cameron C, Tudor-Locke C. CANPLAY pedometer normative reference data for 21,271 children and 12,956 adolescents. Med Sci Sports Exerc. 2013;45(1):123–9.

22. Riddoch C, Edwards D, Page A, et al. The European Youth Heart Study—cardiovascular disease risk factors in children: rationale, aims, study design, and validation of methods. J Phys Act Health. 2005;2:115–29.
23. Riddoch CJ, Andersen LB, Wedderkopp N, Harro M, Klasson-Heggebø L, Sardinha LB, Cooper AR, Ekelund U. Physical activity levels and patterns of 9 and 15-yr-old European children. Med Sci Sports Exerc. 2004;36(1):86–92.
24. Andersen LB, Harro M, Sardinha LB, Froberg K, Ekelund U, Brage S, Anderssen SA. Physical activity and clustered cardiovascular risk in children: a cross-sectional study (the European Youth Heart Study). Lancet. 2006;368(9532):299–304.
25. Ruiz JR, Ortega FB, Martínez-Gómez D, et al. Objectively measured physical activity and sedentary time in European adolescents. Am J Epidemiol. 2011;174(2):173–84.
26. Ness AR, Leary SD, Mattocks C, et al. Objectively measured physical activity and fat mass in a large cohort of children. PLoS Med. 2007;4(3):476–84.
27. Riddoch CJ, Leary SD, Ness AR, et al. Prospective associations between objective measures of physical activity and fat mass in 12–14 year old children: the Avon Longitudinal Study of Parents and Children (ALSPAC). BMJ. 2009;339:b4544.
28. Konstabel K, Veidebaum T, Verbestel V, et al. Objectively measured physical activity in European children: the IDEFICS study. Int J Obes. 2014;38:S135–43.
29. Dunstan DW, Kingwell DA, Larsen R, et al. Breaking up prolonged sitting reduces postprandial glucose and insulin responses. Diabetes Care. 2012;35(5):976–83.
30. Smith L, Ekelund U, Hamer M. The potential yield of non-exercise physical activity energy expenditure in public health. Sports Med. 2015;45(4):449–52.
31. Tremblay MS, Esliger DW, Tremblay A, Colley R. Incidental movement, lifestyle-embedded activity and sleep: new frontiers in physical activity assessment. Appl Physiol Nutr Metabol. 2007;32(Suppl 2E):S208–17.
32. Church TS, Thomas DM, Tudor-Locke C, Katzmarzyk PT, Earnest CP, Rodarte RQ, Martin CK, Blair SN, Bouchard C. Trends over 5 decades in U.S. occupation-related physical activity and their association with obesity. PLoS ONE. 2011;6(5):e19657.
33. Bassett DR, Tremblay MS, Esliger DW, Copeland JL, Barnes JD, Huntington GE. Physical activity and body mass index of children in an old-order Mennonite community. Med Sci Sports Exerc. 2007;39:410–5.
34. Tremblay MS, Esliger DW, Copeland JL, Barnes JD, Bassett DR. Moving forward by looking back: lessons learned from lost lifestyles. Appl Physiol Nutr Metabol. 2008;33(4):836–42.
35. Katzmarzyk PT, Mason C. The physical activity transition. J Phys Act Health. 2009;6:269–80.
36. Ojiambo RM, Easton C, Casajus JA, Konstabel K, Reilly JJ, Pitsiladis Y. Effect of urbanization on objectively measures physical activity levels, sedentary time, and indices of adiposity in Kenyan adolescents. J Phys Act Health. 2012;9:115–23.
37. Onywera VO, Adamo KB, Sheel AW, Waudo JN, Boit MK, Tremblay M. Emerging evidence of the physical activity transition in Kenya. J Phys Act Health. 2012;9:554–62.
38. Esliger DW, Tremblay MS. Physical activity and inactivity profiling: the next generation. Appl Physiol Nutr Metabol. 2007;32(Suppl 2E):S195–207.
39. Tremblay MS, Warburton DER, Janssen I, et al. New Canadian physical activity guidelines. Appl Physiol Nutr Metabol. 2011;36(1):36–46.
40. Murphy MH, Blair SN, Murtagh EM. Accumulated versus continuous exercise for health benefit: a review of empirical studies. Sports Med. 2009;39(1):29–43.
41. Stone M, Rowlands AV, Middlebrooke AR, Jawis MN, Eston RG. The pattern of physical activity in relation to health outcomes in boys. Int J Pediatr Obes. 2009;4(4):306–15.
42. Holman RM, Carson V, Janssen I. Does the fractionalization of daily physical activity (sporadic vs. bouts) impact cardiometabolic risk factors in children and youth? PLoS ONE. 2009;6(10), e25733.
43. Glazer NL, Lyass A, Esliger DW, et al. Sustained and shorter bouts of physical activity are related to cardiovascular health. Med Sci Sport Exerc. 2013;45(1):109–15.

44. Miyashita M, Burns SF, Stensel DJ. Accumulating short bouts of brisk walking reduces postprandial plasma triacylglycerol concentration and resting blood pressure in healthy young men. Am J Clin Nutr. 2008;88(5):1225–31.
45. Bailey RC, Olson J, Pepper SL, Porszasz J, Barstow TJ, Cooper DM. The level and tempo of children's physical activities: an observational study. Med Sci Sports Exerc. 1995;27:1033–41.
46. Lee I-M, Sesso HD, Oguma Y, Paffenbarger Jr RS. The "weekend warrior" and risk of mortality. Am J Epidemiol. 2004;160(7):636–41.
47. Hardman AE. Acute responses to physical activity and exercise. In: Bouchard C, Blair SN, Haskell WL, editors. Physical activity and health. 2nd ed. Champaign, IL: Human Kinetics; 2012.
48. Active Healthy Kids Canada. Don't let this be the most physical activity our kids get after school. 2011 Active Healthy Kids Canada Report Card on Physical Activity for Children and Youth. Toronto, ON: Active Healthy Kids Canada; 2011.
49. Brockman R, Jago R, Fox KR. The contribution of active play to the physical activity of primary school children. Prev Med. 2010;51(2):144–7.
50. Cleland V, Crawford D, Baur LA, Hume C, Timperio A, Salmon J. A prospective examination of children's time spent outdoors, objectively measured physical activity and overweight. Int J Obes. 2008;32:1685–93.
51. Pearce M, Page AS, Griffin TP, Cooper AR. Who children spend time with after school: association with objectively recorded indoor and outdoor physical activity. Int J Behav Nutr Phys Act. 2014;11(1):45.
52. Jankowska MM, Schipperijn J, Kerr J. A framework for using GPS data in physical activity and sedentary behavior studies. Exerc Sport Sci Rev. 2015;43(1):48–56.
53. Levine JA, Eberhardt NL, Jensen MD. Role of non-exercise activity thermogenesis in resistance to fat gain in humans. Science. 1999;283:212–4.
54. Levine JA, Lanningham-Foster LM, McCrady SK, Krizan AC, Olson LR, Kane PH, Jensen MD, Clark MM. Interindividual variation in posture allocation: possible role in human obesity. Science. 2005;307:584–6.
55. Westerterp KR. Pattern and intensity of physical activity. Nature. 2001;410:539.
56. Colley RC, Hills AP, King NA, Byrne AM. Exercise-induced energy expenditure: implications for exercise prescription and obesity. Pat Educ Couns. 2010;79(3):327–32.
57. Westerterp KR, Kester ADM. Physical activity in confined conditions as an indicator of free-living physical activity. Obes Res. 2003;11:865–8.
58. Sedentary Behaviour Research Network. Standardized use of the terms "sedentary" and "sedentary behaviours" [letter to the editor]. Obes Res. 2012;37:540e2.
59. Tremblay MS, Colley R, Saunders TJ, Healy GN, Owen N. Physiological and health implications of a sedentary lifestyle. Appl Physiol Nutr Metab. 2010;35:725–40.
60. Tremblay MS, LeBlanc AG, Kho ME, Saunders TJ, Larouche R, Colley RC, Goldfield G, Connor GS. Systematic review of sedentary behaviour and health indicators in school-aged children and youth. Int J Behav Nutr Phys Act. 2011;8:98.
61. Chaput JP, Carson V, Gray CE, Tremblay MS. Importance of all movement behaviors in a 24 hour period for overall health. Int J Environ Res Public Health. 2014;11:12575–81.
62. Chaput JP. Sleep patterns, diet quality and energy balance. Physiol Behav. 2014;134:86–91.
63. Buman MP, Winkler EA, Kurka JM, et al. Reallocating time to sleep, sedentary behaviors, or active behaviors: associations with cardiovascular disease risk biomarkers, NHANES 2005–2006. Am J Epidemiol. 2014;179:323–34.
64. Leproult R, Van Cauter E. Role of sleep and sleep loss in hormonal release and metabolism. Endocr Dev. 2010;17:11–21.
65. Sadeh A, Sharkey KM, Carskadon MA. Activity-based sleep-wake identification: an empirical test of methodological issues. Sleep. 1994;17(3):201–7.
66. Cole RJ, Kripke DF, Gruen W, Mullaney DJ, Gillin JC. Automatic sleep/wake identification from wrist activity. Sleep. 1992;15(5):461–9.

67. Galland B, Meredith-Jones K, Terrill P, Taylor R. Challenges and emerging technologies within the field of pediatric actigraphy. Front Psychiatry. 2014;5:99.
68. Meltzer LJ, Montgomery-Downs HE, Insana SP, Walsh CM. Use of actigraphy for assessment in pediatric sleep research. Sleep Med Rev. 2012;16:463–75.
69. Sadeh A. The role and validity of actigraphy in sleep medicine: an update. Sleep Med Rev. 2011;15:259–67.
70. Girschik J, Fritschi L, Heyworth J, Waters F. Validation of self-reported sleep against actigraphy. J Epidemiol. 2012;22:462–8.
71. Mekary RA, Willett WC, Hu FB, et al. Isotemporal substitution paradigm for physical activity epidemiology and weight change. Am J Epidemiol. 2009;170:519–27.
72. Panagiotakos DB, Pitsavos C, Polychronopoulos E, et al. Can a Mediterranean diet moderate the development and clinical progression of coronary heart disease? A systematic review. Med Sci Monit. 2004;10:RA193–8.
73. Marshall SJ, Biddle SJH, Sallis JF, et al. Clustering of sedentary behaviors and physical activity among youth: a cross-sectional study. Pediatr Exerc Sci. 2002;14:401–17.
74. Ferrar K, Chang C, Li M, Olds TS. Adolescent time use clusters: a systematic review. J Adolesc Health. 2013;52:259–70.
75. Stone MR, Faulkner GE. Outdoor play in children: associations with objectively-measured physical activity, sedentary behavior and weight status. Prev Med. 2014;65:122–7.
76. Owens J, Adolescent Sleep Working Group and Committee on Adolescence. Insufficient sleep in adolescents and young adults: an update on causes and consequences. Pediatrics. 2014;134:921–32.
77. Schmid SM, Hallschmid M, Schultes B. The metabolic burden of sleep loss. Lancet Diabetes Endocrinol. 2015;3:52–62.
78. Chaput JP, Tremblay A. Insufficient sleep as a contributor to weight gain: an update. Curr Obes Rep. 2012;1:245–56.
79. Tremblay MS, Haskell WL. From science to physical activity guidelines. In: Bouchard C, Blair SN, Haskell WL, editors. Physical activity and health. 2nd ed. Champaign, IL: Human Kinetics; 2012. p. 359–78.
80. Clemes SA, Biddle SJ. The use of pedometers for monitoring physical activity in children and adolescents: measurement considerations. J Phys Act Health. 2013;10(2):249–62.
81. Grant PM, Ryan CG, Tigbe WW, Granat MH. The validation of a novel activity monitor in the measurement of posture and motion during everyday activities. Br J Sports Med. 2006;40:992–7.
82. Cain KL, Sallis JF, Conway TL, Van Dyck D, Calhoon L. Using accelerometers in youth physical activity studies: a review of methods. J Phys Act Health. 2013;10(3):437–50.
83. Pedisic Z, Bauman A. Accelerometer-based measures in physical activity surveillance: current practice and issues. Br J Sports Med. 2015;49:219–23.
84. Watson KB, Carlson SD, Carroll DD, Fulton JE. Comparison of accelerometer cut points to estimate physical activity in US adults. J Sports Sci. 2014;32(7):660–9.
85. Loprinzi PD, Cardinal BJ, Crespo CJ, et al. Differences in demographic, behavioral, and biological variables between those with valid and invalid accelerometry data: implications for generalizability. J Phys Act Health. 2013;10:79–84.
86. Loprinzi PD, Smit E, Cardinal BJ, et al. Valid and invalid accelerometry data among children and adolescents: comparison across demographic, behavioral, and biological variables. Am J Health Promo. 2014;28:155–8.
87. Roth MA, Mindell JS. Who provides accelerometry data? Correlates of adherence to wearing an accelerometry motion sensor: the 2008 health survey for England. J Phys Act Health. 2013;10:70–8.
88. Katzmarzyk PT, Barreira TV, Broyles ST, et al. The international study of childhood obesity, lifestyle and the environment (ISCOLE): design and methods. BMC Public Health. 2013;13:900.
89. Tudor-Locke C, Barreira TV, Schuna Jr JM, et al. Improving wear time compliance with a 24-hour waist-worn accelerometer protocol in the International Study of Childhood Obesity, Lifestyle and the Environment (ISCOLE). Int J Behav Nutr Phys Act. 2015;12:11.

Chapter 7
Can the Epidemiologist Learn more from Sedentary Behaviour than from the Measurement of Physical Activity?

Valerie Carson, Travis Saunders, and Mark S. Tremblay

Abstract Recent research suggests that there are biological mechanisms provoked by sedentary behaviours that result in health consequences even after accounting for the influence of an individual's habitual physical activity. Objective monitors, like accelerometers and inclinometers, allow for the direct measurement of the time spent sedentary in addition to monitoring the extent of physical activity. This allows for epidemiological analyses considering either behaviour or both behaviours in combination. The relatively new field of sedentary physiology and epidemiology allows for creative exploration of correlates, determinants, and consequences of varying levels of sedentary behaviour that can inform novel interventions to promote healthy living.

7.1 Introduction

In the past, many research studies have defined people as sedentary when they are not sufficiently active to meet recognized physical activity guidelines. More recently, the Sedentary Behaviour Research Network (www.sedentarybehaviour.org) recognized that being sedentary is a distinct and separate behaviour, not merely

---

V. Carson (✉)
Faculty of Physical Education and Recreation, University of Alberta, Edmonton, AB, Canada
e-mail: vlcarson@ualberta.ca

T. Saunders
Department of Applied Human Sciences, Faculty of Science, University of Prince Edward Island, Charlottetown, PE, Canada

M.S. Tremblay
Healthy Active Living and Obesity Research Group, CHEO Research Institute, Ottawa, ON, Canada

Department of Pediatrics, University of Ottawa, Ottawa, ON, Canada

R.J. Shephard, C. Tudor-Locke (eds.), *The Objective Monitoring of Physical Activity: Contributions of Accelerometry to Epidemiology, Exercise Science and Rehabilitation*, Springer Series on Epidemiology and Public Health,
DOI 10.1007/978-3-319-29577-0_7

the absence of adequate physical activity, and it defined sedentary behaviour as any waking periods characterized by an energy expenditure ≤1.5 METs while the individual is in a sitting or reclining posture [1]. In contrast, the Network has suggested that authors use the term "inactive" to describe those who are not performing sufficient amounts of moderate- to vigorous-intensity physical activity (MVPA, i.e., not meeting specified physical activity guidelines) [1]. The distinction between inactive and sedentary is important in the context of objective monitoring because devices like accelerometers are able to measure and quantify both sedentary time and the time spent at various intensities of physical activity, allowing epidemiologists to study each behaviour both separately, and together (as discussed in Chap. 6).

To further clarify terminology, specific sedentary *behaviours* (e.g., watching television, travelling in a motorized vehicle reading, eating, or using a computer) are not captured by accelerometers, nor is the context of the behaviour (e.g., at home, at work, in school, or outdoors), and these are important issues to examine in order to develop a full understanding of the relationships between various sedentary behaviours and health (e.g., screen time behaviours vs. non-screen time behaviours; sedentary recreational behaviours vs. sedentary occupational behaviours). Nevertheless, accelerometers are capable of providing reliable estimates of sedentary *time*; this is a very important measure, because the amount of daily time spent sedentary is far greater than the amount of time spent in MVPA. A typical daily behaviour pattern includes <5 % of MVPA and >40 % of sedentary time [2]. In Canada, children are sedentary approximately 8.6 hours/day, while adults spend 9.5 hours/day in sedentary activities [3, 4]. Similar amounts of sedentary time are observed in other developed countries [5, 6]. Esliger and Tremblay summarized many of the indices of sedentary time that accelerometers can capture, and that can be used for epidemiological purposes [7].

In recent years, an increasing number of epidemiological and laboratory based studies have suggested that sedentary behaviour is an independent health risk factor in both children and adults. Multiple systematic reviews and meta-analyses have concluded that sedentary behaviour is associated with an increased risk of cardiovascular disease, diabetes mellitus, certain types of cancer and premature mortality in adults [8–12], as well as an excessive body mass, a low level of fitness and metabolic dysfunction in children and youth [13, 14]. Importantly, the most recent evidence suggests that these relationships are independent of an individual's level of physical activity [8, 12]. For example, a meta-analysis by Biswas et al. concluded that compared to individuals who sit the least, individuals who sit the most are at a 24 % increased risk of all-cause mortality and a 91 % increased risk of diabetes mellitus, independently of physical activity levels. Smaller but statistically significant associations were observed for cancer and cardiovascular disease incidence and mortality [12]. While the authors of this report noted that the associations were weakened among those with high levels of physical activity [12], their results nonetheless support the assertion that sedentary behaviour and physical activity represent unique constructs with independent impacts on chronic disease morbidity and mortality.

7.2 Biology of Sedentary Behaviour

Several biological mechanisms have been suggested which may explain the observed linkage between sedentary behaviours and an increased risk of morbidity and mortality, as seen frequently in epidemiological studies [15–18]. Laboratory-based studies of adults have consistently found that prolonged periods of sedentary behaviour rapidly result in reduced insulin sensitivity and glucose tolerance [18–22]. Peddie and colleagues compared the insulin and glucose response to a test meal following 9 hours of uninterrupted sitting, to the response seen when 9 hours of sitting was interspersed with light intensity walk breaks taken every 30 minutes [20]. They reported that interruptions in sedentary time led to a 39 % reduction in the glucose response to a test meal, and a 26 % reduction in the plasma insulin response, when compared to the findings for uninterrupted sitting. Similarly, Buckley et al. reported that standing for 4 hours reduced the glucose response to a test meal by 43 %, in comparison to sitting [22]. Some studies have also reported that uninterrupted bouts of sedentary behaviour increase triglyceride levels [18, 19], although this finding has been less consistent than the impact upon insulin and glucose levels [20, 21].

In contrast to the consistent associations between prolonged sedentary behaviour and metabolic dysfunction that are seen in adults, uninterrupted sitting appears to have little or no acute impact upon healthy children and youth. In comparison to 8 hours of prolonged sitting, Saunders et al. observed no change in the insulin, glucose, and triglyceride responses to a test meal during 8 hours of sitting interrupted by light intensity walking breaks or structured physical activity [23]. Sisson et al. [24] have reported similar results. Although only a small number of studies have examined the acute impact of prolonged sedentary behaviour in children and youth, the available evidence suggests that any effect is much smaller (or nonexistent) when compared to that observed in adults.

The metabolic responses to uninterrupted sedentary behaviour in adults are likely due to changes occurring at the level of the skeletal muscle. Models of prolonged skeletal muscle inactivity (casting, denervation, and nerve injury) have all led to reduced insulin sensitivity [25–29]. This likely reflects a reduced glucose transport protein (GLUT) content, which is also reduced in these models of sedentary behaviour [25, 27–29]. In support of this view, Latouche et al. reported that breaking up prolonged sitting with a light intensity walking break every 20 minutes led to an increased expression of several genes related to carbohydrate metabolism, including those thought to regulate GLUT activity [30]. Through animal models of sedentary behaviour, Hamilton (Fig. 7.1) et al. have found that prolonged muscle inactivity may also reduce the activity of lipoprotein lipase [17], providing a possible explanation for the relationship that some have observed between sedentary behaviour and dyslipidemia [19]. However, this last mechanism has yet to be investigated in humans.

Available evidence suggests that sedentary behaviour is also likely to impact vascular function and blood pressure [31–34]. In a study of 20 healthy adults,

Fig. 7.1 Marc Hamilton introduced the paradigm of inactivity physiology through his extensive lab-based work on prolonged sedentary time

Hamburg et al. [34] reported that 3 days of bed rest resulted in a 29 % reduction in reactive hyperemia (a measure of vascular health). Changes in vascular function have also been observed in response to more typical periods of sedentary behaviour. Shvartz et al. reported that 5 hours of uninterrupted sitting resulted in increased peripheral resistance, diastolic blood pressure and mean arterial pressure in a group of 13 young men [31]. In a separate study of 12 young men, Thosar et al. found that reactive hyperemia was reduced after just 1 hour of uninterrupted sitting [33]. In contrast, Larsen et al. observed that simply interrupting 5 hours of prolonged sitting with light intensity walking breaks resulted in small but significant (3 mmHg, $p < 0.01$) reductions in both systolic and diastolic blood pressure relative to uninterrupted sitting [32]. Taken together, these results suggest that uninterrupted sedentary behaviour is likely to increase blood pressure and impair vascular function in adults. Cardiovascular relationships have not been studied in children and youth to date.

The final mechanism through which sedentary behaviour is likely to impact health is through an association with other health-related behaviours, in particular food intake and sleep [35–37]. Research suggests that many common screen-based sedentary behaviours (e.g., TV viewing, computer use, and video game use) promote an excessive food intake in both adults and children [30–33]. For example, in a small intervention study, Harris et al. reported that exposure to food advertisements increased food intake by roughly 30 % in adults, and up to 45 % in children [37]. Similarly, screen time is associated with reduced sleep [36, 38], which is in itself a predictor of weight gain [39, 40]. Thus prolonged screen-based sedentary behaviour is likely to have both a direct metabolic impact, as well as a negative influence upon other health-related behaviours.

7.3 Correlates of Sedentary Behaviours and Sedentary Time

The described mechanistic differences between the impact of physical activity and sedentary behaviour on health strongly support the idea that sedentary behaviour is distinct from an inadequate volume of physical activity. From a behavioural standpoint, understanding whether the correlates or factors that influence sedentary behaviour differ from the correlates or factors that influence physical activity can test this idea. Neville Owen (Fig. 7.2) and colleagues developed the Ecologic Model of Sedentary Behavior as a conceptual approach to understanding the correlates of sedentary behaviour [41]. Consistent with other ecological models, this model postulates that sedentary behaviour across all age-groups, is influenced by correlates at multiple levels such as individual, social, environmental, and health policy [41, 42]. Individual or intrapersonal correlates can include demographic, biological, cognitive, and behavioural factors [43, 44]. Social or interpersonal correlates can include family and peer factors [43, 44]. Environmental correlates can include factors related to home, community or neighbourhood [43, 44]. Health policy correlates can include local, provincial/state, and national laws [41]. According to this model, the correlates that influence sedentary behaviour are specific to the setting where the sedentary behaviour occurs [41]. For example, the correlates of sedentary behaviour for children and youth often differ between home and school/daycare settings. Similarly for adults, correlates are likely to differ between home and work settings. Understanding the most important setting-specific multi-level correlates that influence sedentary behaviour, using high-quality measures, should enable the development of more successful health promotional interventions [41].

The majority of research examining correlates of sedentary behaviour in children and youth has focused on one particular type of sedentary behaviour, namely television viewing [42]. The allocation of time to television viewing has typically been measured subjectively with self- or proxy-report questionnaires, logs/diaries,

Fig. 7.2 Neville Owen led the development of the Ecologic Model of Sedentary Behavior

or interviews [45]. A number of reviews have synthesized evidence on the most consistent correlates of television viewing in children and youth from birth to 18 years old [42–44, 46, 47]. The most recent review [42] identified ethnicity (being non-Caucasian) as a consistent positive correlate of television viewing at the individual level, socio-demographic status as a consistent negative correlate, parental television viewing as a consistent positive correlate at the social level; and having a TV set in the bedroom as a consistent positive correlate at the environmental level. It is not surprising that the home setting is a particularly important determinant of television viewing among children and youth.

Although television viewing is a highly prevalent sedentary behaviour among children and youth, evidence suggests that television viewing should not be used as a marker for total sedentary time [48]. Compared to television viewing, considerably less research has examined the correlates of accelerometer-derived total sedentary time in children and youth. Research on this topic is growing, but currently it is difficult to determine whether specific correlates are consistently associated with total sedentary time. The only exception to this generalization is for sex; a number of studies have reported that females engage in more total sedentary time than males [3, 5, 49–53]. For instance, in a large representative sample of Canadian youth aged 15–19 year old, females had larger totals of sedentary time than males of similar age [3]. Similarly, a survey of a large representative sample of American children and youth found that females aged 6–19 years had higher levels of total sedentary time then males in the same age range [5].

In comparison to children, there has been less research focusing on the correlates of adult's sedentary behaviour [41]. However, similar to children, the majority of research that has been conducted has focused on the correlates of subjectively measured television viewing [54].

A recent review of the correlates of sedentary behaviour in adults [54] found that although sociodemographic and behavioral correlates of sedentary behaviour were frequently examined at the individual level, evidence on individual cognitive, social, and environmental correlates were generally lacking. Findings differed depending on the type of sedentary behaviour that was measured. Increasing age, unemployment or retirement, body mass index, a sedentary attitude, and depressive symptoms were consistent positive correlates of television viewing; in contrast, educational status, leisure time physical activity, and life satisfaction were consistent negative correlates. For computer use, educational status and a sedentary attitude were consistent positive correlates and age was a consistent negative correlate. For video game use, sex (being male) was a consistent positive correlate, and leisure time physical activity was a consistent negative correlate of screen-time. For self-reported general sitting, having children in the home was a consistently negative correlate. Only three of the studies included in this review had measured accelerometer-derived total sedentary time, and consistent associations across the studies had not been observed [10, 17, 18].

To date, many of the studies examining the correlates of sedentary behaviour have drawn upon correlates from the physical activity literature [42, 55]. Future research needs to focus upon correlates that are specific to sedentary behaviour, as outlined in the Ecologic Model of Sedentary Behavior [41]. This should include research on the correlates of objectively measured total sedentary time across various age groups. Current research suggests that health promotional interventions may need to consider specific types of sedentary behaviour individually when developing and implementing tactics to reduce sedentary behaviour.

7.4 Measures and Indices of Sedentary Behaviours

Even though the intensity of sedentary behaviour is reasonably fixed, by definition [1], there are multiple characteristics of sedentary behaviours that can be explored to understand there particular impact upon health [7, 16]. Just as physical activity can be described by the total amount and the pattern of frequency, intensity, time (duration), and type (mode) ("FITT" variables), sedentary behaviours can be characterized using their "SITT" characteristics (sitting frequency, interruptions, time, type) [16]. Epidemiological analyses can examine these characteristics as continuous (e.g., total daily sedentary time) or categorical (e.g., by tertiles of sedentary interruption episodes; or TV viewing vs. reading) variables. Much as in the analysis of physical activity, epidemiologists can explore "bouts" of sedentary time of differing duration (e.g., the number of daily bouts of greater than 60 consecutive minutes; or the longest daily bout) as well as temporal patterns within the day (e.g., the proportion of sedentary time occurring in the morning, afternoon, or evening) and between days (e.g., weekdays vs. weekend days; or Mondays vs. Fridays). The universe of sedentary epidemiology is just being discovered, and it merits much further exploration. As discussed in Chap. 6, the most robust and informative epidemiological discoveries will likely emerge from analyses that consider all movement/non-movement behaviours together (i.e., physical activity, sedentary behaviours and sleep) [2]. Table 7.1 provides a summary of objective monitoring devices capable of providing measures of sedentary time and their respective advantages and disadvantages.

7.5 Sedentary Measures: Do They Tell Us Anything New?

Research in physical activity epidemiology has traditionally focused on MVPA or exercise [56]. As previously outlined in this book, participation in regular physical activity, in particular MVPA, brings numerous health benefits across an individual's lifespan. New epidemiological evidence suggests that engaging in excessive sedentary behaviour may be harmful to health, regardless of how active one is [56]. Not only the duration of sedentary behaviour but also its patterns seem important for health; for instance, Healy (Fig. 7.3) and colleagues reported that accumulating sedentary behaviour with more frequent breaks may have less

Table 7.1 Summary of objective measurement devices that can provide an index of sedentary time

Device	Advantages	Disadvantages
Heart rate monitors	• Relatively inexpensive	• Need to distinguish "sedentary" heart rate for individual participants. • Influenced by non-movement exposures (e.g., emotions, caffeine) • Uncomfortable for long periods
Accelerometers	• Accurate measure of overall sedentary time	• Cannot distinguish sitting from standing • Moderate burden to participants • Results are influenced by data processing • Expensive
Inclinometers	• Criterion standard for distinguishing sitting from standing in field-based studies	• Moderate burden to participants • Expensive
Direct observation	• Criterion standard for distinguishing sitting from standing in laboratory studies	• Requires a large time commitment from researchers
Indirect calorimetry	• Precise measurement of energy expenditure	• Cannot distinguish sitting from standing • Possible respondent reactivity • Uncomfortable • Expensive
Sitting pads	• Accurate measurement of sitting time	• Cannot assess other movement behaviours (e.g., standing) • Limited to a single chair

adverse health effects than uninterrupted sitting [57, 58]. However, to date findings have not been consistent across all types of sedentary behaviours and all age groups.

Machado de Rezende et al. [59] synthesized evidence from 27 systematic reviews that were published between 2004 and 2013; these examined relationships between sedentary behaviour and health, primarily though observational studies. A total of 11 of these reviews assessed physical activity as a covariate [59]. On average, statistical adjustment for physical activity was made in 63 % of the studies included. Findings differed depending on the age group and type of sedentary behaviour that was examined. For children and youth, a meta-analysis of moderate-quality randomized controlled trials found that increased television viewing was associated with poorer body composition [14]. A review of longitudinal studies reported that higher television viewing or screen time was associated with poorer physical fitness, independently of physical activity levels [60]. Higher television viewing was also consistently associated with lower academic achievement in school-aged children and youth [14] and in two reviews was linked to poorer cognitive development in the early years [13], but physical activity data

Fig. 7.3 Genevieve Healy provided some of the first observational evidence that frequently interrupting sedentary time is associated with beneficial health outcomes

were not extracted. No consistent evidence was found linking accelerometer-derived total sedentary time with health outcomes in this age group [59]. On the other hand, in adults, a number of reviews reported associations between higher television viewing, screen time, and self-reported total sitting time vs. all-cause mortality and cardiovascular disease mortality [8, 10, 11, 61, 62], and these associations were independent of physical activity levels [8, 10, 61, 62]. Similar findings were reported for cardiovascular disease in two reviews [8, 61] and type 2 diabetes mellitus in three reviews [8, 11, 61]. Additionally, higher television viewing, screen time, self-reported sitting time, and accelerometer-derived total sedentary time were associated with the metabolic syndrome in one review, independently of physical activity levels [63]. Similar findings have been reported in another recent meta-analysis of adult data [12].

None of the 27 reviews synthesized by Machado de Rezende et al. focused specifically on older adults [59]; however, Machado de Rezende published a second review focusing on individuals >60 years of age [64]. This article excluded studies if they did not include physical activity as a covariate. High quality evidence indicated that greater self-reported sitting was associated with an increase of all-cause mortality [65, 66]. Moderate quality evidence indicated higher television viewing was associated with the metabolic syndrome, as well as with a higher waist to hip ratio [67]. Additionally, higher accelerometer-derived total sedentary time was associated with a greater waist circumference and body mass index (BMI) [68]. A very low quality study in older adults also found that greater accelerometer-observed breaks in sedentary time was protective against the metabolic syndrome [69]. Similar findings have been observed in representative samples of Canadian and US adults [58, 70]. More specifically, in Canadian adults longer breaks in sedentary time were associated with a smaller waist circumference, a lower systolic blood pressure, and more favourable HDL-cholesterol, triglycerides, glucose, and insulin values [70], and in US adults were associated with an improved waist circumference and C-reactive protein levels [58]. Similar findings have not been observed in representative samples of Canadian and US children and youth [71, 72]. However, one study by Carson et al. suggests that if more time is spent in longer sedentary bouts, this is associated with a greater body mass in children with lower levels of MVPA, but not in those with higher levels of MVPA [73].

Understanding the relationship between sedentary behaviour and health across the lifespan cannot be determined simply by measuring physical inactivity. Measures of both physical activity and sedentary behaviour are needed in order to understand the impact of the two behaviours on health. For adults in particular, both the duration and the patterns of sedentary behaviour appear to be important. As seen with correlates of sedentary behaviour, associations between sedentary behaviour and specific health indicators may differ both by the type of sedentary behaviour and the age group. This suggests a combination of measurement tools may be needed to gain a full understanding of the health consequences of sedentary behaviour.

7.6 Limitations of Objective Measurements of Sedentary Behaviour

As with other health-related behaviours, measurements of self-reported sedentary behaviour have the potential to introduce large amounts of error and bias [74]. Thus, although instrumentation is not perfect, some observers have argued the need for an increased use of objective measures of sedentary behaviour, or a combination of objective and subjective measurements [15, 74, 75]. At present the most common device for the objective measurement of sedentary time is a hip-worn accelerometer. This method has traditionally been favoured in large epidemiological surveys, including the US National Health and Nutrition Examination Survey [5, 71] (currently transitioning to wrist-worn accelerometers), the Canadian Health Measures Survey [3, 4], and the International Children's Accelerometry Database [76].

Accelerometers provide a measure of movement (typically assessed as "counts" per minute), with periods when movement falls below a critical threshold (typically 100 counts per minute [40, 41] being considered as sedentary time. However, hip- and wrist-worn accelerometers cannot provide a measure of posture/incline. Thus, they are unable to distinguish between sitting/lying down and standing still [77]; Dowd et al. have reported that the Actigraph accelerometer (Actigraph Corporation, Pensacola, USA) displayed 0 % accuracy for distinguishing sitting from standing [78]. However, likely because of the relatively limited amount of time most people spend standing still on a given day, accelerometers appear to provide a valid measure of sedentary time. For example, Ridgers et al. [79] reported that over 6.5 hours of wear-time, sedentary time derived using the Actigraph monitor differed by an average of less than 6 minutes (equivalent to 2.5 % of total sedentary time) when compared to the criterion standard activPAL (PAL Technologies, Glasgow, UK).

The other major limitation of accelerometry is that the results are known to be influenced by data processing, which varies substantially from study to study [15, 77, 80]. Definitions of non-wear time (e.g., the number of minutes with consecutive values of 0 counts per minute), the number of hours required for a day to be considered "valid", the number of valid days required for an individual to be included in an analysis, and even the threshold used to define sedentary behaviour vary across studies [8]. This problem is especially obvious with large public data-sets, such as the US National Health and Nutrition Examination Survey, which are analyzed by multiple authors using different data processing methods [81]. Not surprisingly, differences in data processing techniques not only influence estimates of sedentary time, but also the associations between sedentary time and health outcomes [82]. Although this problem has been widely noted, at present there is no consensus on the "ideal" data-processing steps to be taken when cleaning accelerometer data, suggesting that the problem is likely to persist into the future.

Another objective measure of sedentary behaviour that is beginning to be used more frequently is inclinometry, which allows an assessment of both movement *and* posture. In a study of 25 healthy children, Aminian and Hinckson reported that the activPAL inclinometer showed perfect agreement with direct observation when identifying periods of sitting, lying, standing and walking [83]; similar results were reported by Dowd et al. [78]. However, inclinometers must be worn on the thigh, which may increase participant burden when compared to hip- or wrist-worn accelerometers. Despite their benefits with respect to accuracy, inclinometers have to date been used far less frequently than accelerometers as tools for objectively measuring sedentary time.

Other methods such as direct or video observation and indirect calorimetry are useful in small laboratory studies where precision is paramount, as they offer an extremely accurate measurement of sedentary behaviours. However, given their high cost, high time commitment, and participant burden, such methods of assessment are neither feasible nor appropriate for large, field-based studies. Sitting pads (e.g., cushions that assess the time spent sitting) also provide extremely accurate measures of sitting [84, 85], although they have yet to become widely used. Such devices can only measure the time spent sitting in a given chair, and thus they are not ideal for assessing the total sedentary time throughout the day.

The final limitation common to all objective measurements of sedentary behaviour (aside from direct observation) is their inability to record the type and context of the sedentary behaviour. For example, accelerometers and inclinometers cannot distinguish between sitting at the dinner table, sitting in front of a television, or lying in bed reading a book. This limitation is extremely important, because some behaviours that occur while sitting likely influence concurrent (e.g., incidental snacking) and proximal (e.g., likelihood of getting up soon) behaviours, with an influence upon the health impact of the recorded sedentary behaviour [15]. Screen-based sedentary behaviours are consistently associated with increased health risk [13, 14], while the impact of reading a book may be neutral [86], or even positive

[87]. For these reasons, some have advocated an approach that combines objective with subjective measures of sedentary behaviour whenever possible [15, 75].

7.7 Future Research Directions

With "sedentary epidemiology" in its infancy there is much research yet to be done. Research priorities include:

- Standardization of the methodology for determining sedentary time from accelerometers (e.g., cut-points, non-wear time, valid days required).
- Development of novel, inexpensive, unobtrusive, high resolution, objective measuring devices to record sedentary time and sedentary behaviours.
- Determining the sedentary behaviour variables that are most important to health outcomes and examining the sensitivity and specificity of these indices across age, sex and socio-cultural groups.
- Determination of whether the health impact of sedentary behaviour is consistent across differing levels of physical activity.
- Examination of the health impact of different types and contexts of sedentary behaviour.
- Studies of dose-response characteristics to help inform the establishment of public health guidelines concerning sedentary time and sedentary behaviours.
- Understanding the modifiable correlates of sedentary behaviour across contexts and age groups.
- Development and evaluation of interventions to reduce sedentary behaviour across contexts and age groups.
- Integrating measurements of movement and non-movement across the 24-hour period for analyses of the impact upon health indicators.

7.8 Conclusions

Evidence from epidemiological and laboratory based studies underlines that sedentary behaviour is not simply the absence of MVPA. Though sedentary behaviour research is growing rapidly, it remains approximately 20 years behind studies of MVPA and exercise-focused research [88]. The recent development of objective measuring devices that enable researchers to capture not only the duration of sedentary behaviour but also various patterns of accumulation has contributed to an explosion of research in this area. Objective measures of sedentary behaviour are not without their limitations; further advances in the area of sedentary epidemiology will require the development and refinement of valid and reliable objective measures of sedentary behaviour that can incorporate the type and contextual factors that are needed. This will allow meaningful contributions to our understanding of the most important health effects and the modifiable correlates of sedentary behaviour, as well as facilitating the development of effective interventions for healthy living across the entire human lifespan.

References

1. Sedentary Behaviour Research Network. Standardized use of the terms "sedentary" and "sedentary behaviours". Appl Physiol Nutr Metab. 2012;37(3):540–2.
2. Chaput JP, Carson V, Gray CE, et al. Importance of all movement behaviors in a 24 hour period for overall health. Int J Environ Res Public Health. 2014;11(12):12575–81.
3. Colley RC, Garriguet D, Janssen I, et al. Physical activity of Canadian children and youth: accelerometer results from the 2007 to 2009 Canadian Health Measures Survey. Health Rep. 2011;22(1):15–23.
4. Colley RC, Garriguet D, Janssen I, et al. Physical activity of Canadian adults: accelerometer results from the 2007 to 2009 Canadian Health Measures Survey. Health Rep. 2011;22 (1):7–14.
5. Matthews CE, Chen KY, Freedson PS, et al. Amount of time spent in sedentary behaviors in the United States, 2003–2004. Am J Epidemiol. 2008;167(7):875–81.
6. Ruiz JR, Ortega FB, Martinez-Gomez D, et al. Objectively measured physical activity and sedentary time in European adolescents: the HELENA study. Am J Epidemiol. 2011;174 (2):173–84.
7. Esliger DW, Tremblay MS. Physical activity and inactivity profiling: the next generation. Can J Public Health. 2007;98 Suppl 2:S195–207.
8. Wilmot EG, Edwardson CL, Achana F, et al. Sedentary time in adults and the association with diabetes, cardiovascular disease and death: systematic review and meta-analysis. Diabetologia. 2012;55(11):2895–905.
9. Lynch BM. Sedentary behavior and cancer: a systematic review of the literature and proposed biological mechanisms. Cancer Epidemiol Biomarkers Prev. 2010;19(11):2691–709.
10. Thorp AA, Owen N, Neuhaus M, et al. Sedentary behaviors and subsequent health outcomes in adults a systematic review of longitudinal studies, 1996–2011. Am J Prev Med. 2011;41 (2):207–15.
11. Proper KI, Singh AS, van Mechelen W, et al. Sedentary behaviors and health outcomes among adults: a systematic review of prospective studies. Am J Prev Med. 2011;40(2):174–82.
12. Biswas A, Oh PI, Faulkner GE, et al. Sedentary time and its association with risk for disease incidence, mortality, and hospitalization in adults: a systematic review and meta-analysis. Ann Intern Med. 2015;162(2):123–32.
13. LeBlanc AG, Spence JC, Carson V, et al. Systematic review of sedentary behaviour and health indicators in the early years (aged 0–4 years). Appl Physiol Nutr Metab. 2012;37(4):753–72.
14. Tremblay MS, LeBlanc AG, Kho ME, et al. Systematic review of sedentary behaviour and health indicators in school-aged children and youth. Int J Behav Nutr Phys Act. 2011;8:98.
15. Saunders TJ, Chaput JP, Tremblay MS. Sedentary behaviour as an emerging risk factor for cardiometabolic diseases in children and youth. Can J Diabetes. 2014;38(1):53–61.
16. Tremblay MS, Colley RC, Saunders TJ, et al. Physiological and health implications of a sedentary lifestyle. Appl Physiol Nutr Metab. 2010;35(6):725–40.
17. Hamilton MT, Hamilton DG, Zderic TW. Role of low energy expenditure and sitting in obesity, metabolic syndrome, type 2 diabetes, and cardiovascular disease. Diabetes. 2007;56 (11):2655–67.
18. Saunders TJ, Larouche R, Colley RC, et al. Acute sedentary behaviour and markers of cardiometabolic risk: a systematic review of intervention studies. J Nutr Metabol. 2012;2012 (2012):712435.

19. Duvivier BM, Schaper NC, Bremers MA, et al. Minimal intensity physical activity (standing and walking) of longer duration improves insulin action and plasma lipids more than shorter periods of moderate to vigorous exercise (cycling) in sedentary subjects when energy expenditure is comparable. PLoS One. 2013;8(2):e55542.
20. Peddie MC, Bone JL, Rehrer NJ, et al. Breaking prolonged sitting reduces postprandial glycemia in healthy, normal-weight adults: a randomized crossover trial. Am J Clin Nutr. 2013;98(2):358–66.
21. Dunstan DW, Kingwell BA, Larsen R, et al. Breaking up prolonged sitting reduces postprandial glucose and insulin responses. Diabetes Care. 2012;35(5):976–83.
22. Buckley JP, Mellor DD, Morris M, et al. Standing-based office work shows encouraging signs of attenuating post-prandial glycaemic excursion. Occup Environ Med. 2014;71(2):109–11.
23. Saunders TJ, Chaput JP, Goldfield GS, et al. Prolonged sitting and markers of cardiometabolic disease risk in children and youth: a randomized crossover study. Metabolism. 2013;62(10):1423–8.
24. Sisson SB, Anderson AE, Short KR, et al. Light activity following a meal and postprandial cardiometabolic risk in adolescents. Pediatr Exerc Sci. 2013;25(3):347–59.
25. Megeney LA, Neufer PD, Dohm GL, et al. Effects of muscle activity and fiber composition on glucose transport and GLUT-4. Am J Physiol. 1993;264(4 Pt 1):E583–93.
26. Richter EA, Kiens B, Mizuno M, et al. Insulin action in human thighs after one-legged immobilization. J Appl Physiol (1985). 1989;67(1):19–23.
27. Chilibeck PD, Bell G, Jeon J, et al. Functional electrical stimulation exercise increases GLUT-1 and GLUT-4 in paralyzed skeletal muscle. Metabolism. 1999;48(11):1409–13.
28. Phillips SM, Stewart BG, Mahoney DJ, et al. Body-weight-support treadmill training improves blood glucose regulation in persons with incomplete spinal cord injury. J Appl Physiol (1985). 2004;97(2):716–24.
29. Petrie M, Suneja M, Shields RK. Low-frequency stimulation regulates metabolic gene expression in paralyzed muscle. J Appl Physiol (1985). 2015;118(6):723–31.
30. Latouche C, Jowett JB, Carey AL, et al. Effects of breaking up prolonged sitting on skeletal muscle gene expression. J Appl Physiol (1985). 2013;114(4):453–60.
31. Shvartz E, Gaume JG, White RT, Reibold RC. Hemodynamic responses during prolonged sitting. J Appl Physiol Respir Environ Exerc Physiol. 1983;54(6):1673–80.
32. Larsen RN, Kingwell BA, Sethi P, et al. Breaking up prolonged sitting reduces resting blood pressure in overweight/obese adults. Nutr Metab Cardiovasc Dis. 2014;24(9):976–82.
33. Thosar SS, Bielko SL, Wiggins CC, et al. Differences in brachial and femoral artery responses to prolonged sitting. Cardiovasc Ultrasound. 2014;12:50.
34. Hamburg NM, McMackin CJ, Huang AL, et al. Physical inactivity rapidly induces insulin resistance and microvascular dysfunction in healthy volunteers. Arterioscler Thromb Vasc Biol. 2007;27(12):2650–6.
35. Saunders TJ, Chaput JP. Is obesity prevention as simple as turning off the television and having a nap? Br J Nutr. 2012;108(5):946–7.
36. Vallance JK, Buman MP, Stevinson C, et al. Associations of overall sedentary time and screen time with sleep outcomes. Am J Health Behav. 2015;39(1):62–7.
37. Harris JL, Bargh JA, Brownell KD. Priming effects of television food advertising on eating behavior. Health Psychol. 2009;28(4):404–13.
38. Chaput JP, Visby T, Nyby S, et al. Video game playing increases food intake in adolescents: a randomized crossover study. Am J Clin Nutr. 2011;93(6):1196–203.
39. Chaput JP, Despres JP, Bouchard C, et al. The association between sleep duration and weight gain in adults: a 6-year prospective study from the Quebec Family Study. Sleep. 2008;31(4):517–23.
40. Patel SR, Hu FB. Short sleep duration and weight gain: a systematic review. Obesity (Silver Spring). 2008;16(3):643–53.
41. Owen N, Sugiyama T, Eakin EE, et al. Adults' sedentary behavior determinants and interventions. Am J Prev Med. 2011;41(2):189–96.

42. Salmon J, Tremblay MS, Marshall SJ, et al. Health risks, correlates, and interventions to reduce sedentary behavior in young people. Am J Prev Med. 2011;41(2):197–206.
43. Hinkley T, Salmon J, Okely AD, et al. Correlates of sedentary behaviours in preschool children: a review. Int J Behav Nutr Phys Act. 2010;7:66.
44. Van Der Horst K, Paw MJ, Twisk JW, et al. A brief review on correlates of physical activity and sedentariness in youth. Med Sci Sports Exerc. 2007;39(8):1241–50.
45. Lubans DR, Hesketh K, Cliff DP, et al. A systematic review of the validity and reliability of sedentary behaviour measures used with children and adolescents. Obes Rev. 2011;12 (10):781–99.
46. Gorely T, Marshall SJ, Biddle SJ. Couch kids: correlates of television viewing among youth. Int J Behav Med. 2004;11(3):152–63.
47. Hoyos Cillero I, Jago R. Systematic review of correlates of screen-viewing among young children. Prev Med. 2010;51(1):3–10.
48. Biddle SJ, Gorely T, Marshall SJ. Is television viewing a suitable marker of sedentary behavior in young people? Ann Behav Med. 2009;38(2):147–53.
49. Klitsie T, Corder K, Visscher TL, et al. Children's sedentary behaviour: descriptive epidemiology and associations with objectively-measured sedentary time. BMC Public Health. 2013;13:1092.
50. Steele RM, van Sluijs EM, Sharp SJ, et al. An investigation of patterns of children's sedentary and vigorous physical activity throughout the week. Int J Behav Nutr Phys Act. 2010;7:88.
51. Byun W, Dowda M, Pate RR. Correlates of objectively measured sedentary behavior in US preschool children. Pediatrics. 2011;128(5):937–45.
52. King AC, Parkinson KN, Adamson AJ, et al. Correlates of objectively measured physical activity and sedentary behaviour in English children. Eur J Public Health. 2011;21(4):424–31.
53. Gomes TN, dos Santos FK, Santos D, et al. Correlates of sedentary time in children: a multilevel modelling approach. BMC Public Health. 2014;14:890.
54. Rhodes RE, Mark RS, Temmel CP. Adult sedentary behavior: a systematic review. Am J Prev Med. 2012;42(3):e3–28.
55. Marshall SJ, Ramirez E. Reducing sedentary behavior: a new paradigm in physical activity promotion. Am J Lifestyle Med. 2011;5:518–30.
56. Owen N, Healy GN, Matthews CE, et al. Too much sitting: the population health science of sedentary behavior. Exerc Sport Sci Rev. 2010;38(3):105–13.
57. Healy GN, Dunstan DW, Salmon J, et al. Breaks in sedentary time: beneficial associations with metabolic risk. Diabetes Care. 2008;31(4):661–6.
58. Healy GN, Matthews CE, Dunstan DW, et al. Sedentary time and cardio-metabolic biomarkers in US adults: NHANES 2003–06. Eur Heart J. 2011;32(5):590–7.
59. de Rezende LF, Rodrigues Lopes M, Rey-Lopez JP, et al. Sedentary behavior and health outcomes: an overview of systematic reviews. PLoS One. 2014;9(8):e105620.
60. Chinapaw MJ, Proper KI, Brug J, et al. Relationship between young peoples' sedentary behaviour and biomedical health indicators: a systematic review of prospective studies. Obes Rev. 2011;12(7):e621–32.
61. Grontved A, Hu FB. Television viewing and risk of type 2 diabetes, cardiovascular disease, and all-cause mortality: a meta-analysis. JAMA. 2011;305(23):2448–55.
62. Ford ES, Caspersen CJ. Sedentary behaviour and cardiovascular disease: a review of prospective studies. Int J Epidemiol. 2012;41(5):1338–53.
63. Edwardson CL, Gorely T, Davies MJ, et al. Association of sedentary behaviour with metabolic syndrome: a meta-analysis. PLoS One. 2012;7(4), e34916.
64. de Rezende LF, Rey-Lopez JP, Matsudo VK, et al. Sedentary behavior and health outcomes among older adults: a systematic review. BMC Public Health. 2014;14:333.
65. Pavey TG, Peeters GG, Brown WJ. Sitting-time and 9-year all-cause mortality in older women. Br J Sports Med. 2015;49(2):95–9.

66. Leon-Munoz LM, Martinez-Gomez D, Balboa-Castillo T, et al. Continued sedentariness, change in sitting time, and mortality in older adults. Med Sci Sports Exerc. 2013;45 (8):1501–7.
67. Gao X, Nelson ME, Tucker KL. Television viewing is associated with prevalence of metabolic syndrome in Hispanic elders. Diabetes Care. 2007;30(3):694–700.
68. Gennuso KP, Gangnon RE, Matthews CE, et al. Sedentary behavior, physical activity, and markers of health in older adults. Med Sci Sports Exerc. 2013;45(8):1493–500.
69. Bankoski A, Harris TB, McClain JJ, et al. Sedentary activity associated with metabolic syndrome independent of physical activity. Diabetes Care. 2011;34(2):497–503.
70. Carson V, Wong SL, Winkler E, et al. Patterns of sedentary time and cardiometabolic risk among Canadian adults. Prev Med. 2014;65:23–7.
71. Carson V, Janssen I. Volume, patterns, and types of sedentary behavior and cardio-metabolic health in children and adolescents: a cross-sectional study. BMC Public Health. 2011;11:274.
72. Colley RC, Garriguet D, Janssen I, et al. The association between accelerometer-measured patterns of sedentary time and health risk in children and youth: results from the Canadian Health Measures Survey. BMC Public Health. 2013;13:200.
73. Carson V, Stone M, Faulkner G. Patterns of sedentary behavior and weight status among children. Pediatr Exerc Sci. 2014;26(1):95–102.
74. Saunders TJ, Prince SA, Tremblay MS. Clustering of children's activity behaviour: the use of self-report versus direct measures. Int J Behav Nutr Phys Act. 2011;8:48. author reply 9.
75. Colley RC, Wong SL, Garriguet D, et al. Physical activity, sedentary behaviour and sleep in Canadian children: parent-report versus direct measures and relative associations with health risk. Health Rep. 2012;23(2):45–52.
76. Ekelund U, Luan J, Sherar LB, et al. Moderate to vigorous physical activity and sedentary time and cardiometabolic risk factors in children and adolescents. JAMA. 2012;307(7):704–12.
77. Chinapaw MJ, de Niet M, Verloigne M, et al. From sedentary time to sedentary patterns: accelerometer data reduction decisions in youth. PLoS One. 2014;9(11):e111205.
78. Dowd KP, Harrington DM, Donnelly AE. Criterion and concurrent validity of the activPAL professional physical activity monitor in adolescent females. PLoS One. 2012;7(10):e47633.
79. Ridgers ND, Salmon J, Ridley K, et al. Agreement between activPAL and ActiGraph for assessing children's sedentary time. Int J Behav Nutr Phys Act. 2012;9:15.
80. Mailey EL, Gothe NP, Wojcicki TR, et al. Influence of allowable interruption period on estimates of accelerometer wear time and sedentary time in older adults. J Aging Phys Act. 2014;22(2):255–60.
81. Tudor-Locke C, Camhi SM, Troiano RP. A catalog of rules, variables, and definitions applied to accelerometer data in the National Health and Nutrition Examination Survey, 2003–2006. Prev Chronic Dis. 2012;9:E113.
82. Atkin AJ, Ekelund U, Moller NC, et al. Sedentary time in children: influence of accelerometer processing on health relations. Med Sci Sports Exerc. 2013;45(6):1097–104.
83. Aminian S, Hinckson EA. Examining the validity of the ActivPAL monitor in measuring posture and ambulatory movement in children. Int J Behav Nutr Phys Act. 2012;9:119.
84. Ryde GC, Brown HE, Peeters GM, et al. Desk-based occupational sitting patterns: weight-related health outcomes. Am J Prev Med. 2013;45(4):448–52.
85. Ryde GC, Gilson ND, Suppini A, et al. Validation of a novel, objective measure of occupational sitting. J Occup Health. 2012;54(5):383–6.
86. Sisson SB, Broyles ST, Baker BL, et al. Television, reading, and computer time: correlates of school-day leisure-time sedentary behavior and relationship with overweight in children in the U.S. J Phys Act Health. 2011;8 Suppl 2:S188–97.
87. Wilson RS, Mendes De Leon CF, Barnes LL, et al. Participation in cognitively stimulating activities and risk of incident Alzheimer disease. JAMA. 2002;287(6):742–8.
88. Healy GN, Owen N. Sedentary behaviour and biomarkers of cardiometabolic health risk in adolescents: an emerging scientific and public health issue. Rev Esp Cardiol. 2010;63 (3):261–4.

Chapter 8
New Perspectives on Activity/Disease Relationships Yielded by Objective Monitoring

Roy J. Shephard

Abstract The Hockley Valley Consensus Symposium based most of its conclusions on dose/response relationships between physical activity and disease on subjective questionnaire reports. In this chapter, we summarize the findings from the Hockley Valley meeting, and we examine how far these conclusions have been amplified and/or modified by the use of objective physical activity monitors. Among a wide range of topics, we have included data on objective activity monitoring in relation to all-cause mortality, cardiac death, cardiovascular disease, stroke, peripheral vascular disease, hypertension, cardiac and metabolic risk factors, diabetes mellitus, obesity, low back pain. osteoarthritis, osteoporosis, chronic chest disease, cancer, depression, quality of life and the capacity for independent living. The introduction of objective monitoring has clarified dose/response relationships in a number of areas, allowing us to define relationships in terms of objective metrics (the number of steps taken per day). However, much of the information that is currently available remains cross-sectional in type. In many areas of rehabilitation, the pedometer/accelerometer seems a useful motivating device, providing well-documented increments of weekly activity. However, there remains a need for well-designed longitudinal trials, using objective monitors to follow changes in habitual activity and thus to demonstrate causality in the association between physical activity and good health.

8.1 Introduction

Epidemiologists traced the first outlines of many physical activity/disease relationships using questionnaire self-reports of habitual physical activity. In this chapter, we will summarize briefly the knowledge gained in this fashion, as agreed at the

Roy J. Shephard (✉)
Faculty of Kinesiology & Physical Education, University of Toronto, Toronto, ON, Canada
e-mail: royjshep@shaw.ca

R.J. Shephard, C. Tudor-Locke (eds.), *The Objective Monitoring of Physical Activity: Contributions of Accelerometry to Epidemiology, Exercise Science and Rehabilitation*, Springer Series on Epidemiology and Public Health,
DOI 10.1007/978-3-319-29577-0_8

Hockley Hills, ON, Consensus Conference of 2001 [1], and we will consider how far these conclusions have been amplified and/or modified for each of several health conditions by the introduction of objective monitoring equipment [2]. We will point to further issues that can yet be resolved by objective monitoring, will note the potential to use objective monitoring tools in the stimulation of known increases of exercise behaviour by experimental subjects, and will comment on outstanding problems that will require further developments of methodology.

8.2 Consensus Findings Based upon Physical Activity Questionnaire Data

The Hockley Valley Consensus Conference of 2001 was "evidence-based," in the sense that is sought to consider all of the evidence of associations between physical activity and each of a substantial number of health conditions, weighing the quality of the individual reports, and thus striving to reach a relatively unbiassed conclusion concerning the strength and form of relationships [3, 4].

As Schriger [5] has pointed out, the conclusions reached by the Consensus Conference were not entirely objective, since in applying this information, the individual investigator must decide how far the studies that were considered match the personal characteristics and environment of the patients that he or she is proposing to advise or treat.

Most questionnaire studies of physical activity have attempted to give a description of the type of activity that has been performed (Chap. 1), but often the gross rather than the net energy expenditure has been estimated, and the intensity of effort has commonly been reported in absolute terms rather than relative to the individual's age, sex, physical condition and the duration of activity [6, 7]. Further, any beneficial effects of increased physical activity are likely to be influenced by the nature of the activity undertaken (resistance vs. aerobic exercise, [6]), and discussion remains concerning the relative important of the observed pattern of physical activity versus the attained level of aerobic fitness (Chap. 1, [8], Table 8.1).

8.3 All-Cause Mortality and Cardiac Disease

Objective studies looking at associations between physical activity and all-cause mortality or cardiovascular disease have been relatively few. Surrogate measures of atherosclerosis have included studies of pulse-wave velocity, assessments of coronary arterial calcification and determination of pericardial fat. Pedometers have been used quite extensively both to examine spontaneous levels of physical activity in stroke victims and as a stimulus to both primary and secondary prevention, and rehabilitation following myocardial infarction.

Table 8.1 The relationships between habitual physical activity and various forms of ill-health, as agreed at the Hockley Valley Consensus Symposium of 2001 [1]

Author	Condition	Findings
Lee and Skerrett [9]	All-cause mortality	20–30 % reduction of risk with gross energy expenditure of 4.2 MJ/week. Greater benefit at higher expenditures, but unclear if due to greater volume or greater intensity. Unclear if benefit with expenditures <4.2 MJ/week
Kohl [10]	Cardiovascular disease	Incidence and mortality from cardiovascular disease, particularly ischaemic heart disease, inversely related to habitual physical activity; effects on stroke remain uncertain. More information needed in women
Fagard [11]	Hypertension	Generally a reduction of systolic and diastolic blood pressures, but relationship to pattern of physical activity unclear
Kelley and Goodpaster [12]	Type 2 diabetes mellitus	Physical activity decreases risk of type 2 diabetes mellitus, but unclear if dose/response for exercise volume or intensity
Leon and Sanchez [13]	Blood lipid profile	Data too inconsistent to establish effects of volume and intensity of physical activity
Rauramaa et al. [14]	Haemostasis	Acute exercise appears to activate haemostasis. Chronic effects upon coagulation and thrombosis unclear
Ross and Janssen [15]	Total and regional obesity	Short term reductions of body fat seem dose-related; long-term effects smaller and less consistent. Dose/response unclear for visceral fat
Thune and Furberg [16]	Cancer	Suggestion of graded dose/response for all-cause cancer, colonic cancer and breast cancer. Effect on other cancers less clear
Vuori [17]	Low back pain, osteoarthritis, osteoporosis	Light activity reduces but heavy activity may increase low back pain. Physical activity can be effective in treatment of osteoarthritis. High intensity loading osteogenic, but dose/response effects poorly known
Dunn et al. [18]	Depression and anxiety	Physical activity lessens depression and anxiety, but lack of evidence on dose/response effects
Spirduso and Cronin [18]	Quality of life and independence	Physical activity enhances quality of life and period of independent living, but no dose/response effect of intensity observed

8.3.1 All-Cause Mortality

At the Hockley Valley symposium, I-Min Lee (Fig. 8.1) and Skerrett [9] noted that questionnaire data pointed to a 20–30 % reduction in the risk of all-cause death with an added gross energy expenditure of 4.2 MJ/week. Benefit appeared to increase with larger weekly energy expenditures, but it remained unclear whether this reflected a greater total volume of physical activity or higher intensities of effort

Fig. 8.1 I-Min Lee has extended many of the epidemiological studies of Ralph Paffenbarger on relationships between questionnaire assessments of habitual physical activity and all-cause and cardiovascular mortality

intensity. Although some earlier work had pointed to a threshold intensity for benefit, by 2001, it was still unclear if risk was reduced with weekly expenditures <4.2 MJ.

To answer these questions, it is necessary to follow large samples of subjects for long periods, and perhaps for this reason, there has as yet been little attempt to relate objective measures of habitual physical activity in healthy individuals to their risk of all-cause mortality. However, activity monitors have been used to study relationships in groups with a higher risk of dying, including a large sample of 3027 community-dwelling women with an initial age of 84 years, patients with chronic obstructive pulmonary disease (below, [19]), and individuals receiving thrice-weekly dialysis for renal failure [20]. Although investigators have attempted to include appropriate co-variates in their analyses, some doubt has remained as to whether a low-level of physical activity predisposed to death, or whether the level of activity was low in those individuals with a poorer initial condition.

8.3.1.1 Very Old Women

The study of Tranah et al. [21] looked at the daily peaks of habitual physical activity as measured by a wrist-mounted accelerometer in very elderly women, finding an association between the height of these peaks and the risk of overall, atherosclerotic, coronary heart disease and other cause of death over the next 4 years, with a hazard ratio of 2.16 for overall deaths among women in the least physically active quartile. This data appears to exclude any concept of an activity threshold for all-cause mortality, all-cause, cancer, atherosclerotic, coronary disease and other causes of death. In terms of both the amplitude of the daily circadian variation in activity counts and the daily average (mid-line) count, Tranah et al. [21] saw the main and statistically significant benefit on comparing the least physically active quartile with the next most active quartile (Table 8.2). One focus of this particular study was an age-related deterioration in circadian variations in activity, and a part of the gradient of risk could reflect either the low overall level of physical activity in

Table 8.2 Hazard ratios for quartiles of daily overall physical activity (amplitudes and mesor of circadian rhythm) of elderly women in relation to all-cause, cancer, atherosclerotic, stroke, coronary disease and other causes of death

	All-cause	Cancer	Atherosclerosis	Stroke	Coronary heart disease	Other
Amplitude (counts/minute)						
>4046	1.00	1.00	1.00	1.00	1.00	1.00
>3413	1.02	1.04	0.78	0.54	1.19	1.31
>2743	1.25	1.45	1.20	0.8	1.31	1.19
<2743	**2.16**	1.39	**1.81**	1.57	**2.23**	**3.11**
Daily mesor (counts/minute)						
>2387	1.00	1.00	1.00	1.00	1.00	1.00
>2092	1.14	1.44	0.92	0.74	1.72	1.20
>1796	1.08	1.09	1.01	0.58	2.20	1.14
<1796	**1.71**	1.16	**1.61**	1.12	**2.77**	**2.09**

Significant changes are shown in bold type. Based on the data of Tranah et al. [21]

those with low peaks of activity, or an impairment of the circadian oscillator, with associated adverse effects upon glucose tolerance. In support of the latter hypothesis, the risk of developing cancer and stroke was greater in those individuals showing a peak of physical activity later than 4.30 p.m.

Objective monitoring can cover not only patterns of physical activity, but also the rhythms of sleep, which are also linked to an individual's prognosis. The average sleep duration, as determined by actigraph, is indicative of survival prospects [22]. The relationship is U-shaped, with an optimum prognosis being linked to a relatively short average sleeping time. A 10.5 year follow up of 444 women initially aged 67.6 years found 90 % survival in those who slept for 300–-390 minutes, 61 % survival in those who slept for less than 300 minutes, and 78 % survival for those who slept for longer than 390 minutes per night.

8.3.1.2 Patients with Chronic Obstructive Pulmonary Disease

Waschki et al. [19] followed 170 cases of chronic obstructive pulmonary disease (COPD) for an average of 48 months. The overall mortality during this period was 15 %, giving a rather low total number of deaths [23]. A Sense-wear arm band was used to assess the overall physical activity, total daily energy expenditure being divided by the sleeping energy expenditure to obtain an average physical activity level (PAL). Four-year risks of death for very inactive, sedentary and physically active patients were respectively, 31, 9 and 0 %; as with the elderly women, the main gain seemed to be on moving from the very inactive to the sedentary category. After statistical adjustments for age and sex, an increase in PAL of 0.14 (equivalent to an added energy expenditure of 0.8–1.0 MJ/day) or the taking of an additional 1845 steps/day) was associated with a lower risk of death (respective hazard ratios 0.46 and 0.49); curve fitting did not indicate any significant curvilinearity of this relationship.

Garcia-Rio and associates examined the prognosis of 173 patients with moderate to very severe chronic obstructive pulmonary disease over 5–8 years [24]. In a multivariate model, the cycle ergometer endurance time at 75 % of maximal aerobic power and the accelerometer vector magnitude unit (VMU) score were found to be independent predictors of mortality. Further, those with a high VMU score (average 234 units) had a lower risk of hospitalization than those with an average VMU of 143 units, and the time to the first hospital admission for COPD was longer for the physically more active group. The risk of hospitalization decreased by 10 % for each 10 increase in VMUs. Kaplan-Meier survival curves showed large gains of survival between the first quartile of activity (<130 units), the second quartile (<200 units) and the third quartile (<270 units), but little additional advantage in the fourth quartile (>270 units).

Zanoria and Zuwallack [25] followed 60 patients with COPD for an average of 53 months. A low level of habitual activity (a triaxial accelerometer count <170 units) was a predictor of both all-cause and respiratory hospital admissions over the period of observation.

8.3.1.3 Patients Receiving Renal Dialysis

Matsuzawa et al. [20] followed a group of 202 out-patients who were receiving thrice-weekly dialysis for an average of 45-months; 34 died during this period. A multivariate analysis showed that the hazard ratio for death was reduced by 0.78 for each increase in "activity" of 10 minutes/day [20], "activity" being here defined as a score of >1 (gentle walking or higher) on a Kenz Lifecorder at entry to the study. A Kaplan-Meier plot showed a substantial difference of prognosis between those who were physically active <50 minutes/day, and those active >50 minutes/day (median pedometer counts of 1833 and 4478 steps/day, respectively).

Matsuzawa et al. [26] made further studies on 116 dialysis patients. They demonstrated that the levels of one important cardiac risk factor (HDL cholesterol) were favourably impacted by greater objectively measured physical activity, independently of other co-variates (a positive partial correlation with the daily step count at entry to the study, $r = 0.18$, $p < 0.02$).

8.3.1.4 Activity on Discharge from Hospital and Prognosis

Ostir et al. [23] studied the physical activity patterns of 224 elderly patients admitted to an acute geriatric hospital on the day of admission and the day of discharge, using a "Step Watch" activity monitor. Among those aged 65–84 years, activity was increased by 28 minutes/day immediately prior to discharge (an increase from 478 to 846 steps/day), but in those aged >85 years there was no such increase. Each 100 steps/day increment of activity was associated with a 3 % decrease of mortality over the next 2 years. In contrast, if the step rate had declined by the time of discharge, there was a four-fold increase in the risk of death over the next 2 years.

8.3.1.5 Conclusions

Although objective information on physical activity and all-cause mortality is limited, longitudinal studies on the very old, patients with chronic obstructive lung disease and those undergoing renal dialysis all show graded associations between the volume of physical activity and enhanced prognosis, with much of the benefit seen on reversing total inactivity. Unfortunately, on the basis of the information collected to date, it is not possible to exclude entirely that a poorer initial prognosis leads to difficulty in increasing physical activity rather than the converse.

Future initiatives to overcome these uncertainties might include long-term evaluation of physical activity and health in a substantial sample of older adults who are shown to be healthy when first recruited, and/or discarding the first 2 or 3 years of follow-up to eliminate latent illness (as has sometimes been done in questionnaire-based investigations).

8.3.2 Cardiovascular Disease

Harold W (Bill) Kohl [10] (Fig. 8.2) reported to the Hockley Valley Symposium that occupational classifications, questionnaire data and aerobic fitness measurements had all shown that the incidence and mortality from cardiovascular disease, particularly ischaemic heart disease, were inversely related to habitual physical activity [10], although estimates of benefit varied substantially, 8 of 28 studies did not support a dose-response relationship, and there was no agreement on the type,

Fig. 8.2 Harold W. Kohl has directed many epidemiological studies of the relationships between questionnaire assessments of physical activity, aerobic fitness and health

intensity and pattern of physical activity needed for benefit. Further, many of the available longitudinal studies relied upon a questionnaire assessment of physical activity at entry to the study. The possible effects of physical activity upon the risk of stroke remained uncertain. More information was also needed concerning the cardiovascular benefits of physical activity in women.

The impact of habitual physical activity upon the risk of cardiovascular disease can be examined objectively by following a large sample of a high-risk population for a substantial number of years, but to date all except one group of investigators seem to have been deterred by the logistics. Potential alternatives are to seek some surrogate of cardiovascular disease: a deterioration of vascular distensibility, calcification of the coronary arteries, or the accumulation of pericardial fat.

8.3.2.1 Prospective Study of High-Risk Patients

Yates and associates [27] conducted a 6-year prospective study of a large international sample of 9306 high-risk patients with impaired glucose tolerance and other cardiac risk factors. There were 531 cardiovascular events during the period of observation. The initial pedometer count for this population averaged 5892 steps/day, and there was little change in this average at 12 months (6320 steps/day). However, some patients showed decreases in their habitual activity of 2000 steps/day or more, and others recorded increases of 2000 steps/day or more. The volume of ambulatory activity seen at entry to the study and any increase of activity observed over the trial were both inversely related to the risk of a cardiac incident, respective hazard ratios per 2000 steps/day amounting to 0.90 and 0.92. Moreover, these ratios persisted after an exhaustive adjustment of data for baseline co-variates, including age, sex, body-mass index, smoking status, existing coronary heart disease or cerebrovascular disease, significant ECG abnormalities, peripheral arterial disease, congestive heart failure, chronic obstructive pulmonary disease, renal function and use of antihypertensive medication.

The authors of this report suggested that no other research group has yet conducted a similar trial. However, the study of Tranah and associates [21] demonstrated associations between modest levels of physical activity and a decreased risk of both coronary artery disease and atherosclerosis (above).

8.3.2.2 Pulse-Wave Velocity and Vascular Distensibility

Several studies have used changes in arterial pulse-wave velocity and other measures of vascular distensibility as indices of cardiovascular atherosclerosis (Table 8.3). In the earlier of these studies, habitual physical activity was determined by questionnaire and/or measurements of aerobic fitness [38–40], making it difficult to define the dose of activity needed to optimize vascular characteristics.

Table 8.3 Studies examining arterial distensibility in relation to objective measures of habitual physical activity

Author	Subjects	Findings
Aoyagi et al. [28]	Cross-sectional study of men and women aged 65–84 years	Central but not peripheral arterial distensibility increased with physical activity, largest effect at low intensities
Gando et al. [29]	Cross-sectional study of 538 unfit but otherwise healthy subjects	Carotid/femoral pulse wave velocity inversely correlated with time spent in both light and moderate physical activity
Hawkins et al. [30]	One-year longitudinal study of 274 sedentary and overweight young adults	Brachial arterial pulse wave velocity slowed if activity increased over year
Hayashi et al. [31]	Longitudinal study of 17 sedentary middle-aged men	16 weeks training programme with Polar monitor increased central vascular distensibility
Matsuda [32]	Cross-sectional study of 413 - middle-aged and elderly subjects	Effect plateaus with energy expenditure >0.8–1.2 MJ/day
O'Donovan et al. [33]	51 older adults with hypertension	Applanation tonometry shows significant inverse relation between physical activity and pulse wave velocity
Ried-Larsen et al. [34]	Cross-sectional study of 336 adolescents	Carotid arterial stiffness shows trend approaching significance over physical activity quartiles
Sakuragi et al. [35]	573 children, average age 10 years	Pulse wave velocity negatively correlated with pedometer step count
Sugawara et al. [36]	Cross-sectional study of 103 post-menopausal women	Carotid arterial stiffness inversely related to duration of light, moderate and vigorous physical activity
Tanaka et al. [37]	Longitudinal study of 413 middle-aged and elderly subjects	Three months walking programme, using Polar monitor, showed increase of central arterial distensibility

Tanaka et al. [37] used Polar heart rate monitors to provide objective evidence of physical activity intensity in a study of middle-aged men. After a 3-month aerobic exercise intervention (mainly walking), there was an increase of central arterial compliance (as determined by simultaneous B mode ultra-sound and arterial applanation tonometry).

Sugawara et al. [36] monitored physical activity by having subjects wear a Kenz Lifecorder for 14 days. They found that the carotid arterial stiffness in a group of 103 post-menopausal women was significantly correlated with the duration of both moderate and vigorous physical activity. Tanabe et al. [41] used the same equipment to examine the impact of exercise intensity upon arterial compliance in 413 middle-aged and elderly subjects; vascular distensibility was greater in those taking more than 30 minutes of moderate activity (3–5 METs) per day than in those taking less than this volume. Mitsuo Matsuda (Fig. 8.3) [32] suggested that the increase of arterial distensibility plateaued if the daily energy expenditure, thus

Fig. 8.3 Mitsuo Matsuda has been active in research linking vascular distensibility to pedometer/ accelerometer counts

measured, exceeded 0.8–1.2 MJ/day; this would equate with about 30 minutes of moderate physical activity (3–6 METs).

A cross-sectional study of 336 Scandinavian adolescents divided objective measurements of physical activity into quartiles [34]. A trend to stiffer carotid arteries in the physically less active youth approached statistical significance even at the average age of 15.6 years.

Aoyaji et al. [28] found that after appropriate adjustments for age, sex and blood pressure, central arterial stiffness (as indicated by carotid-femoral and delta brachio-tibial pulse wave velocities) was significantly lower in physically active individuals than in those with lower pedometer/accelerometer readings. In contrast, peripheral vascular stiffness was unrelated to habitual physical activity. Hayashi et al. [31] also found that the main effect of 16 weeks of moderate intensity training, monitored by a Polar heart rate monitor, was upon central vascular distensibility. In contrast, Yamada et al. [40] in their questionnaire-based investigation, concluded that the main effect of habitual activity was upon the peripheral vessels.

Sakuragi et al. [35] examined the correlates of carotid-femoral pulse wave velocity in 5873 healthy 10-year old children. Even at this age, there was a small negative univariate correlation between the pedometer step count and vascular stiffness ($r = -0.08$, $p = 0.046$).

Hawkins et al. [30] strengthened the causal inference behind such associations, demonstrating in a sample of 274 sedentary and overweight young adults that the brachial-arterial pulse wave velocity was reduced over a 1-year lifestyle programme in relation to increases in total accelerometer counts and time spent in moderate physical activity (1952–5724 counts/minute); however, changes in sedentary time and light physical activity were unrelated to changes in pulse wave velocity seen over the course of the study.

Objective monitoring of physical activity with pedometer/accelerometers confirms that in the case of central vascular distensibility, benefit begins with a relatively low dose of physical activity (Chap. 1). In the study of Aoyagi et al. [28], the largest benefit was seen on moving from the least active subject quartile (averaging 3570 steps/day, and 4.8 minutes/day of moderate physical activity), to the next most active quartile (averaging 5838 steps/day, and 12.2 minutes/day of moderate physical activity). Likewise, Sugawara et al. [36] found that carotid arterial stiffness was inversely related to the duration not only of vigorous physical activity (>5–6 METs, depending on age), but also to the duration of moderate (>3–4 METs) and light (<3–4 METs) activity. Likewise, Gando et al. [29] demonstrated that the carotid/femoral pulse wave velocity was correlated with triaxial accelerometer determinations of the time that the older members of a group of 538 unfit but otherwise healthy subjects allocated to moderate (>3 METs, $r = -0.31$), light (<3 METs, $r = -0.39$) and sedentary ($r = 0.44$) activities, but was not correlated with the time spent in vigorous physical activity.

Most of the studies cited in Table 8.3 included some individuals with hypertension, but O'Donovan et al. [33] specifically tested relationships in 53 older adults with hypertension, some 16 members of the group also having the metabolic syndrome. A significant inverse cross-sectional association was still seen between triaxial accelerometer measurements of habitual activity and pulse wave velocity as determined by applanation tonometry ($r = -0.53$).

Further analysis of the Nakanojo data set confirmed the main thrust of dose/response relationships previously set out by Aoyagi et al. [28], although also defining small sex-related differences, probably attributable to lower levels of physical activity and fitness in the female subjects of this sample [42].

8.3.2.3 Vascular Calcification and Pericardial Fat

The amount of coronary artery calcification detected by computed tomography appears directly and strongly related to future cardiac incidents. Several questionnaire-based studies found no association between coronary vascular calcification and habitual physical activity. However, an inverse relationship became apparent with objective monitoring of physical activity.

Gabriel and associates made two electron beam tomographic assessments of coronary calcification in 148 healthy older women, allowing a 12-year interval between assessments [43]. Subjects were classified into three groups (no calcification observed at either test visit, calcification seen at the second visit, and calcification apparent at both visits). Questionnaire responses failed to reveal any inter-group differences of habitual physical activity. However, accelerometer estimates of the times spent in moderate and vigorous physical activity were significantly greater in the first than in the third subject group (81.6 vs. 62.4 minutes/day).

Kristi Storti (Fig. 8.4) and associates [44] also applied the electron beam tomography technique to a cross-sectional study of 173 younger and 121 older women. In the older women (average age 73.9 years), there was a statistically significant association between the likelihood of finding coronary calcification and the daily step count recorded by a Yamax pedometer. The incidence ranged from 31 % in those taking no significant activity (counts <3610 steps/day) to 21 % in

Fig. 8.4 Kristi Storti has carried out a number of studies relating vascular health to objectively monitored physical activity

those taking >5466 steps/day, with no additional benefit in those taking >7576 steps/day; the range 5466–7566 steps/day corresponds to quite a limited volume of physical activity even in the elderly (Chap. 1). Adjusting for covariates, and setting the most active group as the norm, the odds ratios for the four quartiles of physical activity were 1.00, 0.94, 1.08 and 1.31, respectively, only the effect of taking no deliberate physical activity being statistically significant. The younger subjects (average age 56.8 years) showed a similar trend, but this was not statistically significant.

In contrast to these positive conclusions, an accelerometer study of the Whitehall civil servants by Hamer et al. [45] found no relationship between coronary calcification and habitual physical activity. One possible reason for this divergent finding was that some 55 % of the Whitehall sample of 10,308 executive class civil servants were already engaging in at least 30 minutes per day of moderate-to-vigorous physical activity; furthermore, the Whitehall data were not sorted by sex. A subsequent examination of 446 participants in the Whitehall study [46] used electron beam computed tomography to determine pericardial fat; the amount of this fat was inversely correlated with the total physical activity of the individual and the minutes per day spent in moderate-to-vigorous physical activity, even after adjusting for numerous co-variates. This is an important finding, since there is a growing belief that pericardial fat releases pro-inflammatory factors that contribute to coronary atherosclerosis. The association was strongest for minutes per day of moderate physical activity (>3 METs).

8.3.3 Stroke

There have been many objective studies underlining limited physical activity and abnormalities of gait following a stroke, but few investigations have yet evaluated the preventive value of physical activity.

8.3.3.1 Physical Activity and Stroke Prevention

Physical activity might modify the risks of stroke through an action upon either systolic blood pressure or vascular atherosclerosis. To date, the results obtained by objective monitoring are limited; two of three studies have pointed to benefit from light and/or moderate activity, and the third report has shown a non-significant trend in the same sense. However, details of dose/response relationships have yet to be elucidated.

Tranah and associates [21] found no significant association between objectively measured physical activity and the risk of stroke in a sample of elderly women (Table 8.2).

Butler and colleagues [47] examined accelerometer data for the physical activity of 262 individuals in the NHANES study who reported having suffered a stroke. Compared with 524 who had not experienced a stroke, the data showed less vigorous and moderate-to-vigorous physical activity (10.9 vs. 13.0 minutes/day), greater sedentary behaviour (10.0 vs. 9.2 hours/day), and lower rates of counting (190 vs. 227 counts/minute). A large cross-sectional family study of a relatively inactive population conducted by Lee et al. [48] also found an association between a self-report of stroke and a low level of habitual physical activity, respective mean z scores for daily step count, and minutes of moderate and light physical activity being −0.63, −0.32 and −0.42 (Table 8.4).

8.3.3.2 Uses of Objective Activity Monitors Following a Stroke

The wearing of a pedometer has been suggested as a potential method of increasing physical activity following a stroke [49]. Care must be taken in the absolute interpretation of data in patients with an abnormal gait and a slow walking rate [50]. Sense-Wear arm band data show a poor correlation with step counts following

Table 8.4 Findings from a large cross-sectional family study in Hong Kong [48], showing mean z scores for associations between various accelerometer measures of habitual physical activity and medical conditions

Condition	Daily step count	Minutes of moderate-to-vigorous physical activity per day (>3 METs)	Minutes of light physical activity per day (<3 METs)
Self-reported hypertension	−0.22	−0.17	−0.12
Measured hypertension	−0.02	0.01	−0.02
Self-reported cancer	−0.43	−0.05	−0.28
Self-reported stroke	−0.63	−0.32	−0.43
Depressive symptoms on PHQ-9	−0.15	−0.04	−0.10
Physical component subscale of medical outcomes questionnaire	−0.13	−0.05	0.06

a stroke [51]; the problem seems to be with the arm band data or its interpretation, since the arm band estimates of metabolic rate correlate poorly with direct measurements of oxygen consumption by metabolic cart [52]. Nevertheless, a change in data can provide useful motivation, provided that measurement errors remain relatively consistent.

Levels of habitual physical activity are consistently low following a stroke [53, 54]. Moore et al. [54] used a Sense-Wear device to show that the habitual activity of a group of 31 stroke patients with an average age of 73 years remained below the levels recommended for the health of the general population in the 6 months following an acute cerebrovascular incident. In a sample of 49 patients, Mudge and Stott [55] attached a Step Watch infrared monitoring device to the unaffected leg of subjects with hemiplegiua, finding an average count of only 4765 steps/day 6 months following a stroke.

However, prognosis following a stroke is enhanced if patients can be encouraged to achieve a Kenz Lifecorder count of at least 6000 steps/day; the positive and negative predictive values for further episodes with habitual physical activity below this threshold were 38.0 and 91.6 % respectively [56]. In an observer-blinded 24-week controlled trial, 35 of 70 stroke patients were enrolled in a programme that included a heart-rate monitored increase of physical activity and a reduced salt intake. There was an immediate decrease of blood pressure and an increase of HDL cholesterol, and after a follow-up of 2.9 years, there had been only 1 vascular event in the experimental group, compared with 12 events in the control group [57].

Plainly, the risk of a recurrence is relatively high following a stroke, and there is thus scope to study substantial groups of patients prospectively to examine how far the risk of a recurrence is related to the individual's habitual level of physical activity.

8.3.4 Peripheral Vascular Disease

At the Hockley Valley Consensus Conference, it remained unclear whether regular physical activity protected against stroke [10]. Rauramaa et al. [14] further concluded that although there was evidence that acute exercise activated haemostasis, the effects of chronic physical activity upon coagulation and thrombosis remained unclear.

We will discuss objective evidence that inactivity predisposes to a prothrombotic state, and will discuss associations between the development of atheromatous plaques and other manifestations of peripheral vascular disease, finally considering the impact of physical activity levels upon the prognosis in peripheral vascular disease.

8.3.4.1 Inactivity and Prothrombotic State

One possible adverse consequence of physical inactivity is the development of a prothrombotic state (Table 8.5). A substantial number of papers have documented very low levels of objectively measured physical activity in peripheral vascular disease, but it remains uncertain from such observations whether inactivity was the primary problem, or whether the observed low levels of physical activity were secondary to the development of claudicant pain. Objective studies of differences in pulse wave velocity between physically active and inactive individuals (Table 8.3) suggest that at least a part of the difference in lifestyle likely pre-dated the onset of claudication. In a cross-sectional analysis of 106 patients with peripheral arterial disease, Gardner and Killewich [59] demonstrated that plasma levels of plasminogen activator (a serine protease that promotes fibrinolysis) were positively related to daily levels of habitual physical activity. Dividing the sample into low, moderate and high levels of habitual activity, respective levels of tissue plasminogen activator were 1.30, 1.65 and 1.60 IU/mL and the corresponding levels of plasminogen activator inhibitor were 21.4, 16.4 and 17.5 AU/mL. Another study of 188 patients with peripheral arterial disease underlined inverse relationships between accelerometer measurements of habitual physical activity and pro-coagulant factors, the fibrin degradation product D-dimer and inflammatory markers [58, 60] (Table 8.5). In this analysis, the largest change in most variables was seen on moving from the fourth to the fifth quintile of physical activity. Dividing a sample of 111 cases of peripheral arterial disease into activity tertiles, Payvandi et al. also noted a gradient of flow-mediated vasodilatation favouring the individuals who were most physically active (first tertile +4.8 %, second tertile +4.6 %, third tertile +7.2 % dilatation following 4 minutes of vascular occlusion [61]).

8.3.4.2 Inactivity and Femoral Atherosclerotic Plaques

The extent of superficial femoral artery atherosclerotic plaques might seem to provide another objective index of the extent of peripheral vascular disease. However, a study of 454 patients found no significant relationships between quintiles of

Table 8.5 Evidence of a prothrombotic state among 188 patients with peripheral vascular disease, divided into quintiles of arbitrary Caltrac accelerometer-measured physical activity units [58]

	Accelerometer-measured physical activity quintiles				
Measure	<443 units	<634 units	<813 units	<1053 units	>1058 units
D-dimer (mg/mL)	1.44	0.91	1.26	0.80	0.70
C-reactive protein (mg/dL)	0.610	0.663	0.507	0.555	0.274
Fibrinogen (mg/dL)	373	389	375	386	338
Plasminogen activator inhibitor (ng/mL)	19.2	17.6	13.2	18.2	14.7

maximal plaque area, mean plaque area or minimal luminal area and habitual physical activity as measured by a Caltrac accelerometer [62].

8.3.4.3 Inactivity and Peripheral Vascular Disease

The most convincing evidence that inactivity precedes peripheral vascular disease comes from studies of youth with type 1 diabetes mellitus. Many such individuals already present early signs of peripheral vascular disease, particularly an increased intima-media thickness and impaired flow-mediated vasodilatation (4.9 vs. 7.3 %) [63]. Moreover, regular physical activity appears to have protective value; adolescents with type 1 diabetes mellitus who undertook at least 60 minutes per day of moderate to vigorous physical activity maintained a flow-mediated vasodilatation that was at least comparable with that of relatively inactive healthy youngsters.

Many studies of older individuals have shown an association between inactivity and peripheral vascular disease. A comparison of 85 patients who suffered from intermittent claudication and 59 individuals without this symptom [64] showed a substantial inter-group difference in both pedometer counts (4737 vs. 8672 steps/day) and estimated energy expenditures (1.49 vs. 2.56 MJ/day).

McDermott and associates [65] compared 20 patients with peripheral arterial disease against 21 individuals who were free of peripheral arterial disease. Accelerometer measurements indicated a substantial inter-group difference in weekly leisure energy expenditures (3.34 MJ vs. 7.28 MJ/week); despite the magnitude of this effect, two self-report questionnaires failed to detect significant inter-group differences in activity levels. In this study, indices of the brachial-ankle pulse wave velocity showed a clear gradient with tertiles of weekly leisure energy expenditure (0.62, 0.89 and 1.05). Further studies by the same investigators indicated that higher accelerometer counts were associated with more favourable characteristics in the leg muscles [66]; again dividing the sample by activity tertiles, the respective calf muscle cross-sectional areas were 5071, 5612 and 5869 mm^2.

8.3.4.4 Inactivity and Prognosis in Peripheral Vascular Disease

The low level of physical activity seen in many patients with peripheral vascular disease may contribute to the subsequent progression of this condition [67]. A 4-year follow-up of 203 patients with peripheral vascular disease physical found that high levels of daily physical activity were associated with greater initial physical fitness, and a slower rate of decline in both the 6-minute walking distance and the fast-paced 4-metre walking velocity [68].

Garg and associates examined the impact of habitual activity upon prognosis in peripheral vascular disease. They completed a 57-month accelerometer-based follow-up of 460 men and women [69] with an initial average age of 71.9 years; 29 % of the group died over the course of the study. Comparing the physically most

Fig. 8.5 Michael Criqui is a cardiovascular epidemiologist who has studied physical activity patterns in patients with peripheral vascular disease

active with the least active quartile, the hazard ratio was 3.48 for all-cause deaths. Similar hazard ratios were observed for cardiac events and cardiac deaths (Fig. 8.5).

8.3.5 Value of Objective Monitoring in Cardiac Rehabilitation

As in many forms of secondary and tertiary care, pedometers and accelerometers are finding increasing application as a means of objectively determining levels of physical activity during cardiac rehabilitation [70] and of motivating individuals to greater physical activity [71]. Where care has not been taken to supplement formal training sessions with a home-based programme, the increase of weekly energy expenditure has sometimes been much less than intended.

8.3.5.1 Need for Objective Monitoring

Guiraud and colleagues [72] compared responses to a physical activity questionnaire with accelerometer data obtained from a "My Wellness Key actimeter" in a group of 72 patients who had completed a 4-week course of cardiac rehabilitation. The correlation between the subjective and objective measurements was relatively low (for total activity, $r = 0.40$; for reports of sports and leisure activity with objective evidence of moderate and vigorous physical activity, $r = 0.37$), underlining the need for objective monitoring in order to assess adherence to exercise prescriptions.

8.3.5.2 Need to Supplement Rehabilitation Centre Programmes

In an uncontrolled study, Stevenson et al. [73] monitored responses at entry to and on completing a 4–6 week programme of early cardiac rehabilitation. There was only a small increase of total activity from programme entry (194 counts/

observation minute) to programme exit (218 counts/minute), and the time spent in moderate activity also increased only from 13.9 to 18.7 minutes/day. However, total activity was increased on the days patients attended the rehabilitation programme (224 vs. 188 counts per observation minute, 19.7 vs. 12.8 minutes of moderate activity).

Ayabe et al. [74] examined the physical activity patterns of 77 patients who had been enrolled in cardiac rehabilitation programmes 3 days per week for 3 or more months. Lifecorder accelerometer monitoring showed leisure activity averaging 6.5 MJ/week, with 125 minutes of moderate activity (3–6 METs) and 5.7 minutes of vigorous activity (>6 METs) per week. Expenditures certainly met minimum public health recommendations on days when patients attended the rehabilitation centre, but they fell far short on other days of the week (respective expenditures of 1.25 vs. 0.73 MJ/day, 27.8 vs. 10.9 minutes of moderate and vigorous activity), emphasizing the need to supplement formal cardiac rehabilitation classes by home programmes.

Jones et al. [75] reached similar conclusions in their study of a 3-day per week Phase III cardiac programme. Physical activity was much higher on the days when patients attended the rehabilitation centre (10,087 vs. 5287 steps/day). Because of the continued low level of physical activity on intervening days, only 8 % of patients met the minimum recommended level of physical activity for the week as a whole. Those who were successful in this regard were generally undertaking a home programme in addition to the formal classes.

8.3.5.3 Potential of Home-Based Rehabilitation

Home-based rehabilitation, with or without telemonitoring, has the attraction that patients do not spend a large part of their potential exercise time driving to and from a rehabilitation centre.

Reid and colleagues [76] used pedometers to assess physical activity patterns in 141 patients who were unwilling to attend a formal cardiac rehabilitation programme. Comparisons were made between a motivational counselling group and those receiving usual care. Relatively small differences were seen: those receiving motivational counselling increased their weekly walking distance from an initial 21.2 to 38.3 km/week at 12 months, whereas the corresponding figures for those receiving usual care were 20.9 and 34.5 km/week.

Oliveira et al. [77] demonstrated that a useful increase of habitual physical activity could be induced by a well-designed 12-week home-based cardiac programme. The physical activity of the experimental group was almost doubled, from 278 to 526 counts/minute per day, with the duration of moderate physical activity rising from 16.8 to 63.7 minutes/day; however, there were no changes of habitual activity in the control group.

Karjailainen et al. [78] used accelerometers to demonstrate that patients with a combination of type 2 diabetes mellitus and coronary artery disease initially undertook less moderate-to-vigorous and vigorous activity than those with coronary disease alone (4.5 vs. 8.2 minutes/day). However, individually-prescribed home-based exercise programmes were effective in increasing the high intensity habitual activity of both groups (from 2.1 to 6.1 minutes/day and from 5.0 to 9.6 minutes/day).

8.3.5.4 Retention of Interest Following Formal Rehabilitation

Considerations of cost commonly limit the duration of formal cardiac rehabilitation to 12 weeks. How well do patients continue to meet activity recommendations after their formal exercise sessions have ceased, and can the continued objective monitoring of activity contribute to the maintenance of enthusiasm?

Some observers have seen only a limited retention of interest in physical activity. Brändström et al. [79] found pedometer readings averaging only 6719 steps/day 6 months following treatment for myocardial infarction at a rural Swedish hospital (apparently without formal rehabilitation). Only a quarter of this group reached a count of 9000 steps/day, and only a fifth of the group attained 10,000 steps/day.

Clark and associates [80] used pedometers to assess the value of peer buddy motivation following completion of a 12-week cardiac rehabilitation programme. There was a substantial initial difference of step count between those entering the buddy programme (7844 steps/day) and those refusing it (5655 steps/day). Moreover, over the following 12 months, the weekly duration of physical activity for the two categories of patient decreased by 40 vs. 393 minutes/week, respectively.

Izawa et al. [81] demonstrated the positive contribution of objective monitoring in patients who had completed a 6-month cardiac rehabilitation programme. Six months later, there was a striking difference of habitual physical activity between a group that continued to monitor their body mass and physical activity and a control group that simply received usual treatment; the Kenz Lifecorder showed respective counts of 10,459 vs. 6923 steps/day.

8.3.6 Conclusions

Objective monitoring facilitates the long-term monitoring of changes in physical activity pattern, an outstanding need that was recognized by Kohl during the 2001 Consensus Conference [10]. Indeed, Aoyagi and associates have now recorded step counts and energy expenditures in a substantial sample of subjects continuously for 5 years or more [82], allowing associations to be drawn between health manifestations and current rather than initial physical activity patterns. Both cross-sectional and longitudinal objective monitoring have demonstrated that manifestations of coronary arterial disease, arterial distensibility, coronary vascular calcification and pericardial fat are all associated with regular participation in physical activity of moderate intensity and volume. Most of the data to date have been obtained on older individuals, and there remains a need to relate beneficial intensities of activity with the initial fitness levels of subjects. There also remains scope for quasi-experimental longitudinal studies, where objective monitors are used to stimulate known increases of physical activity in a sub-sample of an initially sedentary but healthy population. Finally, objective data have not as yet fully clarified the value of physical activity in the prevention of stroke.

8.4 Hypertension

In his presentation to the Hockley Valley Consensus Symposium, Richard Fagard (Fig. 8.6) concluded that regular exercise generally reduced both systolic and diastolic blood pressures, but relationships to the pattern of physical activity that was undertaken remained unclear [11]. After commenting on the issue of night-time blood pressure measurements and cardiovascular risk, we will review objective cross-sectional data suggesting only weak associations between habitual physical activity and systemic blood pressures, as well as longitudinal data that demonstrate both short-term and longer-term influences.

8.4.1 Night-Time Blood Pressure Determinations and Cardiovascular Risk

Hermida et al. [83] used Actigraph measurements to distinguish blood pressure readings obtained when patients were lying awake from when they were asleep.

Fig. 8.6 Richard Fagard has for many years examined relationships between habitual physical activity and hypertension

They followed 3344 subjects for 5.6 years. In persons without diabetes, the cardiovascular risk increased with waking pressures >130/75 mm Hg, and sleeping values >120/70 mm Hg. However, for those with diabetes, the corresponding thresholds were 120/75 and 105/60 mm Hg. The cardiovascular risk increased more steeply in women than in men when blood pressures exceeded these values [84], and it was best countered by taking one or more blood pressure-lowering medications immediately before retiring [85].

Wuerzner et al. [86] pointed out that loss of the night-time dipping of blood pressure was an important factor contributing to cardiovascular events. In a sample of 103 patients, most of whom were hypertensive, they found that the night-time dipping of blood pressure was positively associated with step counts, even though the daytime blood pressures were unrelated to step counts. Given the limited association between physical activity and daytime pressures, this seems an area that merits furtjer exploration.

8.4.2 Cross-Sectional Association Between Physical Activity and Hypertension

Given the impact of even modest amounts of habitual physical activity upon arterial stiffness (above), one might anticipate finding similar associations between physical inactivity and hypertension. However, in general the observed relationships have been weak (Table 8.6). This might reflect the fact that physical activity reduces central but not peripheral arterial stiffness (Aoyagi et al. [28]), or possibly the association has been weakened by the inclusion of body mass or obesity as a covariate in many of the analyses. Because any effects are quite small, the statistical significance of associations depends in part upon the size of the subject pool.

8.4.2.1 Significant Associations

Five investigations found small inverse associations. A study of 5505 children aged 11–12 years recorded total physical activity (average counts/minute) and minutes of moderate-to-vigorous physical activity per day [92]. Small inverse associations with systolic blood pressure were seen for both the duration of moderate physical activity (−0.66 per 15 minutes) and total daily activity (−0.44 per 100 counts/minute), with the latter variable being dominant in multiple regressions including both variables.

Huffman et al. [91] reported pedometer data from 7118 individuals with cardiovascular disease or impaired glucose tolerance. Small but statistically significant associations were found between pedometer counts and systolic blood pressure, triglycerides and fasting blood glucose.

Table 8.6 Relationship between habitual physical activity and systemic blood pressures, as seen in cross-sectional investigations

Author	Subjects	Findings
Camhi et al. [87]	1391 adults in NHANES	Blood pressures unrelated to time spent in light activity
Chan et al. [88]	182 sedentary adults	Weak inverse association between diastolic pressure and step count (r = −0.21)
Christensen et al. [89]	64 Tarahumara Indians, 18 with hypertension	No significant associations of activity with blood pressures
Healy et al. [90]	169 patients with diabetes	Activity unrelated to blood pressures
Huffman et al. [91]	7118 adults with cardiovascular disease or impaired glucose tolerance	Weak inverse association of pedometer counts with systolic pressures
Leary et al. [92]	5505 children aged 11–12 years	Inverse associations with duration of moderate-to-vigorous activity (β = −0.66 per 15 minutes) and total activity (β = −0.44/100 counts)
Lee et al. [48]	Large cross-sectional study of Hong Kong families	Self-reported hypertension shows weak association with daily step count and minutes of moderate activity
Luke et al. [93]	NHANES sample	Self-reported hypertension but not measured blood pressures inversely related to mean activity and time spent in moderate activity
Natali and associates [94]	1384 asymptomatic subjects aged 30–60 years	No relation of blood pressure to actigraph data

A cross-sectional survey of 182 sedentary Canadian employees [88] also found a weak association between daily step count and diastolic blood pressure (r = −0.21).

Cerebral palsy often leads to low levels of habitual physical activity, and Ryan et al. [95] demonstrated that in a sample of 90 such children aged 6–17 years, there was a negative association between their habitual activity and hypertension. Despite a low prevalence of obesity, hypertension was commonly observed. Logistic regression analyses showed significant correlations with sedentary activity (0.064), moderate activity (0.389), vigorous activity (−0.501), and mean activity count (−0.277).

Finally, a large cross-sectional family study found a weak relationship between habitual physical activity and self-reported hypertension [48], but this association was not confirmed when pressures were measured in the laboratory.

8.4.2.2 Negative Reports

Other investigators have found no significant association between habitual physical activity and blood pressures. Camhi et al. [87] examined NHANES data for 1391 adults; blood pressures were unrelated to Actigraph measurements of the time spent and the step count during light physical activity. A further analysis of NHANES data for 2003–2006 was conducted by Luke et al. [93]. This emphasized the wide gap between questionnaire responses and objective measurements of habitual physical activity. HDL-cholesterol, plasma glucose, and self-reported hypertension and diabetes were significantly correlated with the mean level of physical activity (counts/minute) and the time spent in bouts of moderate-to-vigorous activity (>2000 counts/minute) lasting longer than 1 minute; however, these associations were weakened after co-varying for levels of obesity, and no significant associations were seen for measured blood pressures or total cholesterol levels.

The Family project also found self-reported hypertension was significantly related to step count, but the relationship was not confirmed by direct measurements of blood pressure [48].

Natali and associates [94] divided 1384 asymptomatic subjects aged 30–60 years into three blood pressure categories (optimal, normal and high normal). Within this categorisation, neither the men nor the women showed any significant inter-group differences in uniaxial Actigraph measurements of habitual physical activity.

Christensen and colleagues [89] used a combined accelerometer and heart rate sensor to determine activity levels in a sample of Tarahumara Indians, a group once famed for their long-distance running, but now afflicted with a high prevalence of diabetes; 28 % of 64 middle-aged Tarahumaran adults had hypertension, but the associations of systolic and diastolic blood pressures with habitual physical activity within this population were weak and statistically non-significant.

Finally, a study of 169 Australian diabetics [90] found no significant associations between accelerometer data and systolic or diastolic blood pressures.

8.4.3 Longitudinal Studies

Despite the limited evidence of cross-sectional associations, a number of objectively monitored longitudinal studies have demonstrated reductions of blood pressure of a similar order to those reported in questionnaire studies.

A single session of walking reduced the blood pressure of elderly individuals by 12/4 mm Hg [96].

A quasi-experimental longitudinal study was conducted on 47 test and 94 control employees of the Merrill Lynch organization [97]. At control sites, subjects received blood pressure and weight measurements plus some health education; at experimental sites, they were also given pedometers and were specifically encouraged to increase their daily physical activity. Over 1-year of observation, the test subjects showed decreases of both blood pressure (−10.6/−7.4 vs. −6.1/−3.1 mm Hg) and body mass index (−1.0 vs. +0.2 kg/m^2) relative to the controls; at the end of the year, 38 % of the experimental group reported taking vigorous physical activity three or more times per week, although only 28 % used the pedometer on a regular basis.

Moreau et al. encouraged a small group of post-menopausal women with borderline or stage 1 hypertension (pressures 130–159/85–99 mm Hg) to increase their physical activity by walking an additional 3 km/day, monitored by a pedometer [98]. Relative to a control group who did not increase their daily activity, the systolic pressure was decreased by 6 mm Hg at 12 weeks, and by 11 mm Hg at 24 weeks, although diastolic pressures remained unchanged. The body mass of the experimental group also decreased by an average of 1.3 kg, but this change appeared unrelated to the alteration of blood pressures.

Iwane et al. [99] persuaded 83 of 730 middle-aged industrial workers to adopt a daily pedometer count >10,000 steps/day; 30 of the 83 individuals were initially hypertensive (pressures >140/90 mm Hg), and they showed an average decrease of pressures from 149/99 to 139/90 mm Hg over a 12-week period when their pedometer counts averaged 13,510 steps/day. However, pressures remained unchanged, both in those who were initially normotensive and in those who did not increase their physical activity.

Zoellner et al. [100] enrolled 269 middle-aged adults, mostly African Americans, in an uncontrolled pedometer-motivated exercise programme. Over 6-months, they observed a substantial decrease in average blood pressures (−6.4/−4.6 mm Hg)

Hultquist and associates [101] compared the merits of a 10,000 steps/day target with the alternative of walking 30 minutes/day in a group of middle-aged women, most of whom were initially normotensive. Over 4 weeks of observation, both approaches reduced blood pressure, with the timed walk being somewhat more effective (respective decreases of 2/2 vs. 4/3 mm Hg).

Swartz et al. [102] persuaded a group of 18 post-menopausal women with a family history of diabetes to increase their pedometer count from an initial 4972 to 9213 steps/day. Initial blood pressures averaged 138/88 mm Hg, but reductions of 8/5 mm Hg were seen over an 8-week programme.

There remains scope to extend these longitudinal investigations, with careful documentation of the patterns of increased physical activity that are adopted.

8.5 Cardiac and Metabolic Risk Factors

In their presentation to the Hockley Valley Consensus Symposium, Art Leon (Fig. 8.7) and Sanchez [13] concluded that questionnaire data on blood lipids were too inconsistent to establish the effects of volume and intensity of physical activity.

There are five main metabolic risk factors [103]: (1) a body mass index >25 kg/m^2; (2) a fasting serum triglyceride level >1.7 mmol/L; (3) a fasting HDL-cholesterol <1.0 mmol/L in men or <1.3 mmol/L in women; (4) a systolic blood pressure >130 mm Hg or a diastolic pressure >85 mm Hg; and (5) a fasting blood glucose >6.1 mmol/L or a haemoglobin A1c >5.5 %. Some authors have commented on individual risk factors, for instance hypertension (above), whereas others have looked at increases in the number of subjects with risk factors at various levels of daily physical activity.

Fig. 8.7 Art Leon has led many investigations examining relationships between habitual physical activity and blood lipids

A lack of physical activity is associated with metabolic risk factors not only in the middle-aged and the elderly, but also in children and university students. Abnormal sleep patterns also modify risks.

8.5.1 Middle-Aged and Elderly Adults

Cross-sectional research on 182 people from Prince Edward Island found that sedentary employees with one or more self-reported components of the metabolic syndrome took fewer steps/day than those individuals with no manifestations of this syndrome [88]. Other studies of middle-aged and older adults have consistently replicated this finding.

Scheers et al. [104] used a Sense-Wear arm band to study relationships between various measures of habitual physical activity and individual metabolic risk factors in a sample of 370 middle-aged Flemish adults (Table 8.7). The strongest relationships were with waist circumference, where statistically significant effects were seen from sedentary time and the duration of light activity, but the largest effects were with moderate physical activity and the average energy expenditure over the entire day (PAL). Relationships with serum triglycerides and HDL-cholesterol were also seen, but as would be anticipated from the previous section of this text, partial correlations with systolic and diastolic blood pressures were relatively weak, and there were no significant relationships with fasting blood glucose. The authors concluded from their analysis that activity of at least moderate intensity may be important to the control of the metabolic syndrome.

Henson et al. [105] studied 878 adults with risk factors for type 2 diabetes mellitus. They found that plasma glucose, triacylglycerol and HDL cholesterol levels were all significantly associated with sedentary time. However, only adiposity was associated with total physical activity and the number of breaks in sedentary time, suggesting that excessive sedentary time is an important variable to modify in the control of metabolic risk factors.

Table 8.7 Partial correlation coefficients between habitual physical activity as monitored by SenseWear arm band and metabolic risk factors in 370 middle-aged Flemish adults, after adjustment for co-variates (only statistically significant associations are shown; no associations with fasting blood glucose were significant)

Variable	Waist circumference	Triglycerides	HDL-cholesterol	Systolic blood pressure	Diastolic blood pressure
Average physical activity (METs)	−0.61	−0.27	0.39	−0.13	−0.23
Sedentary time (hours/day)	0.34	0.16	−0.24		0.18
Length of sedentary bout (minutes)	0.31	0.15	−0.21	0.11	0.12
Breaks in sedentary bouts (n/day)	−0.18				
Light activity (<3 METs, hours/day)	0.16				
Moderate activity (<6 METs, hours/day)	−0.50	−0.21	0.33		−0.17
Vigorous activity ((>6 METS, hours/day)	−0.34	−0.19	0.29		−0.17
Moderate + vigorous (hours/day)	−0.52	−0.23	0.35		−0.18
Moderate + vigorous in bouts >10 minutes (hours/day)	−0.54	−0.24	0.36		−0.19

Based on data of Scheers et al. [104]

Camhi et al. [87] analyzed data from a NHANES survey of 1371 adults. They also found the main impact was upon the lipids. Blood pressures were unrelated to Actigraph measurements of the time spent and the step count during light physical activity (an average 113 minutes, 3554 steps/day). However, a greater time spent in light physical activity and a greater number of steps taken at a light intensity of effort were associated with a lower odds of high triglycerides, a low LDL-cholesterol, a large waist circumference and a lower risk of the metabolic syndrome. No associations were found between habitual physical activity and blood pressure or blood sugar.

Bankoski et al. [106] also examined NHANES data. People with the metabolic syndrome spent a longer percentage of the day sitting (67.3 vs. 62.2 %), had longer sedentary bouts (17.7 vs. 16.7 minutes), had a lower intensity of activity while sitting (14.8 vs. 15.8 counts/minute) and had fewer sedentary breaks (82.3 vs. 86.7) than those not affected by the metabolic syndrome.

Ekelund et al. [107] examined 258 adults aged 30–50 years with a family history of diabetes mellitus. They established that after controlling for age, sex and obesity, the time spent in moderate and vigorous physical activity was associated with metabolic risk, but in their study no associations were seen for light activity and sedentary time. Total body movement (counts/day) was independently associated with triglycerides, insulin, HDL-cholesterol and summed metabolic risk factors.

Table 8.8 Standardized regression coefficients linking components of physical activity profile and metabolic risk factors in 160 Australian diabetics studied by Healy and Associates [90]

Variable	Sedentary duration	Light intensity duration	Moderate to vigorous intensity duration	Mean activity intensity
Waist circumference	**0.22**	**−0.20**	**−0.16**	**−0.27**
Triglycerides	**0.19**	−0.14	**−0.23**	**−0.24**
HDL cholesterol	−0.12	0.11	0.07	0.05
Systolic blood pressure	0.06	−0.07	0.00	−0.10
Diastolic blood pressure	0.03	−0.05	0.06	−0.01
Fasting plasma glucose	0.13	−0.12	−0.08	**−0.18**
Metabolic risk score	**0.23**	**−0.20**	**−0.17**	**−0.23**

Bold type indicates statistically significant associations

In contrast, a study of 169 Australian diabetics [90] found significant associations between accelerometer estimates of time spent sitting, and engagement in light and moderate physical activity (Table 8.8). Independently of the time spent in moderate and vigorous physical activity, there were significant associations between the times spent in sedentary and light intensity activities and waist circumference and other metabolic risk factors.

Park et al. [103] examined daily pedometer/accelerometer records for 220 Japanese subjects aged 65–84 years. Data were related to each of the five main metabolic risk factors. Those having a high blood pressure, a high blood glucose and three or more of the metabolic risk factors all showed significantly lower step counts and significantly shorter daily periods of exercise at an intensity >3 METs than their peers who did not have these manifestations. The odds ratio of presenting three or more metabolic risk factors was increased 4.3-fold in those in the lowest step count quartile, and 3.3-fold for those in the lowest quartile of activity at an intensity >3 METs. For those in the second quartile of step counts (average 5581 steps/day), the odds ratio was 2.58, but no significant advantage was seen with a further increase in the daily step count.

8.5.2 Children and Young Adults

Alhassan and Robinson [108] sought the optimum level of habitual physical activity discriminating the presence of two or more cardiac risk factors in a sample of 261 girls with an average age of 9.4 years. The clearest discrimination came with an Actigraph record showing >16.6 minutes of physical activity per day at an intensity >2000 counts/minute. Interestingly, this threshold is substantially lower than the minimum daily recommendation derived from questionnaire studies.

An association between low levels of daily physical activity and cardiac risk factors was demonstrated in a group of 462 adolescent Australian girls [109]. The mean pedometer count was 9617 steps/day, but comparing the physically most active with the least active quartile, the latter group were more likely to be obese (odds ratio 4.70) and to have a low level of cardiorespiratory fitness (odds ratio 3.27). All 47 girls who had three or more cardiac risk f actors had a count of less than 7400 steps/day.

Morrell et al. [110] examined 1610 university undergraduates; their habitual physical activity ranged from <7500 to >12,500 steps/day, and across this range there was a progressive decrease in the average number of metabolic syndrome criteria from 0.94 in the physically least active to 0.73 in the most active students.

8.5.3 *Sleep Quality*

Accelerometers have also been used to examine sleep quality and fragmentation in relation to manifestations of cardiac risk factors.

Petrov et al. [111] used a wrist actigraph to examine 503 middle-aged adults with no previous history of cardiovascular disease. The 10-year change in lipid concentrations was associated with sleep patterns. Each additional hour of sleep per night was associated with higher levels of total cholesterol (5.2 mg/dL) and LDL cholesterol (3.4 mg/dL). However, sleep quality and sleep fragmentation were unrelated to blood lipids.

Johannson and associates [112] also used an actigraph (Actiwatch-16) to compare sleep quality between 57 patients with stable coronary arterial disease and 47 controls. They found that among those with coronary artery disease, overall sleep quality was poorer in the men, and in the women sleep duration was shorter, with a higher fragmentation index, longer periods of wakefulness and more frequent naps.

Yngman-Uhlin et al. [113] compared sleep quality between patients undergoing peritoneal dialysis ($n = 28$), those with established coronary artery disease ($n = 22$) and the general population ($n = 18$). The Actiwatch-L records showed greater sleep fragmentation and poorer sleep quality in those undergoing dialysis.

8.5.4 *Primary, Secondary and Tertiary Prevention*

Primary preventive programmes are initiated before clients have shown any manifestations of metabolic risk. Secondary programmes are begun after the appearance of such risk factors, and tertiary programmes are followed after patients have developed clinical manifestations of disease such as a myocardial infarction.

8.5.4.1 Primary and Secondary Prevention

A group of 762 sedentary Australian employees volunteered to participate in a broadly-based primary preventive workplace health programme that included provision of a pedometer and entry of the data into a web-site diary [114]. The reported pedometer counts were boosted by 10,000 steps for each 6.4 km of cycling that had been undertaken. At entry, the employees were physically somewhat more active than their peers at the same companies (average count 11,453 steps/day, compared with 10,676 steps/day). Over a 4-month period, 79 % of the volunteers reported adhering to the prescribed programme, and there was a 6.5 % increase in those reporting that they met daily physical activity guidelines (from 40.9 to 47.5 %). There was also some decrease in reported daily sitting time (0.6 hours/day) among participants, and the pedometer count increased by 640 steps/day. Small decreases of systolic and diastolic blood pressure (1.8 mm Hg) and waist circumference (average 1.6 cm) were seen, but these could have arisen from aspects of the programme other than the observed increase in daily walking.

A study of alternative methods of primary prevention in the U.S. National Guard compared the benefit obtained from a pedometer-based intervention with that seen after enrolment in a traditional high-intensity fitness programme. Participants were soldiers who had initially failed a 2-mile-run test [115]. The intervention lasted for 12 weeks, and was followed by 12 weeks of maintenance and observation. The pedometer scores of the first group improved substantially over the intervention period, but regressed during the 12 weeks of maintenance, whereas there was little change with the traditional programme (respective scores for the pedometer programme, 6415, 8027 and 7702 steps/day; for the traditional fitness programme 7300, 6593 and 7478 steps/day). Cardiac risk factors changed only minimally over the period of observation.

Tully et al. [116] carried out a small controlled 12-week trial that examined changes in cardiac risk factors in response to 12 weeks of pedometer-monitored walking (30 minutes per day, 5 days per week). Relative to control subjects, there were significant reductions in systolic and diastolic blood pressures and stroke risk as computed by the Framingham risk score.

Simmons et al. [117] further studied 365 people with a family history of diabetes; energy expenditures were measured by both the flex heart rate method and by accelerometry. Small increases in both measures of physical activity were observed over 1 year, and this change was associated with a decrease in a clustered metabolic risk score based upon all five metabolic risk factors.

8.5.4.2 Tertiary Prevention

Richardson et al. followed a group of 324 sedentary adults who were overweight and had type 2 diabetes mellitus or coronary artery disease for a period of 16 weeks [118]. During this time they were able to log on to a web-site that provided individually-tailored motivational feedback, and this increased their daily pedometer count by an average of 1888 steps/day. A half of the group also had access to a chat room where they could discuss progress with other study participants; this increased study adherence, but did not further boost step counts relative to the remaining participants. Others are currently looking at the merits of combining pedometer data with use of the Smart Phone as a means of augmenting daily physical activity [119].

Bäck et al. [120] studied cardiac risk factors in 65-year-old patients 6 months after recovery from a cardiac event in relation to pedometer data. The average count in a sample of 332 patients was 7027 steps/day. After adjusting for confounders, significant correlations were found between daily activity and HDL cholesterol ($r = 0.19$), triglycerides ($r = -0.19$), glucose tolerance ($r = -0.23$), body mass index ($r = -0.21$), night-time heart rate, 24-hour average heart rate and muscular endurance.

8.5.5 Conclusions

Objective studies of habitual physical activity have to date confirmed the impression from questionnaire studies that there is an association between inadequate physical activity and development of the metabolic syndrome. In general, the strongest association has been with body fat and body lipids, although the study of Park et al. is an exception in showing associations with blood pressure and blood glucose [103]. The majority of objective studies with the exception of Ekelund et al. [107] point to benefit with relatively low volumes and intensities of effort, but further longitudinal studies are needed to confirm a causal relationship.

8.6 Diabetes Mellitus

Kelley and Goodpaster [12] advised the Hockley Valley Consensus Symposium that although physical activity decreased the risk of type 2 diabetes mellitus, it was unclear from questionnaire studies whether the volume or the intensity of exercise was the protective factor. One complicating factor, more important for type I than for type II diabetes, is that physical activity may have been restricted because of the disease. To date, most objective studies have focussed upon metabolic risk factors (as discussed in the previous section of this chapter), fasting insulin levels, and the development of insulin resistance.

8.6.1 Type 1 Diabetes Mellitus

A study of children and youth aged 6–17 years found that (perhaps because of parental over-protection) habitual daily physical activity was lower in

32 individuals with type-1 diabetes mellitus than in 42 healthy control subjects (567 vs. 695 counts/minute) [63]. In consequence, the vasculature of students with type I diabetes showed a higher intima-media thickness, and less flow-mediated vasodilation than the controls (4.9 vs. 7.3 %).

8.6.2 Type 2 Diabetes Mellitus

Until recently, type 2 diabetes mellitus has been seen mainly in middle-aged and older adults, but with the increasing inactivity of children and adolescents, the prevalence of this condition is progressively increasing in younger individuals.

8.6.2.1 Middle-Aged and Older Adults

The one negative report is from Camhi et al. [87]. They examined NHANES data for 1391 adults, finding that fasting blood sugar levels were unrelated to Actigraph measurements of the time spent and the step count during light physical activity. Possibly, their findings were negative because they evaluated only light activity, but not moderate-to vigorous activity or total physical activity. Others, also, have found a lack of effect of physical activity upon the fasting blood glucose, but associations have been detected with insulin levels and insulin sensitivity.

A cross-sectional study of 346 men and 455 women aged 30–60 years measured insulin sensitivity by means of a hyperinsulinaemic euglycaemic clamp and habitual physical activity by accelerometer [121]. Insulin sensitivity was positively associated with total activity, the time spent in sedentary and light activities and the average intensity of activity undertaken. Insulin sensitivity increased with an increase of the average activity count per minute from 150 to 550 counts/minute, but apparently plateaued at the highest average counts (>600/minute).

Gando et al. [122] obtained 28 days of accelerometer measurements on 807 healthy Japanese men and women. The time spend in light activity (<3 METs) was inversely associated with insulin resistance, even when the data were adjusted for participation in moderate and vigorous physical activity.

Ekelund et al. examined 192 adults with a family history of diabetes mellitus [123]; as in their study of metabolic risk factors (above), they found that whereas the time spent in moderate and vigorous physical activity predicted fasting insulin and insulin resistance, glucose regulation was unrelated to the time spent in sedentary and light activities.

Leheminant and Tucker [124] examined 264 middle-aged women. Sedentary time, and the time spent in moderate and vigorous physical activity were all correlated with insulin resistance in this sample, but relationships became non-significant after adjusting for the percentage of body fat, suggesting that it may be necessary to consider body composition as an intermediary in the effect of physical activity upon glucose regulation.

Fig. 8.8 Richard Telford has undertaken longitudinal studies of physical activity and insulin resistance in young children

Fig. 8.9 Andrea Kriska has been involved in many epidemiological investigations of objectively measured physical activity and health, including studies of diabetes mellitus

8.6.2.2 Children and Adolescents

A 2-year longitudinal study by Richard Telford (Fig. 8.8) and associates [125], examined 241 boys and 257 girls who were initially aged 8.1 years; in the boys, they found significant relationships between pedometer indications of change in habitual physical activity and change of insulin resistance over the 2-year interval, although no association was apparent in the initial cross-sectional data. In the girls, insulin resistance was unrelated to the pedometer data.

Thomas et al. [126] studied a small group of 32 teenagers aged 12–18 years; intravenous glucose tolerance was significantly correlated with total physical activity, moderate physical activity, and the number of 5-minute bouts of moderate physical activity.

Andrea Kriska (Fig. 8.9) and associates [127] evaluated 699 adolescents aged 10–17 years, with a history of type 2 diabetes lasting for <2 years. These youngsters spent 56 minutes longer per day sitting than even the obese adolescents of the NHANES survey, and they showed extremely low levels of moderate and vigorous physical activity.

8.6.3 *Fatty Infiltration of the Liver*

Fatty infiltration of the liver is another metabolic phenomenon associated with inadequate physical activity [128].

In a large sample of 1307 non-diabetic patients, the extent of fatty infiltration of the liver was correlated with the level of habitual physical activity as measured by an Actigraph [129]. There was a gradient of physical activity from 35×10^4 counts/day in those with the least fatty infiltration to 28×10^4 counts/day in those most affected. Inactivity was also associated with other coronary and metabolic risk factors, including low density lipoprotein cholesterol, systolic blood pressure, intima media thickness and reduced insulin sensitivity.

8.6.4 *Pedometers and Exercise Promotion in Type 2 Diabetes Mellitus*

As in other chronic disease conditions, pedometers and accelerometers have played an important role in documenting and encouraging programmes for the primary, secondary and tertiary treatment of diabetes.

8.6.4.1 Critique of Pedometer Monitoring

Tudor-Locke et al. [130, 131] have used monitoring by inexpensive pedometers as a means to encourage patients with type 2 diabetes mellitus to engage in an incremental walking programme. When wearing such devices, walking was increased by an average of 34 minutes/day, with clients taking an additional 3000 steps/day, and much of this increase of physical activity was retained 2 months following conclusion of the formal exercise programme.

In Australia [132], a brief intervention using a pedometer and a step-recording diary was effective in increasing the 2-week walking distance relative to a comparison group (223 vs. 164 minutes), and 63.5 % of the experimental group achieved the recommended volume of moderate intensity physical activity, as compared with 41.8 % in the control group. However, another study of 57 - Australians with an average age of 62 years found that use of a pedometer had no motivational advantage relative to exercise counselling alone [133].

As in other clinical conditions, activity counts have shown that relative to the provision of exercise information, a detailed exercise consultation was a more effective method of increasing the physical activity of diabetic patients. Over a 4-week period, an experimental group showed a 4 % increase in weekly activity counts, whereas there was a 9 % decline in the controls [134].

Bjorgaas et al. [135] emphasized that the wearing of a pedometer was helpful in establishing realistic perceptions of the volume of daily activity that was being undertaken. Nevertheless, some patients at high risk of diabetes have criticized reliance upon pedometer monitoring, in part because such instruments are only accurate when recording walking activity, and in part because the commonly demanded target of 10,000 steps/day has seemed too rigorous to some of them [136].

8.6.4.2 Response of Glucose Regulation to Pedometer-Based Programmes

Prospective studies have commonly shown an improvement of glucose regulation in response to pedometer-based exercise programmes.

A group of patients with impaired glucose tolerance was divided between those receiving a structured pedometer-based educational programme and two alternative approaches [137]. At 12 months, those using the pedometer showed significant decreases in fasting and 2-hour post glucose challenge values (−0.32 and −1.31 mmol/L), whereas no changes were seen with traditional educational programmes.

Swartz et al. [102] encouraged women with type 2 diabetes mellitus who began with a pedometer count averaging 4972 steps/day to move towards a walking target of 10,000 steps/day. The actual count rose to an average of 9213 steps/day; this was associated with a reduction of blood pressure and a decreased area under the curve on a glucose tolerance test, even though the programme did not result in a decrease in body mass.

Andersen et al. [138] studied 150 physically inactive Pakistani men living in Oslo. A 5-month programme that included use of an accelerometer aimed to enhance the lifestyle of this population, which is at high risk of diabetes. In consequence, physical activity was increased 15 % relative to a control group. At the follow-up examination, there were no inter-group differences of fasting or post-prandial glucose, but the intervention group showed 27 % lower insulin levels following an oral glucose tolerance test.

In Japan, 2205 patients with established type 2 diabetes mellitus were involved in an intensive lifestyle programme that included the provision of pedometers [139]. Over a period of 2 years, small but statistically improvements in HbAC1 were seen relative to control subjects. Likewise, Sone et al. evaluated a 3-year pedometer-monitored lifestyle programme in 2205 patients with established type 2 diabetes mellitus [139]. Small but statistically significant differences in HbAc1 levels were seen after 2 years of participation, whereas control subjects showed no change.

Yates et al. [140] confirmed that the gains of physical activity and resulting improvements in glucose regulation realized in their pre-diabetes programme were sustained at 24 months.

8.6.4.3 Impact of Habitual Physical Activity upon Long-Term Prognosis

The institution of pedometer-monitored activity programmes has a favorable impact, not only upon glucose regulation, but also the likelihood of developing the various complication of diabetes. However, in some of these studies there remains a need to confirm whether activity has averted complications, or whether activity was low because of complications.

Data on 224 men and 103 women with type 2 diabetes mellitus showed that habitual physical activity, as measured by pedometer, was linked with low levels of inflammatory markers and less evidence of arterial stiffness [141].

A cross-sectional study of 100 patients with type 2 diabetes mellitus found negative correlations of habitual physical activity with peripheral neuropathy and muscular strength; those with neuropathy took 1967 fewer steps/day, and those with muscular weakness took 1782 fewer steps/day [142].

The intermuscular adipose tissue volume was inversely correlated with the daily step count in 22 patients with type 2 diabetes mellitus and peripheral neuropathy ($r = -0.44$), and lesser fat deposits were associated with a better muscular performance [143].

8.7 Obesity

Robert Ross (Fig. 8.10) and Ian Janssen reported to the Hockley Valley Consensus Symposium that short-term reductions of body fat were related to the dose of physical activity that was undertaken, but long-term effects were small and inconsistent; further, dose/response relationships were unclear for the reduction of visceral fat [15].

Fig. 8.10 The research of Robert Ross focusses on the role of physical activity in the management of obesity

Objective data on the relationship between physical activity and obesity are discussed in Chap. 9. In brief, pedometer/accelerometer monitoring has documented rather precisely a deficit of physical activity in the obese, amounting to around 2000 steps/day, or 15–20 minutes/day of moderate and/or vigorous physical activity relative to peers of normal weight. The difference in daily physical activity is quite small, underlining that the build-up of fat depots has usually occurred over the course of several years.

There is also good evidence that the physical activity of an obese adult can be augmented by 2000–3000 steps/day, and that this initiates a slow but consistent loss of total body fat (0.05–0.1 kg/week). There is scope to link objective activity monitoring to the loss of visceral fat, but this has not as yet been attempted.

8.8 Skeletal Conditions

Ilka Vuori [17] (Fig. 8.11) reported to the Hockley Valley Consensus Symposium on the issues of low back pain, osteoarthritis and osteoporosis. He concluded that light activity reduced low back pain, but that heavy activity had a detrimental effect. Moderate physical activity could also be an effective treatment for osteoarthritis. High intensity loading had an osteogenic effect, but dose/response effects were poorly understood in the context of preventing osteoporosis.

8.8.1 Low Back Pain

A prospective study showed that high levels of physical activity during childhood (as assessed at age 9 years) protected against back pain (neck, mid-back and low back) in early adolescence (age 12 years) [144]. Relative to the physically most

Fig. 8.11 Ilka Vuori has for many years examined relationships between habitual physical activity and health

active children, the least active group had a multivariate odds ratio of 3.3 for low back pain and 2.7 for mid-back pain.

Wedderkopp et al. [145] underlined that subjective reports were a most inaccurate method of determining the habitual physical activity of children and adolescents. Pain variables showed no relationship to objective measurements of habitual physical activity as determined with an accelerometer.

A retrospective analysis of findings in 82 adults aged 55–79 years found that pain severity had an adverse impact upon daily physical activity in terms of step counts, distance walked and walking velocity, suggesting that back pain could be a significant barrier to rehabilitation programmes in this age group [146].

van Weering et al. [147] found no difference in the volume of habitual physical activity recorded by triaxial accelerometer between 29 patients with chronic low back pain and 20 controls; however, those with back pain tended to take their activity earlier during the day.

Verbrunt et al. used triaxial accelerometry to demonstrate that habitual physical activity was no lower in 13 patients with chronic non-specific low back pain than in age and sex-matched controls [148].

A study was undertaken 6 months following lumbar disc surgery; subjective reports suggested a negative relationship between pain and physical activity on the day of questioning, but this relationship was not seen in 8-hour accelerometer data [149].

8.8.2 *Osteoarthritis*

In terms of the early prevention of osteo-arthritis, there remains a need to carry out the critical experiment of documenting details of the volume and intensity of habitual activity from an early age, with any associated trauma, in relation to the development of osteoarthritis in middle age. Once the condition is established, many people become physically very inactive. A light intensity and moderate volume of physical activity may slow disease progression, but excessive physical activity (>10,000 steps/day) appears to have a negative impact upon prognosis. Objective monitoring can be helpful in setting an appropriate regimen.

8.8.2.1 Osteoarthritis and Inactivity

Following the development of osteoarthritis, many people become quite inactive physically. A study of 1111 adults with osteoarthritis of the knee found only 12.9 % of men and 7.7 % of women meeting physical activity recommendations; 40.1 % of the men and 56.5 % of the women took no moderate activity during a week of observation [150]. A related cross-sectional accelerometer study of 1908 adults with or at risk of symptomatic osteoarthritis found that only 13 % of participants met current daily physical activity recommendations, and 45 % were "inactive," with no periods of moderate-to-vigorous physical activity [151].

A further Sense-Wear arm-band study of 51 patients with an average age of 68 years again found that individuals with severe arthritis of the knee or the hip joint took much less activity than their peers (6632 vs. 8576 steps/day), pointing to the need to stimulate the daily activity of such patients in order to avoid other adverse consequences of physical inactivity [152].

8.8.2.2 Physical Activity and Disease Progression

Dunlop and associates [150] examined disability onset in 1680 community-dwelling adults aged 49 years or older, and studied disability progression (marked by a growing inability to perform the basic activities of daily living) in a further 1814 individuals with symptomatic osteoarthritis. They observed an inverse association between participation in light physical activity and both the onset and the progression of disability [150]. Dividing study participants into activity quartiles, the respective hazard ratios for disability onset were 1.00, 0.62, 0.47 and 0.58, and for disease progression 1.00, 0.59, 0.50, and 0.53.

Ng and associates [153] evaluated a programme that included progressive walking and the administration of glucosamine sulphate to patients with symptomatic osteoarthritis of the hip or knee. They were able to increase activity by 3000 steps/day in the first 6 weeks of this regimen, and by 6000 steps/day in the second 6 weeks. Symptoms were reduced, particularly over the period 6–12 weeks, with further small improvements over the following 12 weeks.

Feinglass and colleagues [154] conducted a 6-month exercise programme for 226 individuals with knee arthritis or rheumatic arthritis. Contrary to their initial hypothesis, the greatest gains of step count and distance walked in response to exercise counselling by physical activity coaches or physicians were seen among that quartile of their subjects who initially were most limited by osteo-arthritis.

Nevertheless, in those with established osteoarthritis, the usually recommended volume of exercise can be excessive. A 2.7 year prospective study of 405 - community-dwelling adults found that in those with knee joint problems, the attainment of a daily pedometer count of >10,000 steps/day was associated with increased cartilage damage (relative risk 1.36) [155]. In those with a large initial cartilage volume, a high step count was protective, but in those with a low initial cartilage volume, taking >10,000 steps/day increased the extent of cartilage loss.

Questionnaires fail to give a useful assessment of the intensity of physical activity in osteoarthritis [156, 157], but an accelerometer is a useful tool in teaching the patient moderate patterns of physical activity that will minimize discomfort [158, 159].

8.8.2.3 Role of Resistance Activity

Current body sensors in general do not detect resistance activity, although such activity is known to be important in the management of patients with osteo-arthritis. There is thus a need to combine accelerometer data with self-reports when examining physical activity patterns in this class of patient.

Farr and colleagues compared the effectiveness of resistance training with self management in 171 patients with osteoarthritis of the knee [160]. Over a 3-month period, the patients allocated to both groups increased their moderate-to-vigorous physical activity from a baseline that averaged 26 minutes/day, by 18 and 22 % respectively. However, at 9 months, the increase was better sustained in those receiving resistance training (10 %) than in those with self-management (2 %). These data show that resistance training, important in the management of osteoarthritis, can be undertaken without compromising involvement in aerobic activity.

8.8.3 Osteoporosis

As yet, there has been only limited use of objective monitoring devices in examining relationships between habitual physical activity and osteoporosis, although cross-sectional and longitudinal studies from our laboratories have clarified dose/response relationships somewhat (Table 8.9). The required dose of activity is moderate, and in young Swedish children who were taking commonly recommended levels of physical activity, active commuting to and from school did not further enhance bone health. In addition to resolving issues of osteoporosis,

Table 8.9 Longitudinal study of adults initially aged 65–83 years, showing the risk that osteosonic bone density would fall below fracture threshold (−1.5 SD of population norm) over 5-year follow-up, in relation to quartiles of habitual physical activity. Significant relative risks are shown in bold type

Activity quartile	Average pedometer count (steps/day)	Risk relative to most active quartile	Average duration of activity >3 METs (minutes)	Risk relative to most active quartile
Men				
Quartile 1	4071	**2.63**	4.1	**3.33**
Quartile 2	6233	**1.75**	9.9	**2.51**
Quartile 3	7654	1.01	19.4	1.12
Quartile 4	10,588	1	34.0	1
Women				
Quartile 1	4427	**2.77**	4.1	**3.94**
Quartile 2	5731	**1.99**	9.6	**1.87**
Quartile 3	7436	1.00	17.4	0.99
Quartile 4	9986	1	31.6	1

Based on Cox proportional hazard analysis of Shephard et al. [161] for the Nakanojo population

physical activity monitors have been exploited to check the energy balance of professional ballet dancers, and to regulate high impact exercise programmes for older women.

8.8.3.1 Cross-Sectional Studies

Our cross-sectional and longitudinal objective studies of adults initially aged 65–83 years have helped to clarify dose response relationships for the prevention of osteoporosis. Cross-sectional observations on 172 free-living adults related pedometer/accelerometer data to an ultrasonic index reflecting bone stiffness in the calcaneus [162]. Linear and exponential models suggested a progressive increase of bone health to pedometer counts as high as 14,000 steps/day. An increased risk of osteoporotic fractures was assumed with a sonic index more than 1.5 SD below the population average for a young adult. Classifying physical activity levels by quartiles, in men the risk was significantly increased relative to the most active quartile (odds ratio 2.23) in the least active quartile (those taking an average of only 4.7 minutes of activity per day at an intensity >3 METs). In women, the effect was larger. Relative to the quartile that was the most active physically, odds ratios for the two least active quartiles were 8.35 and 4.94 in terms of step count (averages of 3523 and 6165 steps/day), and 3.53 and 2.83 in terms of duration of activity at an intensity >3 METs (averages of 4.4 and 11.9 minutes/day).

Given that at least moderate physical activity is needed to strengthen bone, it is not surprising that in a 4.1-year prospective study of 3027 women with an average initial age of 84 years, bone quality was associated with the height of circadian peaks of physical activity [21].

Alwis et al. [163] examined the impact of active commuting to school on bone health in Swedish children (97 girls and 133 boys aged 7–9 years). In this study, the children who walked or cycled to school had no advantage over those travelling by car or bus in terms of either bone width or bone mineral accrual, but accelerometer records showed no overall difference of daily physical activity between the two groups, and both were getting at least 60 minutes of moderate to vigorous physical activity per day.

8.8.3.2 Longitudinal Studies

A 6-year prospective study on the Nakanojo population of seniors initially aged 65–83 years [161] related objective annual physical activity data to the likelihood that the sonic index of bone density for an individual would drop below the critical fracture threshold. Kaplan-Meier curves and Cox proportional hazard analyses of the data set the minimum required daily level of activity somewhat higher than appeared from our cross-sectional analyses, above. For the men, the relative risk of a fracture was increased significantly in the lower two quartiles of step count (relative to the most active quartile by ratios of 2.63 and 1.75, respectively) and ratios for the lower two durations of physical activity >3 METs were also significantly increased, at 2.77 and 1.91. For the women, the corresponding ratios were for step count 3.33 and 2.51, and for the duration of moderate activity >3 METs 3.94 and 1.87. The implication is that for good bone health in this age group, subjects should maintain their level of physical activity at or above the level seen in the third population quartile (for the men, an average of 7650 steps/day, 19.4 minutes of moderate activity, and for the women an average of 7440 steps/day, 17.4 minutes of moderate activity). However, even this requirement is somewhat less than the amount of physical activity recommended in widely circulated questionnaire-based public health recommendations.

8.8.3.3 Other Applications of Objective Monitoring

Hoch and associates combined information from accelerometers with dietary records, assessing energy balance in a group of 22 professional ballet dancers [164]. Seventeen of the group showed an inadequate energy intake relative to their daily activity, and five had a low bone density (here implying 1 SD rather than 1.5 SD below the population norm).

Niu and colleagues have also used accelerometers to monitor and regulate high impact exercise programmes for post-menopausal women [165].

8.9 Chronic Chest Disease

The Consensus Conferences of 1988 and 1992 included sessions on exercise and chronic chest disease [166, 167], but perhaps because the benefits of physical activity were small, this topic was not discussed at the Hockley Valley Symposium.

The development of chronic obstructive pulmonary disease (COPD) is commonly linked to smoking, exposure to industrial dusts, and innate problems such as asthma and cystic fibrosis, with habitual physical activity playing little role in its aetiology. Rehabilitation programmes generally yield small gains of performance. However, it is difficult to sustain patient compliance, and the basis of benefit is as yet unclear; possibly mechanisms include a strengthening of skeletal or chest muscles, and a habituation to sensations of dyspnoea.

Habitual physical activity is typically low in patients with COPD, and is inversely related to disease severity, hyperinflation, markers of inflammatory response, quality of life, risk of hospitalization and mortality. Such data certainly encourage the prescription of walking programmes for patients with COPD, although more longitudinal studies are needed to exclude the possibility that an initial lower level of disease permits the attainment of a greater step count with associated health benefits. The available objective data point to small gains of step count in response to rehabilitation, although the final level of physical activity remains very low in most of the affected patients.

8.9.1 Habitual Physical Activity in Chronic Chest Disease

Because walking is usually the main activity of patients with chronic chest disease, there is a close correspondence between estimates of whole body and lower limb movement [168].

Pitta et al. [169] used a Dynaport monitor to compare activity patterns between 50 patients with COPD and 25 healthy individuals of similar age. The patients showed a shorter walking time (44 vs. 81 minutes/day), a shorter standing time (99 vs. 295 minutes/day), and a lower movement intensity when walking (1.8 vs. 2.4 m/s^2).

Watz and associates [170] applied a Sense-Wear arm band to patients with COPD; they found a progressive decrease of daily step count with disease severity, as assessed using the clinical classification of the Global Initiative for Chronic Obstructive Lung Disease, from around 8000 steps/day and 150 minutes at an intensity >3 METs for those in category I to around 2000 steps/day and about 50 minutes/day at an intensity >3 METs for those in category 4.

Moy et al. [171] observed a similar gradation of SAM (StepWatch Activity Monitor) scores with disease severity in a sample of 127 patients with stable COPD; the average count for their patients was 5680 steps/day.

8.9.2 Physical Activity and Pathological Concomitants of Chronic Chest Disease

Garcia-Rio et al. [172] noted a strong negative correlation between the daily step count and the tendency to hyperinflation of the lungs during exercise.

Moy et al. [173] found a dose/response relationship between inflammatory markers and step count in a cross-sectional study of 171 patients with chronic obstructive pulmonary disease. Each additional 1000 steps/day as recorded by a pedometer was associated with an 0.96 mg/L decrease in c-reactive protein and an 0.96 pg/L decrease in IL-6 concentrations, suggesting that an increase of walking distance was associated with a decrease of inflammatory response.

8.9.3 Physical Activity and Prognosis

Observations in a group of 17 male patients with COPD showed that a higher step count was associated with a greater 6 minute walking distance and a higher quality of life [174].

Garcia-Rio et al. [175] measured the habitual physical activity of 173 patients with moderate to severe COPD, using a triaxial accelerometer. In their study, movement patterns were expressed as vector magnitude units (VMUs). The risk of hospitalization over a 5–8 year follow-up showed a dose/response relationship with the individual's VMU, and the mortality risk was also lower for those with a high VMU.

In an average 4-year follow-up of 170 out-patients with COPD, mortality averaged 15.4 % [19]. A Sense-Wear arm-band assessment of daily physical activity proved the strongest predictor of all-cause mortality, with a hazard ratio of 0.46 for death linked to an 0.14 increase in PAL, here defined as the ratio of total-day energy expenditure to whole night sleeping energy expenditure.

A study of 36 pediatric oncology patients found that many had sleeping problems, including decreased total sleep time, increased awakenings and a lower efficiency of sleep, but that increased physical activity was positively associated with total sleep and sleep efficiency [176].

8.9.4 Response to Rehabilitation

Devices such as the Actigraph seem an effective approach to examining the small increases of physical activity likely to occur during the rehabilitation of patients with chronic obstructive pulmonary disease (COPD) [177].

The Omron pedometer was used to provide weekly feedback to patients enrolled in a rehabilitation programme organized by Moy and colleagues [178]. The end-result of the use of their programme was an increase of habitual physical activity from an initial step count of 2908 to 4171 steps/day at 3 months. Although encouraging, this gain must be set against the observations of DePew et al. [179], who found that a count of <4600 steps/day was associated with severe inactivity (an overall activity level <1.4 METs).

8.10 Cancer

Inger Thune (Fig. 8.12) and Furburg [16] indicated to the Hockley Valley Consensus Symposium that there were suggestions of graded dose/response relationships for all-cause cancer, colonic cancer and breast cancer, but that the effect of physical activity on the risk of developing cancer at other sites was less clear.

A major problem in epidemiological studies of cancer is the long lag period between carcinogenic change and the appearance of clinical disease. Most patients have considerable difficulty in recalling the patterns of physical activity that they have adopted over their life course. In principle, the impact of physical activity could be determined in a very long-term study, where objective measurements of were made annually, beginning perhaps as a young adult. However, such a demanding investigation has not yet been initiated.

Cross-sectional comparisons have demonstrated objective associations with physical activity and a reduced risk of cancer, and an impact has also been shown

Fig. 8.12 Inger Thune has made major contributions to elucidating relationships between physical activity and cancer

on one surrogate marker (prostate serum antigen levels). Most reports have suggested a low level of physical activity following a diagnosis of cancer, in part because of the side-effects of chemotherapy; but in a few instances recommended levels of physical activity have been maintained. The level of physical activity following treatment offers insight into the overall health status of the patient. As in other conditions, objective monitors offer a useful stimulus to greater physical activity, this being particularly helpful when the patient is living in a remote location.

8.10.1 Role of Physical Activity in Cancer Prevention

A number of cross-sectional studies suggest the role of a physically active lifestyle in the prevention of cancer, reflecting the general association between current physical activity patterns and a lifetime interest in exercise, but longitudinal objective data that would resolve dose/response issues for the various forms of cancer have yet to be collected.

One community marked by a high level of habitual physical activity is the Old Order Amish of Ohio and Southern Ontario. Members of this sect show a substantially lower incidence of cancer than the general population, and also much higher pedometer counts; the average reading for the Amish male in rural Ohio is 11,447 steps/day, compared with 7605 steps/day in the average person, and in Amish farmers (who use no mechanized equipment) the count rises to 15,278 steps/day [180]. There are many other differences of lifestyle between the Amish people and the general population, and further research is needed to ascertain how far their high level of physical activity contributes to their low cancer rates.

Lee et al. [48] conducted a large cross-sectional family study of a relatively inactive population; they noted that a self-reported history of cancer of any type was significantly and inversely correlated with daily step count, and with the daily duration of moderate and light physical activity (respective mean z scores, −0.43, −0.05 and −0.28), the first of these associations being the strongest.

Dallal et al. [181] compared accelerometer data for 996 cases of breast cancer with 1164 controls. The odds ratio for cancer among women in the highest quartile of moderate-to-vigorous physical activity was 0.38 relative to those in the lowest activity quartile; likewise, the odds ratio for sedentary behaviour (quartile 1 vs. quartile 4) was 1.81. Lynch and associates [182] examined NHANES data; they also found negative associations of breast cancer risk with time spent in light physical activity and positive associations with sedentary time.

One early indicator of prostatic cancer is a rise in prostatic serum antigen (PSA) levels. Among participants in the NHANES surveys of 2003–2006, PSA levels were inversely associated with accelerometer measurements of habitual physical activity [183]. For every hour increase of sedentary behaviour, there was a 16 % increase in the odds of showing an elevated PSA, and for every hour increase of light physical activity these odds were diminished by 18 %.

8.10.2 Current Physical Activity of Cancer Patients

Most cancer survivors are relatively inactive physically, in part because of the side-effects of treatment, but possibly also ante-dating the appearance of clinical disease. On the other hand, if they can be persuaded to become more active, this reduces the risk of a recurrence of the neoplasm, cancer mortality and associated morbidity, as well as reducing pain and other side-effects of the cancer [184].

Aznar et al. [185] undertook a pilot study of children aged 4–7 years who were undergoing treatment for acute lymphoblastic leukaemia None of them met public health recommendations for daily physical activity; the average time that they spent in moderate to vigorous physical activity was 328 minutes/week, compared with 506 minutes/week in age-matched healthy children.

Tillmann and colleagues [186] examined 28 survivors of acute lymphoblastic leukaemia aged 5.7–14.7 years who had been treated by means other than radiotherapy; they found that 5 years later, accelerometer counts were still only about two thirds of those in their age-matched peers. Moreover, bone mineral densities were low, especially in children with low accelerometer counts.

Tan and associates [187] evaluated a sample of 38 children aged 3–12 years old who were undergoing chemotherapy for acute leukaemia. They again showed much lower physical activity levels than age-matched controls (26 vs. 192 counts/minute) and they spent more of their day in sedentary pursuits (1301 vs. 1020 minutes).

Vermaete et al. [188] followed lymphoma patients before, during and following chemotherapy, using a Dynaport MiniMode accelerometer. Physical activity was initially only 86 % of that in healthy individuals, and during and following treatment of the cancer a vicious cycle of inactivity, deconditioning and fatigue commonly developed.

Other studies of adults have also found low levels of habitual physical activity. Among survivors of prostate cancer, waist circumference was inversely associated with participation in moderate to vigorous physical activity as determined by

Actigraph accelerometer; a 13.7 cm difference was found between top and bottom activity quartiles [189]. However, the body fat of these patients showed no apparent relationship with the duration of light physical activity or sedentary time.

Granger et al. [190] obtained accelerometry data from 50 patients with non-small cell lung carcinoma; again, 30 of the 50 were not meeting recommended levels of daily physical activity, and the group showed poor levels of strength and health-related quality of life relative to healthy controls.

If physical activity is not encouraged, breast cancer survivors are commonly extremely sedentary. Data from the U.S. NHANES survey found that the average breast cancer patient was sedentary for 66 % of their day, and light activity (accelerometer counts <1950/minute) occupied the remaining 33 % of their day [191]. Another analysis of the same data set concluded that only 13 % of cancer survivors met current minimum physical activity requirements, and that lack of activity was associated with obesity [192]. The time spent in moderate to vigorous activity was negatively correlated with measures of adiposity, while body fat content was positively associated with sedentary time.

Broderick and associates [193] measured physical activity 6 weeks, 6 months, and 1 year following chemotherapy for breast cancer. The initial low levels of physical activity persisted throughout the year, with an associated gain in body mass.

However, a low level of habitual physical activity is not an inevitable consequence of cancer treatment. Ruiz-Casado et al. [194] reported that 94 % of a group of 204 Spanish cancer survivors met international recommendations for the daily duration of moderate physical activity, and in Australia, 16 of 19 children who had been treated for acute lymphoblastic anaemia 6 months to 5 years previously also met the recommended levels of physical activity for their age group [195].

8.10.3 Physical Activity as a Measure of Health Status

One of the problems in physical activity epidemiology is that an inactive lifestyle can predispose to disease, but ill-health can also reduce habitual physical activity. This is particularly true following cancer, Objective measurements of physical activity can thus provide a useful measure of the impact of cancer and of chemotherapy upon overall physical function and the extent of cachexia [196], also providing a surrogate measure of the individual's quality of life [197].

Accelerometer measurements collected for 7 days post-operatively allowed Inoue et al. [198] to compare the merits of laparoscopic colorectal surgery with traditional open abdominal surgery. The period of convalescence required to reach 90 % of pre-operative levels of physical activity was much shorter for laparoscopy (an average of 3.4 days) than for open surgery (an average of 6.8 days). Laparoscopy for gastric resection was also associated with a faster recovery of physical activity than that seen with open abdominal surgery [199].

Accelerometer data have also proven helpful in assessing the curtailment of physical activity following pulmonary resection [200] and periods of intensive chemotherapy [201]. Tonosaki and Ishikawa [202] demonstrated that in the first week of adjuvant chemotherapy, the pedometer counts of breast cancer patients dropped to an average of 3841 steps/day, with a physical activity level of only 1.43 METs; during the second week, there was a small recovery, to 4058 steps/day.

8.10.4 Physical Activity Following Cancer

Social cognitive approaches, using the theory of planned behaviour [203], offer a helpful framework for the encouragement of physical activity following treatment of cancer [204, 205]. Objective monitors play an important role in patient motivation, and in general substantial increases of daily physical activity can be demonstrated, with beneficial effects on the accumulation of fat and bone density.

Vallance and colleagues [206] compared methods of motivation in a sample of 377 breast cancer survivors (Fig. 8.13). Moderate to vigorous physical activity was increased by 30 minutes/week in those receiving only the usual treatment, by 70 minutes/week in those receiving detailed print materials, by 89 minutes/week in those given a pedometer, and by 87 minutes/week in those receiving both a pedometer and print materials. The last group also showed an enhanced quality of life and less fatigue, relative to the usual treatment group. A follow-up at 6 months [207, 208] found continuing increases of 9, 39, 69 and 56 minutes/week of moderate to vigorous physical activity in the four patient groups.

Irwin et al. [209] allocated 75 post-menopausal breast cancer survivors between usual treatment and a 6-month exercise programme that comprised 150 minutes/week of gym-based and home-based exercise. The moderate to vigorous activity of the experimental group increased by an average of 129 minutes/week, compared with only 44 minutes/week in the usual care group.

Fig. 8.13 Kerry Courneya has made extensive study of exercise programmes for survivors of breast cancer

Matthews et al. arranged a 12-week walking programme for breast cancer survivors; this was completed by 34 of 36 patients [210]. Accelerometer data showed that their walking activity was increased by 11.9 MET-h/week, compared with an increase of only 1.7 MET-h/week in usual treatment controls.

Nikander et al. [211] provided breast cancer survivors with a session of step aerobics or circuit training during alternate weeks, with 2 days of supplementary home training per week [211]. Adherence to the formal sessions was 78 %. Based on accelerometer readings, they deduced that patients were developing reaction forces six times body mass, sufficient to enhance bone mass.

The on-line monitoring of pedometer scores is a particularly useful tactic to encourage physical activity among cancer survivors living in rural areas [212].

The feedback of pedometer data also reduces fatigue scores when adult survivors of childhood cancer participate in home-based exercise programmes [213].

8.11 Depression and Anxiety

At the Hockley Valley Consensus Symposium, Dunn et al. [18] concluded that physical activity lessened anxiety and depression, although there was a lack of evidence on dose/response effects. A recent review of activity monitoring in depression concluded that during the daytime, depressed patients were physically less active than their peers, but over the course of treatment, patients showed an increase of daytime activity and a decrease of night-time activity [214].

As yet, most objective studies of physical activity and depression have been cross-sectional in type, and it has been difficult to determine whether sedentary behaviour caused depression, or depression reduced the desire for physical activity. Two longitudinal studies support the first of these hypotheses. However, there remains a need for more longitudinal studies of those initially in good psychological health to establish cause and effect and to define the dose of physical activity that is most effective in prevention of depression. Various medical conditions cause depression, but in a number of instances, this problem has been helped by an increase of physical activity.

8.11.1 Cross-Sectional Studies of Physical Activity and Depression

In addition to confirming inferences from questionnaire-based studies, cross-sectional studies of physical activity and depression have provided relatively consistent information on dose/response relationships, pointing to benefit from relatively light physical activity.

8.11.1.1 Cross-Sectional Associations Between Activity and Depression

Vallance et al. [215] examined 385 men aged 55 years or older. When daily step counts were arranged as quartiles, physical, mental and global health-related quality of life were all significantly higher for the first two quartiles relative to the physically least active quartile. Depression was also significantly less in the third quartile relative to those in the least active quartile.

A large cross-sectional family study of Hong Kong residents found a weak but statistically significant relationship between depressive symptoms as reported on the PHQ-9 questionnaire and the daily step count [48]. However, z scores were not significant for the times spent in moderate and light physical activity (Table 8.4).

The multivariate analysis of Drieling et al. [216] also showed a significant inverse relationship between a 7-day pedometer recording of habitual physical activity and symptoms of depression among Latino-American immigrants.

Aronen et al. [217] examined relationships in children using a belt-worn actigraph (Basic Mini Motionlogger). They found that lower levels of physical activity were a significant concomitant of depression, although in their comparison of 22 depressed children and 22 controls, differences in 24-hour activity score were relatively small.

McKercher et al. [218] discussed the psychological profiles of young adults with depression, comparing those who remained physically active with those who were inactive. Information from 4-day use of a Yamax digiwalker pedometer was combined with data from the International Physical Activity Questionnaire. The physically active men were less likely to report insomnia, fatigue, and thoughts of suicide, while among the women who were active, reports of hypersomnia, irrational guilt, vacillating thoughts and suicide were less common.

8.11.1.2 Dose/Response Relationships

Several studies have suggested that quite modest levels of physical activity are associated with a lesser likelihood of depression relative to those who have a sedentary lifestyle.

Kazuhiro Yoshiuchi (Fig. 8.14) and colleagues were able to define dose/response relationships between psychological variables and habitual physical activity a little more closely, using pedometer/accelerometer data from the Nakanojo study [219]. In a sample of 184 Nakanojo residents aged 65–85, only 8 showed evidence of depression on the Hospital Anxiety and Depression Scale (marked by a D score >11). However, 7 of these 8 were extremely sedentary, taking less than 4000 steps/day. The one exception took 8067 steps/day, but at a very low intensity (only 6.6 minutes/day being spent at an intensity >3 METs). In contrast, all of the subjects who were physically active (step count >10,000 steps/day, >30 minutes/day at an intensity >3 METs) had low D scores, generally below 5. Within the sample as a whole, there were weak but statistically significant negative

Fig. 8.14 Kazahiro Yoshiuchi has studied relationships between objective measurements of physical activity and psychological health

Table 8.10 Relationship between a questionnaire assessment of depression and duration of daily physical activity as seen in NHANES data for 2005–2006

Class of activity	Minimal depression	Mild depression	Moderate or severe depression
Sedentary (minutes/day)	451.6	444.5	445.5
Light (<3 METs, minutes/day)	147.1	135.2	124.9
Moderate (>3 METs, minutes/day)	26.0	21.8	19.7
Vigorous (>6 METs, minutes/day)	6.7	6.4	5.5

Based on data of Song et al. [221]

correlations of D score with both daily step count ($r = -0.206$) and duration of physical activity at an intensity >3 METs ($r = -0.214$). These observations suggest that the likelihood of depression is reduced in seniors taking at least 4000 steps/day, and exercising at an intensity of >3 METs for 5 or more minutes per day.

Lopinski [220] took NHANES data for 2003–2006, finding that for every 60 minutes/day of light physical activity, survey participants were 20 % less likely to be depressed; in their study, moderate to vigorous physical activity was also associated with a lower risk of depression. A further study of the same data set by Song et al. [221] looked at levels of depression in 4058 men and women over the age of 20 years, using Patient Health Questionnaire 9; 13.9 % of participants showed mild depression, and 5.6 % moderate to severe depression, and these individuals engaged in less light and moderate physical activity than their peers (Table 8.10). A third analysis of the NHANES data [222] looked at the records of 2862 adults, finding depression in 6.8 % of this group. Dividing the duration of moderate to vigorous physical activity (MVPA) into quartiles, the odds ratios for depression in the four quartiles were 1.00, 0.55, 0.49 and 0.37, with the main benefit occurring on moving from quartile 1 (MVPA <8.5 minutes/day) to quartile 2 (MVPA 8.5–19.2 minutes/day).

McKercher et al. [223] studied the relationship of depression as assessed by a diagnostic interview to objective measurements of physical activity in 1995 young adults. Women taking >7500 steps/day had a 50 % lower prevalence of depression than those who were sedentary (<5000 steps/day); as little as 1.25 hours/week of leisure activity reduced the prevalence of depression by 45 %. However, high durations of physical activity at work (>2 hours/week) doubled the prevalence of depression. In their study, depression among the men appeared unrelated to objective measurements of physical activity.

Ewald and associates [224] also found that in a older Australians, most of the benefit in terms of an enhanced mood state on the SF-36 questionnaire occurred in the lower half of the daily step count distribution (<8000 steps/day).

8.11.2 Prospective Trials of Exercise and Depression

The available prospective trials give some support to the view that physical activity causes an enhanced mood state. Further support is found in the treatment of depression associated with various medical conditions.

Patel et al. [225] conducted a prospective trial in 225 older adults with an initial low level of physical activity. After 9 months of pedometer-based motivation that increased moderate leisure time activity by 68 minutes/day and total activity by 55 minutes/day, depressive symptomatology as assessed by the Geriatric Depression Scale and the SF-36 mental functioning scale was significantly decreased, D scores dropping from an average of 1.8 to 1.4, and SF-36 mental functioning scores rising from 84.9 to 87.7.

Prohaska et al. [226] followed 572 elderly Latino subjects who were enrolled in an exercise programme for a period of 24 months. Initial levels of activity were low (2743 steps/day in the 158 who were depressed, 3257 steps/day in the 413 who were not). Adherence to the exercise programme was poorer in those who were depressed but increases of physical activity over the 24 months of observation were similar, whether subjects initially showed signs of depression or not.

8.11.3 Physical Activity and Disease-Related Depression

Depression is a common accompaniment of various medical conditions, including pregnancy, heart disease, diabetes, osteoarthritis, cancer, multiple sclerosis, fibromyalgia and schizophrenia.

8.11.3.1 Pregnancy and Depression

Loprinzi et al. [227] examined the relationship between depression and physical activity in a sample of 141 pregnant women from the 2005–2006 NHANES data.

Some 19 % of the group manifested depression, and they were less active than their peers (8.8 vs. 14.5 minutes of moderate to vigorous physical activity, 4.8 vs. 20.9 % meeting recommended levels of daily physical activity).

8.11.3.2 Cardiac Conditions and Depression

Alosco et al. [228] scored patients with cardiac failure on the Beck Depression Inventory (BDI). Accelerometer data showed an average of 587 minutes of sedentary time, and only 0.3 minutes of vigorous activity per day; moreover, the number of items reported on the BDI was inversely related to the individual's level of physical activity.

Horne et al. [229] found that depression both before and following cardiac surgery was associated with low levels of objectively measured physical activity. In a multivariate analysis, the odds ratio of showing depression pre-surgery was 2.03 for those patients who were inactive, and throughout the study, those who were depressed were taking 50 minutes/week less physical activity.

The depression associated with congestive heart failure sometimes causes patients to under-estimate the volume of their physical activity; Skotzko et al. [230] examined 33 elderly patients with congestive heart failure. Although those who were depressed reported reduced physical activity, this was not confirmed by objective monitoring.

8.11.3.3 Diabetes Mellitus and Depression

A 12-month pedometer-motivated rehabilitation programme for 291 patients with type 2 diabetes mellitus [231] increased their physical activity from an initial average of 3226 to 4499 steps/day, and at the same time reduced their Beck Depression Inventory scores from 26.7 to 14.2, with an enhanced quality of life (particularly an increase of the SF-12 mental composite score from 36.8 to 44.7).

Another pedometer-motivated exercise programme followed 1650 medically under-served patients with diabetes mellitus [232] for 5 years; it showed a slower rate of decline of physical activity in this group than in those receiving usual care, and this was accompanied by less evidence of depression.

8.11.3.4 Osteoarthritis and Depression

White and colleagues [233] studied the walking activity of 1018 patients with osteoarthritis of the knee, using a Step-Watch monitor. They found that depression was associated with a lesser volume of walking among those with painful knees, and a positive affect with associated with a greater volume of walking.

8.11.3.5 Cancer and Depression

Sabiston and associates [234] used a Yamax digiwalker and a pedometer diary to demonstrate that in a sample of patients following treatment of breast cancer, physical activity was a significant partial mediator of the relationships between pain and depression and between pain and positive affect.

8.11.3.6 Multiple Sclerosis and Depression

Physical activity as determined by a uniaxial Actigraph accelerometer showed an inverse association ($r = -0.25$) with depression in the early stages of multiple sclerosis [235], and a weak positive association with the quality of life of these patients ($r = 0.07$) [236].

8.11.3.7 Fibromyalgia and Depression

McLoughlin et al. [237] reported that in 39 patients with fibromyalgia, levels of physical activity as measured both by Actigraph acclerometer and questionnaire were low relative to controls (counting rates of 224 vs. 294/minute). In these patients, the time spent in moderate activity was inversely correlated with depression scores ($r = -0.37$), whereas sitting time was positively correlated with depression ($r = 0.49$).

8.11.3.8 Schizophrenia and Depression

Wichniak et al. [238] found low levels of Actigraph determined levels of physical activity in schizophrenic patients who were being treated with olanzapine or risperidone. Moreover, the low 24-hour levels of physical activity were linked to depressive symptoms ($r = 0.23$), often with metabolic side effects from the low levels of activity.

8.12 Quality of Life and Physical Activity

Questionnaire-based studies have generally shown a greater quality of life in physically more active individuals. Spirduso and Cronin [18] concluded at the Hockley Valley Consensus Symposium that physical activity increased both the quality of life and the duration of independent living, but no dose/response effect was observed.

We will now examine objective evidence from cross-sectional and longitudinal studies, and will consider interactions between physical activity and the health-related quality of life (HRQOL) in a variety of chronic clinical conditions.

8.12.1 Cross-Sectional Studies of Physical Activity and Quality of Life

Objective measurements of habitual physical activity have confirmed the association between the amount of activity and the quality of life and have in general pointed to a dose/response relationship where the main effect is seen on moving from a very sedentary lifestyle to quite low levels of physical activity.

8.12.1.1 Dose/Response Relationships Seen in Nakanojo

In the Nakanojo study, Yasunaga et al. [239] tested the relationship between quality of life as assessed by a Japanese language version of the SF-36 questionnaire and objective pedometer/accelerometer data on habitual physical activity in a sample of 181 elderly Japanese aged 65–85 years. Associations with physical activity were fairly uniform across the various dimensions of the SF-36, so an overall HRQOL was assessed by summing scores across the 8 SF-36 scores. Whether objectively measured physical activity was assessed in terms of steps/day or minutes of activity at an intensity >3 METs, the largest improvement of health-related quality of life was seen on moving from the first to the second physical activity quartile, both in men and in women. Not even the physically most active subjects attained the theoretical maximal HRQOL of 100 units (Table 8.11).

Table 8.11 Relationship between habitual physical activity and health-related quality of life (sum of 8 SF-36 scores) in 181 Japanese men (M) and women (F) aged 65–85 years

Sex	Activity quartile (steps/day)			
	Quartile 1 (M 3944, F 3411 steps/day)	Quartile 2 (M 6546, F 5202 steps/day)	Quartile 3 (M 8329, F 6907 steps/day)	Quartile 4 (M 11,233, F 10,036 steps/day)
Men	73.2	84.0	86.2	88.9
Women	68.4	82.6	84.2	80.1
	Activity quartile (minutes of activity >3 METs)			
	Quartile 1 (M 5.2, F 5.2 minutes/day)	Quartile 2 (M 15.5, F 11.7 minutes/day)	Quartile 3 (M 24.5, F 17.4 minutes/day)	Quartile 4 (M 41.4, F 32.0 minutes/day)
Men	70.1	85.4	88.2	88.5
Women	70.7	78.4	85.1	81.1

Based on data of Yasunaga et al. [239]

However, the gain from the first to the second quartile of physical activity was of clinical significance, whereas further gains in the upper two quartiles were clinically unimportant.

A further analysis of the Nakanojo data set looked at potential interactions between the volume and the intensity of physical activity. After controlling for step count, the SF-36 score was higher in those who spent more of their time at an intensity >3 METs; moreover, engagement in physical activity at an intensity >3 METs was associated more closely with the physical than with the mental components of the SF-36 score [240].

A third report on the Nakanojo study [241] noted that both the age-adjusted number of stressful life events and their cumulative severity were negatively correlated with daily step counts and with the number of minutes spent exercising at an intensity >3 METs.

8.12.1.2 Other Cross-Sectional Studies in Adults

Other cross-sectional objective physical activity data have confirmed the association with HRQOL. While not adding much to the understanding of dose/response relationships, some reports have emphasized a greater association with physical than with mental health.

Data obtained on a representative sample of 5537 English adults aged 40–60 years [242] emphasized the importance of using objective measures of physical activity. The partial regression coefficients with HRQOL as assessed by the EUROQOL-5D scale were substantially larger for objective (0.072) than for subjective (0.026) measures of physical activity.

Withall et al. [243] examined 228 elderly adults (average age 78 years). Both step count and the daily duration of moderate-to-vigorous physical activity were positively associated with physical well-being ($r = 0.50$. 0.55 respectively), but were less strongly linked to mental well-being ($r = 0.15$, 0.15). Moreover, well-being was not significantly related to sedentary time.

Jepsen et al. [244] carried out a cross-sectional analysis on 49 adults who were severely obese. Accelerometer measures of habitual physical activity were positively associated with life satisfaction ($p = 0.024$) and physical functioning ($p = 0.025$), but were unrelated to mental health.

Vallance et al. [245] divided 297 post-menopausal women into those achieving a pedometer count of at least 7500 steps/day, and those who did not; the group who were physically more active achieved better scores on both the physical and the mental components of the HRQOL. Vallance et al. [215] used the Rand health status inventory to examine the quality of life in 385 men aged >55 years. Pedometer counts were divided into quartiles, with the two highest quartiles showing higher scores for physical, mental and global health than the two lowest quartiles.

A study of 754 adults older than 66 years [246] found that self-rated health was higher in those who were physically more active; however, the benefits of such activity did not differ whether the activity was performed out-of-doors or indoors.

8.12.1.3 Cross-Sectional Studies in Children

Studies in children again show an association between habitual physical activity and quality of life, but as yet there are insufficient data to establish whether dose/response relationships are similar to those seen in adults.

Herman et al. [247] evaluated the self-rated health of children aged 8–10 years. In the boys (but not the girls), the odds of reporting less than excellent self-rated health were increased two-fold among those not meeting public health physical activity recommendations, and the odds rose six-fold in those assigned to the lowest tertile of moderate to vigorous physical activity.

A cross-sectional comparison between 132 normal weight and 107 obese children aged 10–13 years [248] found respective average actigraph counts of 27,037 and 32376/hour; the obese children also showed lower HRQOL scores (74.4 vs. 87.1). The lower QOL in the obese could possibly relate in part to appearance and self image, although in a group of 177 overweight and obese children aged 8–12 years, Shoup et al. [249] found that psychosocial and total QOL were positively related to habitual activity, irrespective of the individual's body mass.

8.12.2 Longitudinal Studies of Physical Activity and Quality of Life

There have now been quite a number of longitudinal studies where habitual activity has been increased, and an enhanced quality of life has been observed (Table 8.12). Such observations help to establish the causal nature of the relationship, although often the longitudinal programmes have included advice on diet and weight loss that could have contributed to the enhanced quality of life. In many of these studies, the increase in daily step count associated with a gain in the HRQOL has been in the range 2000–3000 steps/day.

8.12.2.1 Longitudinal Studies in Adults

An Australian worksite health programme used pedometers to encourage the 762 participants to reach and/or maintain a goal of 10,000 steps/day [250]. At 4 months, the WHO-Five well-being index was increased by an average of 3.5 units, and the score remained increased by an average of 3.4 units after cessation of the formal programme. Those who initially had a poor sense of well-being on

Table 8.12 Longitudinal studies of increased physical activity and associated gains in the health-related quality of life

Author	Subjects	Study duration	Increase of physical activity	Findings
Freak-Poli et al. [250]	762 Australian workers	4 months	Step counts >10,000 steps/day realized	Increase of WHO-5 well-being index
Mutrie et al. [251]	37 primary care patients >65 years of age	12 weeks	Step counts increased to 6743/day, compared with 5671/day in controls	SF-36 scores all increased in experimental group, largest in physical domain
Fortuno et al. [252].	131 elderly adults	6 months	Activity increased by 2661 steps/day	Increased EUROQoL scores
Hawkes et al. [253]	22 near-relatives of colon cancer victims	6 weeks	Moderate to vigorous activity increased 151 minutes/week	Increase in both physical and mental components of SF-36
Long et al. [254]	89 sedentary women	16 weeks	2750 steps/day increase	Improved QOL on Dartmouth questionnaire
Morgan et al. [255]	107 overweight and obese subjects	6 months	1600–2000 steps/day increase	Increased score on physical component of SF-12
Fitzsimons et al. [256]	79 inactive adults	12 weeks	2388 steps/day increase	Gains in mood and EUROQoL scores
Wallmann et al. [257]	153 middle-aged adults	15 weeks	3240 steps/day increase	Improvements in subjective QOL

average tended to a lower step count than those with positive well-being (11,223 vs. 12,066 steps/day). Moreover, a half of workers who initially had a poor sense of well-being moved into the positive well-being category as their step count increased. Unfortunately, the statistical significance of conclusions from this study was limited by a high initial step count in many of those volunteering for the study. A further analysis of this data set [258] found that (contrary to some of the cross-sectional studies) there was an increase in the mental component of the quality of life among those workers who increased their physical activity.

A small 12-week study engaged 37 primary care patients aged >65 years in a pedometer-motivated exercise programme [251]. It increased daily pedometer counts to 6743 steps/day, as compared with 5671 steps/day in control subjects. Scores on all SF-36 scores increased in the experimental group, and as in the cross-sectional data, the largest changes were seen in the physical health component.

In Spain, 131 elderly adults completed a 6-month once-weekly supervised exercise programme [252]. This induced an increase in activity of 2661 steps/day, mainly in the men, and there was also an improvement in EUROQoL scores.

Hawkes et al. [253] instituted a 6-week pedometer-motivated lifestyle programme for 22 near relatives of patients who had suffered colo-rectal cancer.

Moderate to vigorous activity was increased by 150.7 minutes/week, and in addition to a reduction in body fat content, there were significant improvements in both the physical and the mental components of the SF-36 HRQOL score.

A 16 week exercise and lifestyle intervention in 89 sedentary women increased the pedometer count of these subjects from 6800 steps/day to around 9550 steps/day, with little change of physical activity in control subjects [254]. The quality of life was assessed by the Dartmouth general health questionnaire; a 4.4 decrease of score from an initial value of 19.7 indicated an increase in the quality of life for the experimental group.

A prospective trial in 107 overweight and obese subjects [255] used one of two pedometer-based motivational programmes to increase physical activity by 1600–2000 steps/day from an initial count of 6900 steps/day over a 6-month period. In addition to a 2.4 % reduction of body fat mass, there was a 1.9–2.5 increase of score on the physical component of the SF-12 HRQOL.

A small trial of pedometer-motivated exercise in 79 relatively sedentary adults found that pedometer counts increased from an initial 6941 to 9327 steps/day at 12 weeks and 8804 steps/day at 24 weeks [256]. There were associated gains in mood and health-related quality of life as measured on the EUROQoL scale.

A German campaign urged 153 middle-aged subjects to increase their activity by 3000 steps/day [257]. Activity was in fact increased from 6646 to 9886 steps/day over a 15-week period, with a decrease of body mass index and improvements in the subjective quality of life.

8.12.2.2 Longitudinal Studies in Children

Two small-scale longitudinal studies induced small increases of physical activity in the experimental group, but in neither case was this sufficient to improve the HRQOL.

A parent-motivated programme increased the physical activity of 39 children aged 5 years. Relative to control schools, the children concerned had less sedentary time, and increased their accelerometer readings by 5.6 counts/minute, with a 4 minutes per day increase in the time allocated to moderate-to-vigorous physical activity. However, the greater physical activity induced only marginal changes in the quality of life or the general perceived health of the children relative to control students [259].

Half of a sample of 107 obese children aged 7–11 years were enrolled in a 6-month exercise and lifestyle programme designed to reduce body fat content [260]. Weight gain was significantly smaller (2 kg) in the intervention group, and there were small increases in physical activity, but no significant changes in the quality of life.

8.12.3 *Quality of Life in Medical Conditions*

Exercise programmes have been prescribed for many chronic conditions, including cardiac disease, stroke, chronic obstructive pulmonary disease, diabetes mellitus, cancer, osteoarthritis and rheumatoid arthritis, multiple sclerosis and renal dialysis. In general, the HRQOL has improved as activity has increased, but in most studies other components of therapy could have contributed to this outcome.

8.12.3.1 Quality of Life in Cardiac Conditions

Müller and colleagues [261, 262] examined the quality of life in 330 adult patients with congenital heart disease, using triaxial accelerometers. Most of the group were relatively active as adults, with 76 % of the group meeting the minimum weekly recommendation for physical activity; on average, the group engaged in at least 59 minutes of moderate activity per day. Perhaps because most of this initial level of activity and a good initial quality of life as assessed by the SF-36 questionnaire, the correlation between quality of life and objective measures of habitual physical activity was quite limited ($r = 0.030$–0.258).

Izawa and associates [263] used the SF-36 questionnaire to assess the quality of life in patients 6 months following formal cardiac rehabilitation; 90 of the 109 patients had continued exercising, with a pedometer count of 9252 steps/day, and on all of the SF-36 measures they had a much higher quality of life than the 19 individuals who (despite apparently identical clinical characteristics) had ceased exercising (a continuing pedometer count of only 4246 steps/day).

A personally tailored exercise programme for 115 patients with coronary heart disease continued for 6 months [264]. Significant gains of physical activity were realized relative to usual care controls (7079 vs. 6186 steps/day, 201 vs. 163 minutes/day of moderate to vigorous physical activity at 6 months, 7392 vs. 6750 steps/day, 201 vs. 170 minutes/day 6 months after completing the programme), and there were associated small but statistically significant improvements on the emotional and the physical scales of the HRQOL relative to the controls.

A pedometer-based socio-cognitive approach enhanced the physical activity of coronary patients in the first year following a heart attack [265], and there were small associated gains in the quality of life relative to usual treatment controls. At 12 months post-infarct, over 80 % of the experimental group were still taking >7500 steps/day whereas in the control group less than 60 % met this standard.

A small controlled trial applied an individually-designed exercise programme (two times per week for 3 months) to 33 patients with congestive heart failure. There was no increase in daily energy expenditure as assessed by accelerometer, and the perceived quality of life remained unchanged [266], emphasizing the potential for HRQOL to change independently of physical activity levels.

Gottlieb and associates [267] evaluated a 6-month graded exercise programme in 33 elderly patients with congestive heart failure. Although maximal oxygen intake increased by an average of 2.4 ml/(kg minute), there was no increase in daily energy expenditure as assessed by accelerometer, and the perceived quality of life remained unchanged.

8.12.3.2 Quality of Life and Stroke

Rand and associates [268] initiated a daily physical activity programme following stroke in 40 adults with a mean age of 66 years. After controlling for the severity of the initial cerebrovascular impairment, the amount of physical activity undertaken explained 10–12 % of the variance in the physical component of the SF-36 score; however, the mental component of the SF-36 was unrelated to physical activity patterns.

8.12.3.3 Quality of Life and Chronic Obstructive Pulmonary Disease

In some studies of patients with chronic obstructive pulmonary disease there has been a dissociation between changes of habitual physical activity and gains in the HRQOL.

Jehn and associates [269] studied the quality of life in 107 patients with chronic obstructive pulmonary disease (COPD), using an SF-36 and a respiratory questionnaire. Age and fast walking were independent predictors of quality of life in these patients, with the latter showing an 0.30 correlation on multivariate analysis.

HajGhanbari et al. [270] emphasized that in patients with COPD, pain could have a negative impact upon both quality of life and habitual physical activity as measured by a triaxial accelerometer.

Mador et al. [271] found improvements in both exercise capacity and quality of life in 24 patients with COPD following rehabilitation, but this did not translate into increased levels of habitual physical activity as measured with an RT3 triaxial accelerometer.

Nguyen and associates [272] also found that a coaching-induced increase of pedometer-monitored physical activity in a small sample of patients with chronic obstructive pulmonary disease did not increase their quality of life relative to a self-monitoring group who showed little change in habitual physical activity.

A comparison between pedometer-based counselling and usual care in 35 patients with COPD [273] showed a small increase of pedometer count in the first group (from 7087 to 7872 steps/day), whereas in the second group, counts decreased (from 7539 to 6172 steps/day); the first group also showed increases in HRQOL, larger on the St. George's Respiratory Questionnaire than on the SF-36 score.

DeBlok et al. [274] compared the effects of pedometer-motivated activity counselling with a standard rehabilitation programme in 21 patients with COPD;

the gains of pedometer count were greater in the first group (1430 vs. 455 steps/day), but HRQOL did not differ between the two groups.

8.12.3.4 Quality of Life in Type 2 Diabetes Mellitus

In patients with type 2 diabetes mellitus, programmes that increased habitual activity also boosted the HRQOL.

A group of 292 patients with a combination of type 2 diabetes mellitus and depression received monthly counselling on depression management and walking [231]. The pedometer count at 12 months was increased from an initial value of 3226 to 4499 steps/day, and there were also gains in the mental composite score on the SF-12 test, values rising from 36.8 to 44.7.

The value of an exercise consultation was compared to usual treatment in a small group of 28 patients with type 2 diabetes mellitus [275]. After a period of 5 weeks, physical activity counts as measured by a CSA accelerometer were 4 % higher in the experimental group, compared with 9 % lower in the control group. The experimental group also showed advantages over the controls in terms of HRQOL.

8.12.3.5 Quality of Life in Cancer During and Following Chemotherapy

Both physical activity and quality of life tend to decline during the period of active chemotherapy, but if activity is subsequently encouraged, the HRQOL may rise.

In lymphoma patients, the DynaPort Minipod pedometer count averaged 6150 steps/day before chemotherapy, 5118 steps/day during treatment, and 5934 steps/day after treatment [188]. The global quality of life diminished from an initial score of 71 to 67 and then 63, with a particular deterioration on the fatigue sub-scale.

At diagnosis, triaxial accelerometer data showed that patients with non-small cell lung cancer were physically less active than healthy individuals, with 60 % failing to meet public health recommendations for daily physical activity. Moreover, physical activity declined further over the next 6 months, as treatment continued, with an associated decline in HRQOL [190].

After the completion of treatment, measurements with an Actigraph GT3X+ accelerometer showed that the quality of life of colon cancer survivors was positively related to their involvement in moderate to vigorous physical activity, with associated advantages of physical function and well-being [276]; however, in this study HRQOL was unrelated to sedentary time.

Likewise, in those treated successfully for breast cancer [277], physical activity as indicated by accelerometer data influenced quality of life, probably indirectly (through increases of self efficacy and health status).

Vallance et al. studied 377 breast cancer survivors [207]; four differing types of behavioural change interventions including provision of a pedometer sought to increase habitual physical activity. Gains of self-reported moderate to vigorous activity at 6 months ranged from 9 to 69 minutes/week, but there were no significant inter-group differences in the SF-36 HRQOL. However, a second study of the same patient-sample [206], using the "Functional assessment in Cancer Therapy Breast (FACT-B) scale," did find a significant gain of HRQOL in the groups receiving the strongest motivation and increasing the duration of their moderate to vigorous activity by 87–89 minutes/week, whereas there was little change in those receiving only the standard treatment, with only a 30 minutes/week increase of physical activity.

In a group of 27 patients who had received chemotherapy for ovarian cancer [278], exercise counselling increased activity by an average of 73 minutes/day, with an increase of pedometer count from around 4200 to 5600 steps/day; in these patients, physical well-being was increased concomitantly with the increase in moderate and vigorous physical activity ($r = 0.48$).

8.12.3.6 Quality of Life in Osteoarthritis and Rheumatoid Arthritis

Prioreschi et al. noted that accelerometry measures of physical activity were lower in 50 patients with rheumatoid arthritis than in matched controls, and they also spent more time sitting [279]. However, within the patient group, the health-related quality of life was correlated with physical activity levels, the more active patients scoring higher on every component of the SF-36 questionnaire.

Despite reporting improvements in pain, function, quality of life, and self-reported levels of physical activity, patients may not actually increase their physical activity following arthroplasty as measured by Actigraph accelerometer; objective monitoring and a plan to increase activity are needed if patients are to meet recommended levels of exercise following surgery [280].

Brandes et al. [281] examined patients using the DynaPort ADL monitor. Thery found that in the first year after total knee arthroplasty, the gait cycle of 53 patients increased on average from a pre-operative 4993/day to 5932/day; this was associated with a large increase in the SF-36 HRQOL (from an average of 43.1–82.5 units).

8.12.3.7 Quality of Life in Multiple Sclerosis

In a 6-month study of 292 adult patients with multiple sclerosis, Moti and associates [236, 282] found that where accelerometer data showed an increase of physical activity, there was an associated small increase in the quality of life, probably through pathways that include changes in pain, fatigue and the sense of self-efficacy. Another study of 46 cases of multiple sclerosis from the same laboratory found that average daily step counts were significantly correlated with self-esteem ($r = 0.30$).

8.12.3.8 Renal Dialysis

Katayama et al. [283] found that on non-dialysis days the health-related quality of life of patients receiving thrice weekly dialysis was positively associated with habitual physical activity >4 METS as assessed by triaxial accelerometers.

8.12.4 Conclusions

Cross-sectional studies suggest that quality of life is augmented by a small increase in habitual physical activity. This conclusion is generally supported by longitudinal studies. However, in patients with clinical conditions, an increase of physical activity has many effects, and it is difficult to discern which of these are responsible for any increase in the HRQOL.

8.13 Independence and Physical Activity

In old age, one of the most important objectives is to conserve sufficient functional capacity to allow independent living. In terms of peak aerobic power, the critical threshold is about 15 ml/(kg minute) [284], and there are parallel thresholds in terms of muscular strength and joint flexibility [285]. Data from the Nakanojo study provide some insights into the minimum levels of physical activity that are required to meet these needs.

The peak walking speed, a correlate of peak aerobic power, and the peak knee extension torque were both correlated with the daily step count ($r = 0.28, 0.20$), and the duration of activity at an intensity >3 METs ($r = 0.27, 0.21$). Dividing the data into quartiles, maximal walking speed (1.81, 2.04, 2.11, 2.25 m/second; 1.80, 2.07, 2.16, 2.18 m/second) showed a significant increase from the first to the third quartile, whether classified by daily step count (3382, 5690, 7589, 10,400) or duration of moderate activity (4.6, 12.4, 20.4, 34.6 minutes). Thus, to optimize aerobic power, the elderly should maintain a step count >7600 steps/day, spending at least 20 minutes/day in moderate physical activity [286].

A whole body dual x-ray absorptiometry scan assessed muscle mass, with the threshold for sarcopaenia set at 1 SD below the norms for young adults [287]. The year averaged step-count and the duration of activity >3 METs were both correlated with the appendicular muscle mass ($r = 0.28, 0.34$, respectively). The odds ratio of demonstrating sarcopaenia (Table 8.13) showed a significant increase in the quartile with the lowest step count, and in the two quartiles with the shortest durations of activity at an intensity >3 METs.

Table 8.13 Odds ratio for demonstrating an appendicular muscle mass in the sarcopaenic range in relation to habitual physical activity. Significant odds ratios are shown in bold type

	Men		Women	
Quartile	Step count	Odds ratio	Step count	Odds ratio
Quartile 1	3427 steps/day	**2.00**	3048 steps/day	**2.66**
Quartile 2	6171	1.20	4999	1.57
Quartile 3	7972	0.79	6942	1.02
Quartile 4	10,545	1.00	9974	1.00
	Activity >3 METs	Odds ratio	Activity >3 METs	Odds ratio
Quartile 1	6.7 minutes/day	**3.39**	5.9 minutes/day	**4.55**
Quartile 2	14.7	**2.03**	10.1	**3.15**
Quartile 3	21.6	1.05	18.5	1.23
Quartile 4	33.5	1.00	31.1	1.00

Based on data from the Nakanojo study of seniors aged 65–85 years [287]

8.14 Other Applications of Objective Physical Activity Monitoring

Among many other potential applications of the objective monitoring of physical activity, we may highlight briefly the estimation of maximal aerobic power, the validation of questionnaire assessments of physical activity patterns, and the examination of sleep patterns.

8.14.1 Prediction of Maximal Oxygen Intake

There has been considerable interest in possible methods of predicting maximal oxygen intake that do not require either expensive laboratory equipment or the willingness of the patient to engage in all-out maximal exercise. Cao et al. [288, 289] suggested that the pedometer/accelerometer provides a simple tool that can be used for this purpose. In 87 Japanese women aged 20–69 years, the correlation between the daily step count and the maximal oxygen intake as determined on a cycle erometer was relatively low ($r = 0.40$). However, a predictive equation based also upon age and body mass index yielded a correlation of 0.81, with a standard error of prediction of 5.3 mL/(kg min), rather similar to the error of many submaximal exercise prediction procedures.

A second report [288] evaluated other possible combinations of physical activity (durations of moderate and vigorous physical activity) and replacement of the BMI term by waist circumference; the optimal equation reduced the standard error of predictions to 3.0 ml/(kg min):

$$\dot{V}O_{2max} = 51.5 - 0.363\,(\text{age, year}) - 0.393\,(\text{waist circ., cm}) + 0.316\,\left(10^{-3}\text{steps/day}\right) + 0.290\,(\text{vigorous activity, min/day})$$

However, critics of this approach have noted that as in many other predictions that avoid all-out exercise, the answer obtained is heavily biassed by body fat content.

Novoa et al. used equations of this type as a simple method to estimate maximal oxygen intake in patients following lobectomy or pneumonectomy [290].

8.14.2 Development and Validation of Subjective Questionnaires

Some authors have used pedometers and accelerometers to help in the development of subjective physical activity questionnaires and as a means of testing their validity. Often, the correlation is poor. In younger adults, a part of the problem is that pedometers and accelerometers are also fallible, giving limited information about some common types of physical activity such as cycling, swimming, resisted exercise, and the climbing of hills. However, even in the elderly whose main form of activity is walking there are often large discrepancies between questionnaires and objective measurements of habitual physical activity.

In the Nakanojo study [291], the correlation between scores on the Physical Activity Questionnaire for elderly Japanese and pedometer/accelerometer data was quite low ($r = 0.41$ for step count, $r = 0.53$ for duration of activity >3 METs). Moreover, the subjective questionnaire exaggerated the absolute levels of physical activity as much as three-fold relative to the objective measurements.

The situation was no better in patients who were undergoing walking-based rehabilitation. Orrell et al. [292] noted that the correlation between questionnaire reports and accelerometer scores in cardiac patients was quite low ($r = 0.21–0.38$ for various measures); commonly, the subjective responses again greatly over-estimated the amount of physical activity that had been undertaken.

8.14.3 Documentation of Sleep Patterns and Sleep Disturbances

Accelerometers have also found application in documenting patterns of disturbed sleep among individuals with depression [293], providing much more accurate information than commonly used sleep questionnaires [294]. Disturbed sleep patterns due to obstructive sleep apnoea are frequently associated with low daytime levels of physical activity and an impaired quality of life [295], as well as impaired survival prospects (see Sect. 8.3.3.1, above).

8.15 Conclusions

The Hockley Valley Consensus Symposium based most of its conclusions on dose/response relationships between physical activity and disease on subjective reports of habitual physical activity. In this chapter, we summarize the findings from the Hockley Valley meeting, and we examine how far these conclusions have been amplified and/or modified by the introduction of objective activity monitors. Among the wide range of topics that are considered, we have included data on objective activity monitoring in relation to all-cause mortality, cardiac death, cardiovascular disease, stroke, peripheral vascular disease, hypertension, cardiac and metabolic risk factors, diabetes mellitus, obesity, low back pain, osteoarthritis, osteoporosis, chronic chest disease, cancer, depression, quality of life and the capacity for independent living. Objective monitoring has clarified a number of dose/response relationships, but much of the information that is currently available remains cross-sectional in type. In many areas of rehabilitation, the pedometer/accelerometer seems a useful motivating device, providing well-documented increments of weekly activity, and offering the prospect of quasi-experimental long-term studies of the effects of graded doses of exercise. However, there remains a need for well-designed longitudinal trials using objective monitors to follow changes in habitual activity; such research is needed to establish causal relationships between increased physical activity and greater health.

References

1. Kesaniemi YA, Danforth E, Jensen MD, et al. Dose-response issues concerning physical activity and health: an evidence-based symposium. Med Sci Sports Exerc. 2001;33(6): S351–8.
2. Lamonte MJ, Ainsworth BE. Quantifying energy expenditure and physical activity in the context of the dose response. Med Sci Sports Exerc. 2001;33(6 Supp):S370–8.
3. Oxman AD, Sackett DL, Guyatt GH. User's guides to the medical literature. 1. How to get started. The evidence-based medical working group. JAMA. 1993;270:2093–5.
4. Rothman KJ, Greenland S. Modern epidemiology. Philadelphia, PA: Lippincott-Raven; 1998.
5. Schriger DL. Analyzing the relationship of exercise and health: methods, assumptions and limitations. Med Sci Sports Exerc. 2001;33(6 Suppl):S359–63.
6. Howley ET. Type of activity: resistance, aerobic and leisure versus occupational physical activity. Med Sci Sports Exerc. 2001;33(6 Suppl):S364–9.
7. Shephard RJ. Absolute versus relative intensity of physical activity in a dose-response context. Med Sci Sports Exerc. 2001;33(6 Suppl):S400–18.
8. Blair SN, Cheng Y, Holder JS. Is physical activity or fitness more important in defining health benefits? Med Sci Sports Exerc. 2001;33(6 Suppl):S379–99.
9. Lee I-M, Skerrett PJ. Physical activity and all-cause mortality: what is the dose-response relationship? Med Sci Sports Exerc. 2001;33(6 Suppl):S459–71.
10. Kohl HW. Physical activity and cardiovascular disease: evidence for a dose-response. Med Sci Sports Exerc. 2001;33(6 Suppl):S472–83.

11. Fagard RH. Exercise characteristics and the blood pressure response to dynamic training. Med Sci Sports Exerc. 2001;33(6 Suppl):S484–92.
12. Kelley DE, Goodpaster BH. Effects of exercise on glucose homeostasis in Type 2 diabetes mellitus. Med Sci Sports Exerc. 2001;33(6 Suppl):S495–501.
13. Leon AS, Sanchez OA. Response of blood lipids to exercise training alone or combined with dietary intervention. Med Sci Sports Exerc. 2001;33(6 Suppl):S502–15.
14. Rauraamaa R, Li G, Vaisänen SB. Dose-response and coagulation and hemostatic factors. Med Sci Sports Exerc. 2001;33(6 Suppl):S516–20.
15. Ross R, Janssen I. Physical activity, total and regional obesity: dose-response considerations. Med Sci Sports Exerc. 2001;33(6 Suppl):S521–7.
16. Thune I, Furberg A-S. Physical activity and cancer risk: dose response and cancer, all-sites and specific. Med Sci Sports Exerc. 2001;33(6 Suppl):S530–50.
17. Vuori IM. Dose-response of physical activity and lower back pain, osteoarthritis, and osteoporosis. Med Sci Sports Exerc. 2001;33(6 Suppl):S551–86.
18. Dunn AL, Treivedi MH, O'Neal HA. Physical activity dose-response effects on outcomes of depression and anxiety. Med Sci Sports Exerc. 2001;33(6 Suppl):S587–97.
19. Washki B, Kirsten A, Holz O, et al. Physical activity is the strongest predictor of all-cause mortality in patients with COPD: a prospective cohort study. Chest. 2011;140(2):331–42.
20. Matsuzawa R, Matsunaga A, Wang G, et al. Habitual physical activity measured by accelerometer and survival in maintenance hemodialysis patients. Clin J Am Soc Nephrol. 2012;7 (12):210–6.
21. Tranah GJ, Blackwell T, Ancoli-Israel S, et al. Circadian activity rhythms and mortality: the study of osteoporotic fractures. J Am Geriatr Soc. 2010;58(2):282–91.
22. Kripke DF, Langer RD, Elliott JA, et al. Mortality related to actigraphic long and short sleep. Sleep Med. 2011;12(1):28–33.
23. Ostir GV, Berges IM, Kuo Y-F, et al. Mobility activity and its value as a prognostic indicator of survival in hospitalized older adults. J Am Geriatr Soc. 2013;61(4):551–7.
24. Garcia-Rio F, Rojo B, Casitas R, et al. Prognostic value of the objective measurement of daily physical activity in patients with COPD. Chest. 2012;142(2):338–46.
25. Zanoria SJ, ZuWallack R. Directly measured physical activity as a predictor of hospitalizations in patients with chronic obstructive pulmonary disease. Chron Respir Dis. 2013;10 (4):207–13.
26. Matsuzawa R, Matsunaga A, Kutsuna T, et al. Association of habitual physical activity measured by an accelerometer with high-density lipoprotein cholesterol levels in maintenance hemodialysis patients. Sci World J. 2013, 780783.
27. Yates T, Haffner DM, Schulte PJ, et al. Association between change in daily ambulatory activity and cardiovascular events in people with impaired glucose tolerance (NAVIGATOR trial): a cohort analysis. Lancet. 2014;383(9922):1059–66.
28. Aoyagi Y, Park H, Kakiyama T, et al. Yearlong physical activity and regional stiffness of arteries in older adults: the Nakanojo study. Eur J Appl Physiol. 2010;109:455–64.
29. Gando Y, Yamamoto K, Murakami H, et al. Longer time spent in light physical activity is associated with reduced arterial stiffness in older adults. Hypertension. 2010;56:540–6.
30. Hawkins M, Gabriel KP, Cooper J, et al. The impact of a change in physical activity on change in arterial stiffness in overweight or obese sedentary young adults. Vasc Med. 2014;19(4):257–63.
31. Hayashi K, Sugawara J, Komine H, et al. Effects of aerobic exercise training on the stiffness of central and peripheral arteries in middle-aged sedentary men. Jpn J Physiol. 2005;55 (4):235–9.
32. Matsuda M. Effects of exercise and physical activity on prevention of arteriosclerosis—special reference to arterial distensibility. Int J Sport Health Sci. 2006;4:316–24.
33. O'Donovan C, Lithander F, Rafteryet T, et al. Inverse relationship between physical activity and arterial stiffness in adults with hypertension. J Phys Act Health. 2014;11(2):272–7.

34. Ried-Larsen M, Grøntved A, Froberg K, et al. Physical activity intensity and subclinical atherosclerosis in Danish adolescents: the European Youth Heart Study. Scand J Med Sci Sports. 2013;23:e168–77.
35. Sakuragi S, Anbhayaratna K, Gravenmaker KJ, et al. Influence of adiposity and physical activity on arterial stiffness in healthy children. The life of our kids study. Hypertension. 2009;53:611–6.
36. Sugawara J, Otsuki T, Tanabe T, et al. Physical activity duration, intensity, and arterial stiffening in postmenopausal women. Am J Hypertens. 2006;19(10):1032–6.
37. Tanaka H, Dinenno F, Monahan KD, et al. Aging, habitual exercise and dynamic arterial compliance. Circulation. 2000;102:1270–5.
38. Boreham CA, Ferreira I, Twisk JW, et al. Cardiorespiratory fitness, physical activity, and arterial stiffness: the Northern Ireland Young Hearts Project. Hypertension. 2004;44 (5):721–6.
39. Kakiyama T, Matsuda M, Koseki S. Effect of physical activity on the distensibility of the aortic wall in healthy males. Angiology. 1998;49(9):749–57.
40. Yamada S, Inaba M, Goto H, et al. Associations between physical activity, peripheral atherosclerosis and bone status in healthy Japanese women. Atherosclerosis. 2006;188:196–202.
41. Tanabe T, Maeda S, Sugawara J, et al. Effect of daily physical activity on systemic arterial compliance in middle-aged and elderly humans: special references in amount and intensity of physical activity. Int J Sport Health Sci. 2006;4:489–98.
42. Aoyagi Y, Shephard RJ. Sex differences in relationships between habitual physical activity and health in the elderly: practical implications for epidemiologists based on pedometer/accelerometer data from the Nakanojo study. Arch Gerontol Geriatr. 2013;56(2):327–38.
43. Gabriel KP, Matthews KA, Pérez A, et al. Self-reported and accelerometer-derived physical activity levels and coronary artery calcification progression in older women: results from the Healthy Women Study. Menopause. 2013;20(2):152–61.
44. Storti KL, Pettee Gabriel KK, Underwood DA, et al. Physical activity and coronary artery calcification in two cohorts of women representing early and late postmenopause. Menopause. 2010;17(6):1146–51.
45. Hamer M, Venuraju SM, Lahiri A, et al. Objectively assessed physical activity, sedentary time, and coronary artery calcification in healthy older adults. Arterioscler Thromb Vasc Biol. 2012;32:500–5.
46. Hamer M, Venuraju SM, Urbanova L, et al. Physical activity, sedentary time, and pericardial fat in healthy older adults. Obesity. 2012;20:2113–7.
47. Butler EN, Evenson KR. Prevalence of physical activity and sedentary behavior among stroke survivors in the United States. Top Stroke Rehabil. 2014;21(3):246–55.
48. Lee PH, Nan H, Yu Y-Y, et al. For non-exercising people, the number of steps walked is more strongly associated with health than time spent walking. J Sci Med Sport. 2013;16:227–30.
49. Sullivan JE, Espe LE, Kelly AM, et al. Feasibility and outcomes of a community-based, pedometer-monitored walking program in chronic stroke: a pilot study. Top Stroke Rehabil. 2014;21(2):101–10.
50. Carroll SL, Greig CA, Lewis SJ, et al. The use of pedometers in stroke survivors: are they feasible and how well do they detect steps? Arch Phys Med Rehabil. 2012;93(3):466–70.
51. Manns PJ, Haennel RG. SenseWear armband and stroke: validity of energy expenditure and step count measurement during walking. Stroke Res Treat. 2012, 247165.
52. Kuys SS, Clark C, Morris NR. Portable multisensor activity monitor (SenseWear) lacks accuracy in energy expenditure measurement during treadmill walking following stroke. Int J Neurol Rehabil. 2014;1:101.
53. Field MJ, Gebruers N, Shanmuga T, et al. Physical activity after stroke: a systematic review. ISRN Stroke. 2013, 464176.

54. Moore SA, Hallsworth K, Plötz T, et al. Physical activity, sedentary behaviour and metabolic control following stroke: a cross-sectional and longitudinal study. PLoS One. 2013;8(1), e55263. Special section.
55. Mudge S, Stott S. Timed walking tests correlate with daily step activity in persons with stroke. Arch Phys Med Rehabil. 2009;90:296–300.
56. Kono Y, Kawajin H, Kamisaka K, et al. Predictive impact of daily physical activity on new vascular events in patients with mild ischemic stroke. Int J Stroke. 2015;10(2):219–23.
57. Kono Y, Yamada S, Yamaguchi J, et al. Secondary prevention of new vascular events with lifestyle intervention in patients with noncardioembolic mild ischemic stroke: a single-center randomized controlled trial. Cerebrovasc Dis. 2013;36:88–97.
58. McDermott MM, Greenland P, Guralnick J, et al. Inflammatory markers, D-dimer, pro-thrombotic factors, and physical activity levels in patients with peripheral arterial disease. Vasc Med. 2004;9:107–15.
59. Gardner AW, Killewich LA. Association between physical activity and endogenous fibrinolysis in peripheral arterial disease: a cross-sectional study. Angiology. 2002;53(4):367–74.
60. Craft LL, Guralnick J, Ferrucci L, et al. Physical activity during daily life and circulating biomarker levels in patients with peripheral arterial disease. Am J Cardiol. 2008;102 (9):1263–8.
61. Payvandi L, Dyer A, McPherson D, et al. Physical activity during daily life and brachial artery flow-mediated dilation in peripheral arterial disease. Vasc Med. 2009;14(3):193–201.
62. McDermott MD, Liu K, Carroll TJ, et al. Superficial femoral artery plaque and functional performance in peripheral arterial disease: walking and leg circulation study (WALCS III). JACC Cardiovasc Imaging. 2011;4(7):730–9.
63. Trigona B, Aggoun Y, Maggio A, et al. Preclinical noninvasive markers of atherosclerosis in children and adolescents with type 1 diabetes are influenced by physical activity. J Pediatr. 2010;157(4):533–9.
64. Sieminski DJ, Gardner AW. The relationship between free-living daily physical activity and the severity of peripheral arterial occlusive disease. Vasc Med. 1997;2(4):286–91.
65. McDermott MM, Liu K, O'Brien E, et al. Measuring physical activity in peripheral arterial disease: a comparison of two physical activity questionnaires with an accelerometer. Angiology. 2000;51(2):91–100.
66. McDermott MM, Guralnick J, Ferrucci L, et al. Physical activity, walking exercise, and calf skeletal muscle characteristics in patients with peripheral arterial disease. J Vasc Surg. 2007;46(1):87–93.
67. McDermott MD, Greenland F, Ferrucci L, et al. Lower extremity performance is associated with daily life physical activity in individuals with and without peripheral arterial disease. J Am Geriatr Soc. 2002;50(2):247–55.
68. Garg PV, Liu K, McDermott MD. Physical activity during daily life and functional decline in peripheral arterial disease. Circulation. 2009;119(2):251–60.
69. Garg PV, Tian L, Criqui MH, et al. Physical activity during daily life and mortality in patients with peripheral arterial disease. Circulation. 2006;114:242–8.
70. Chase JA. Systematic review of physical activity intervention studies after cardiac rehabilitation. J Cardiovasc Nurs. 2011;26(5):351–8.
71. Bravata DM, Smith-Spangler C, Vandana G, et al. Using pedometers to increase physical activity and improve health: a systematic review. JAMA. 2007;298(19):2296–304.
72. Guiraud T, Granger R, Bousquet M, et al. Validity of a questionnaire to assess the physical activity level in coronary artery disease patients. Int J Rehabil Res. 2012;35:270–4.
73. Stevenson TG, Riggin K, Nagelkirk PR, et al. Physical activity habits of cardiac patients participating in an early outpatient rehabilitation program. J Cardiopulm Rehabil Prev. 2009;29(5):299–303.
74. Ayabe M, Brubaker PH, Dobrosieski D, et al. The physical activity patterns of cardiac rehabilitation program participants. J Cardiopulm Rehabil Prev. 2004;24:80–6.

75. Jones NL, Schneider PL, Kaminski LA, et al. An assessment of the total amount of physical activity of patients participating in a phase III cardiac rehabilitation program. J Cardiopulm Rehabil Prev. 2007;27:81–5.
76. Reid RD, Morrin LI, Higginson LAJ, et al. Motivational counselling for physical activity in patients with coronary artery disease not participating in cardiac rehabilitation. Eur J Prev Cardiol. 2011;19(2):161–6.
77. Oliveira J, Ribeiro F, Gomes H. Effects of a home-based cardiac rehabilitation program on the physical activity levels of patients with coronary artery disease. J Cardiopulm Rehabil Prev. 2008;28:392–6.
78. Karjailainen JJ, Kiviniemi AM, Hautala AJ, et al. Effects of exercise prescription on daily physical activity and maximal exercise capacity in coronary artery disease patients with and without type 2 diabetes. Clin Physiol Funct Imaging. 2012;32(6):445–54.
79. Brändström Y, Brink E, Grankvist G, et al. Physical activity six months after a myocardial infarction. Int J Nurs Pract. 2009;15:191–7.
80. Clark AM, Munbday C, McLaughlin D, et al. Peer support to promote physical activity after completion of centre-based cardiac rehabilitation: evaluation of access and effects. Eur J Cardiovasc Nurs. 2012;11(4):388–95.
81. Izawa KP, Watanabe S, Omiya K, et al. Effect of the self-monitoring approach on exercise maintenance during cardiac rehabilitation. A randomized, controlled trial. Am J Phys Med Rehabil. 2005;84:313–21.
82. Shephard RJ, Park H, Park S, et al. Objectively measured physical activity and progressive loss of lean tissue in older Japanese adults: longitudinal data from the Nakanojo study. J Am Geriatr Soc. 2013;61(11):1887–93.
83. Hermida RC, Ayala DE, Mojón A, et al. Ambulatory blood pressure thresholds for diagnosis of hypertension in patients with and without type 2 diabetes based on cardiovascular outcomes. Chronobiol Int. 2013;30(1–2):132–44.
84. Hermida RC, Ayala DE, Mojón A, et al. Differences between men and women in ambulatory blood pressure thresholds for diagnosis of hypertension based on cardiovascular outcomes. Chronobiol Int. 2013;30(1–2):221–32.
85. Ayala DE, Hermida RC, Mojón A, et al. Cardiovascular risk of resistant hypertension: dependence on treatment-time regimen of blood pressure-lowering medications. Chronobiol Int. 2013;30(1–2):340–52.
86. Wuerzner G, Bochud M, Zweiacker C, et al. Step count is associated with lower nighttime systolic blood pressure and increased dipping. Am J Hypertens. 2013;26(4):527–34.
87. Camhi SM, Sisson SB, Johnson WD, et al. Accelerometer-determined moderate intensity lifestyle activity and cardiometabolic health. Prev Med. 2011;52(5):358–60.
88. Chan CB, Spoangler E, Valcour J, et al. Cross-sectional relationship of pedometer-determined ambulatory activity to indicators of health. Obes Res. 2003;11(12):1563–70.
89. Christensen DL, Alcala-Sanchez I, Leal-Berunen I, et al. Physical activity, cardio-respiratory fitness, and metabolic traits in rural Mexican Tarahumara. Am J Hum Biol. 2012;24 (4):558–61.
90. Healy GN, Wijndaele K, Dunstan DW, et al. Objectively measured sedentary time, physical activity and metabolic risk. Diabetes Care. 2008;31(2):369–71.
91. Huffman KM, Sun J-L, Thomas L, et al. Impact of baseline physical activity and diet behavior on metabolic syndrome in a pharmaceutical trial: results from NAVIGATOR. Metabolism. 2014;63(4):554–61.
92. Leary SD, Ness AR, Smith GD, et al. Physical activity and blood pressure in childhood: findings from a population-based study. Hypertension. 2008;51(1):92–8.
93. Luke A, Dugas LR, Durazo-Arvizu RA, et al. Assessing physical activity and its relationship to cardiovascular risk factors: NHANES 2003–2006. BMC Public Health. 2011;11:387.
94. Natali A, Muscelli E, Casolaro A, et al. Metabolic characteristics of prehypertension: role of classification criteria and gender. J Hypertens. 2009;27(12):2394–402.

95. Ryan JM, Hensey O, McLoughlin B, et al. Reduced moderate-to-vigorous physical activity and increased sedentary behavior are associated with elevated blood pressure values in children with cerebral palsy. Phys Ther. 2014;94(8):1144–53.
96. Lima LG, Moriguti JC, Ferriolli E, et al. Effect of a single session of aerobic walking exercise on arterial pressure in community-living elderly individuals. Hypertens Res. 2012;35:457–62.
97. Gemson DH, Commisso R, Fuente J, et al. Promoting weight loss and blood pressure control at work: impact of an education and intervention program. J Occup Environ Med. 2008;50 (3):272–81.
98. Moreau KL, Degarmo R, Langley J, et al. Increasing daily walking lowers blood pressure in postmenopausal women. Med Sci Sports Exerc. 2001;33(11):1825–31.
99. Iwane M, Arita M, Tomimoto S, et al. Walking 10,000 steps/day or more reduces blood pressure and sympathetic nerve activity in mild essential hypertension. Hypertens Res. 2000;23(6):573–80.
100. Zoellner J, Connell C, Madson MB, et al. HUB city steps: a 6-month lifestyle intervention improves blood pressure among a primarily African-American community. J Acad Nutr Diet. 2014;114:4603–12.
101. Hultquist CN, Albright C, Thompson DL, et al. Comparison of walking recommendations in previously sedentary women. Med Sci Sports Exerc. 2005;37(45):676–83.
102. Swartz AM, Strath DJ, Bassett DR, et al. Increasing daily walking improves glucose tolerance in overweight women. Prev Med. 2003;37(4):356–62.
103. Park S, Park H, Togo F, et al. Year-long physical activity and metabolic syndrome in older Japanese adults: cross-sectional data from the Nakanojo study. J Gerontol A Biol Sci Med Sci. 2008;63(10):1119–23.
104. Scheers T, Philippaerts R, Lefevre J. SenseWear determined physical activity and sedentary behavior and metabolic syndrome. Med Sci Sports Exerc. 2013;45(3):481–9.
105. Henson J, Yates T, Biddle SJH, et al. Associations of objectively measured sedentary behaviour and physical activity with markers of cardiometabolic health. Diabetologia. 2013;56(5):1012–20.
106. Bankoski A, Harris TB, McClain JJ, et al. Sedentary activity associated with metabolic syndrome independent of physical activity. Diabetes Care. 2011;34(2):497–503.
107. Ekelund U, Griffin SJ, Wareham NJ. Physical activity and metabolic risk in individuals with a family history of type 2 diabetes. Diabetes Care. 2007;30(2):337–42.
108. Alhassan S, Robinson TN. Defining accelerometer thresholds for physical activity in girls using ROC analysis. J Phys Act Health. 2010;71(1):45–53.
109. Schofield G, Schofield L, Hinckson EA, et al. Daily step counts and selected coronary heart disease risk factors in adolescent girls. J Sci Med Sport. 2009;12:148–55.
110. Morrell CS, Cook SB, Carey GB. Cardiovascular fitness, activity, and metabolic syndrome among college men and women. Metab Syndr Relat Disord. 2013;11(5):370–6.
111. Petrov M, Kim Y, Lauderdale D, et al. Longitudinal associations between objective sleep and lipids: the CARDIA study. Sleep. 2013;36(11):1587–95.
112. Johansson A, Svanborg E, Edéll-Gustafsson U. Sleep-wake activity rhythm and health-related quality of life among patients with coronary artery disease and in a population-based sample—an actigraphy and questionnaire study. Int J Nurs Pract. 2013;19(4):390–401.
113. Yngman-Uhlin P, Johansson A, Fernström A, et al. Fragmented sleep: an unrevealed problem in peritoneal dialysis patients. Scand J Urol Nephrol. 2011;45(3):206–15.
114. Freak-Poli R, Wolfe R, Backholer K, et al. Impact of a pedometer-based workplace health program on cardiovascular and diabetes risk profile. Prev Med. 2011;53:162–71.
115. Talbot LA, Metter EJ, Morrell CH, et al. A pedometer-based intervention to improve physical activity, fitness and coronary heart disease risk factors in National Guard personnel. Mil Med. 2011;176(5):592–600.
116. Tully MA, Cupples ME, Chan WS, et al. Brisk walking, fitness, and cardiovascular risk: a randomized controlled trial in primary care. Prev Med. 2005;41:622–8.

117. Simmons RK, Griffin SJ, Steele R, et al. Increasing overall physical activity and aerobic fitness is associated with improvements in metabolic risk: cohort analysis of the ProActive trial. Diabetologia. 2008;51(5):787–94.
118. Richardson CR, Buis LR, Janney AW, et al. An online community improves adherence in an internet-mediated walking program. Part 1: results of a randomized controlled trial. J Med Internet Res. 2010;12(4), e71.
119. Glynn LG, Hayes PS, Casey M, et al. SMART MOVE—a smartphone-based intervention to promote physical activity in primary care: study protocol for a randomized controlled trial. Trials. 2013;14:157.
120. Bäck C, Cider A, Gillström J, et al. Physical activity in relation to cardiac risk markers in secondary prevention of coronary artery disease. Int J Cardiol. 2013;168(1):478–83.
121. Balkau B, Mhamdi L, Oppert J-M, et al. Physical activity and insulin sensitivity: the RISC study. Diabetes. 2008;57(10):2613–8.
122. Gando Y, Murakami H, Kawakami R, et al. Light-intensity physical activity is associated with insulin resistance in elderly Japanese women independent of moderate- to vigorous-intensity physical activity. J Phys Act Health. 2014;11(2):266–71.
123. Ekelund U, Brage S, Griffin SJ, et al. Objectively measured moderate- and vigorous-intensity physical activity but not sedentary time predicts insulin resistance in high-risk individuals. Diabetes Care. 2009;32(6):1081–6.
124. Lecheminant JD, Tucker LA. Recommended levels of physical activity and insulin resistance in middle-aged women. Diabetes Educ. 2011;37(4):573–80.
125. Telford RD, Cunningham RB, Shaw JE, et al. Contrasting longitudinal and cross-sectional relationships between insulin resistance and percentage of body fat, fitness, and physical activity in children-the LOOK study. Pediatr Diabetes. 2009;10(8):500–7.
126. Thomas AS, Greene LF, Ard JD, et al. Physical activity may facilitate diabetes prevention in adolescents. Diabetes Care. 2009;32(1):9–13.
127. Kriska A, Delahanty L, Edelstein S, et al. Sedentary behavior and physical activity in youth with recent onset of type 2 diabetes. Pediatrics. 2013;131(3):e850–6.
128. Shephard RJ, Johnson N. Effects of physical activity upon the liver. Eur J Appl Physiol. 2015;115(1):1–46.
129. Gastaldelli A, Kozakova M, HØjlund K, et al. Fatty liver is associated with insulin resistance, risk of coronary heart disease, and early atherosclerosis in a large European population. Hepatology. 2009;49(5):1537–44.
130. Tudor-Locke CE, Myers A, Bell RC, et al. Preliminary outcome evaluation of the First Step Program: a daily physical activity intervention for individuals with type 2 diabetes. Patient Educ Couns. 2002;47(1):23–8.
131. Tudor-Locke C, Bell RC, Myers RC, et al. Controlled outcome evaluation of the First Step Program: a daily physical activity intervention for individuals with type II diabetes. Int J Obes Relat Metab Disord. 2004;28(1):113–9.
132. Furber S, Monger C, Franco L, et al. The effectiveness of a brief intervention using a pedometer and step-recording diary in promoting physical activity in people diagnosed with type 2 diabetes or impaired glucose tolerance. Health Promot J Austr. 2008;19 (3):189–95.
133. Engel L, Lindner H. Impact of using a pedometer on time spent walking in older adults with type 2 diabetes. Diabetes Educ. 2006;32(1):98–107.
134. Kirk AF, Higgins LA, Hughes AR, et al. A randomized controlled trial to study the effect of exercise consultation on the promotion of physical activity in people with Type 2 diabetes: a pilot study. Diabet Med. 2001;18:877–82.
135. Bjorgaas M, Viuk KT, Saeterhaug A, et al. Relationship between pedometer-registered activity, aerobic capacity and self-reported activity and fitness in patients with type 2 diabetes. Diabetes Obes Metab. 2005;7(6):737–44.
136. Korkiangas EE, Alahuta MA, Husman PM, et al. Pedometer use among adults at high risk of type 2 diabetes, Finland, 2007–2008. Prev Chronic Dis. 2010;7(2):A37.

137. Gorely T, Bull F, Khunti K. Effectiveness of a pragmatic education program designed to promote walking activity in individuals with impaired glucose tolerance: a randomized controlled trial. Diabetes Care. 2009;32(8):1404–10.
138. Andersen E, Hoastmark AT, Holme I. Intervention effects on physical activity and insulin levels in men of Pakistani origin living in Oslo: a randomised controlled trial. J Immigr Minor Health. 2013;15(1):101–10.
139. Sone H, Katagiri A, Ishibashi S, et al. Effects of lifestyle modifications on patients with type 2 diabetes: the Japan Diabetes Complications Study (JDCS) study design, baseline analysis and three year-interim report. Horm Metab Res. 2002;34(9):509–15.
140. Yates T, Davies MJ, Sehmi S, et al. The pre-diabetes risk education and physical activity recommendation and encouragement (PREPARE) programme study: are improvements in glucose regulation sustained at 2 years? Diabet Med. 2011;28(10):1268–71.
141. Jennersjö P, Ludvigsson J, Länne T, et al. Pedometer-determined physical activity is linked to low systemic inflammation and low arterial stiffness in type 2 diabetes. Diabet Med. 2012;29 (9):1119–25.
142. van Sloten TT, Savelberg HHCM, Duimet-Peeters IG, et al. Peripheral neuropathy, decreased muscle strength and obesity are strongly associated with walking in persons with type 2 diabetes without manifest mobility limitations. Diabetes Res Clin Pract. 2011;91(1):32–9.
143. Tuttle LJ, Sinacore DR, Cade DR, et al. Lower physical activity is associated with higher intermuscular adipose tissue in people with type 2 diabetes and peripheral neuropathy. Phys Ther. 2011;91(6):923–30.
144. Wedderkopp N, Kjaer P, Hestbaek L, et al. High-level physical activity in childhood seems to protect against low back pain in early adolescence. Spine J. 2009;9(2):134–41.
145. Wedderkopp N, Leboeuf-Yde C, Andersen B, et al. Back pain in children: no association with objectively measured levels of physical activity. Spine. 2003;28(17):2019–24.
146. Yamakawa K, Tsai CK, Haig AJ, et al. Relationship between ambulation and obesity in older persons with and without low back pain. Int J Obes Relat Metab Disord. 2004;28(1):137–43.
147. van Weering MGH, Vollenbroek-Hutten MMR, Tonis TM, et al. Daily physical activities in chronic lower back pain patients assessed with accelerometry. Eur J Pain. 2009;13 (6):649–54.
148. Verbunt JA, Westerterp KR, VanderHeijden GJ, et al. Physical activity in daily life in patients with chronic low back pain. Arch Phys Med Rehabil. 2001;82(6):726–30.
149. Hasenbring MI, Plaas H, Fischbein B, et al. The relationship between activity and pain in patients 6 months after lumbar disc surgery: do pain-related coping modes act as moderator variables? Eur J Pain. 2006;10(8):701–9.
150. Dunlop DD, Song J, Semanik PA, et al. Relation of physical activity time to incident disability in community dwelling adults with or at risk of knee arthritis: prospective cohort study. BMJ. 2014;348, g2472.
151. Sun K, Song J, Lee J, et al. Relationship of meeting physical activity guidelines with health-related utility. Arthritis Care Res. 2014;66(7):1041–7.
152. Holsgaard-Larsen A, Roos EM. Objectively measured physical activity in patients with end stage knee or hip osteoarthritis. Eur J Phys Rehabil Med. 2012;48(4):577–85.
153. Ng NTM, Heesch KC, Brown WJ. Efficacy of a progressive walking program and glucosamine sulphate supplementation on osteoarthritic symptoms of the hip and knee: a feasibility trial. Arthritis Res Ther. 2010;12(1):R25.
154. Feinglass J, Song J, Semanik P, et al. Association of functional status with changes in physical activity: insights from a behavioral intervention for participants with arthritis. Arch Phys Med Rehabil. 2012;93(1):172–5.
155. Dore DA, Winzenberg TM, Ding C, et al. The association between objectively measured physical activity and knee structural change using MRI. Ann Rheum Dis. 2013;72(7):1170–5.
156. Svege I, Kolle E, Risberg MA. Reliability and validity of the Physical Activity Scale for the Elderly (PASE) in patients with hip osteoarthritis. BMC Musculoskelet Disord. 2012;13:26.

157. Winter CC, Brandes M, Muller C, et al. Walking ability during daily life in patients with osteoarthritis of the knee or the hip and lumbar spinal stenosis: a cross sectional study. BMC Musculoskelet Disord. 2010;11:233.
158. Murphy SL, Smith DM, Lyden AK. Type of activity pacing instruction affects physical activity variability in adults with symptomatic knee or hip osteoarthritis. J Phys Act Health. 2012;9(3):360–6.
159. Murphy SL, Lyden A, Smith DM, et al. Effects of a tailored activity pacing intervention on pain and fatigue for adults with osteoarthritis. Am J Occup Ther. 2010;64(6):869–76.
160. Farr JN, Going SB, McKnight PE, et al. Progressive resistance training improves overall physical activity levels in patients with early osteoarthritis of the knee: a randomized controlled trial. Phys Ther. 2010;90(3):356–66.
161. Shephard RJ, Park H, Park S et al. Objectively measured physical activity and calcaneal bone health in older Japanese adults: dose/response relationships in longitudinal data from the Nakanojo study. Calc Tissue Internat. 2014 (in press).
162. Park H, Togo F, Watanabe E, et al. Relationship of bone health to yearlong physical activity in older Japanese adults: cross-sectional data from the Nakanojo study. Osteporos Int. 2007;18(3):285–93.
163. Alwis G, Linden C, Dencker M, et al. Bone mineral accrual and gain in skeletal width in pre-pubertal school children is independent of the mode of school transportation—one-year data from the prospective observational pediatric osteoporosis prevention (POP) study. BMC Musculoskelet Disord. 2007;8:66.
164. Hoch AZ, Papanek P, Szabo A, et al. Association between the female athlete triad and endothelial dysfunction in dancers. Clin J Sport Med. 2011;21(2):119–25.
165. Niu K, Ahola R, Guo H, et al. Effect of office-based brief high-impact exercise on bone mineral density in healthy premenopausal women: the Sendai Bone Health Concept Study. J Bone Miner Metab. 2010;28(5):568–77.
166. Jones NL, Killian KJ. Exercise in chronic airway obstruction. In: Bouchard C, Shephard RJ, Stephens T, et al., editors. Exercise, fitness and health. Champaign, IL: Human Kinetics; 1990. p. 547–60.
167. Whipp BJ, Casaburi R. Physical activity, fitness and chronic lung disease. In: Bouchard C, Shephard RJ, Stephens T, editors. Physical activity, fitness and health. Champaign, IL: Human Kinetics; 1994. p. 749–61.
168. Walker PP, Burnett A, Flavahan PW, et al. Lower limb activity and its determinants in COPD. Thorax. 2008;63(8):683–9.
169. Pitta F, Troosters MA, Spruit VS, et al. Characteristics of physical activity in daily life in chronic obstructive pulmonary disease. Am J Respir Crit Care Med. 2005;171:972–7.
170. Watz H, Washki B, Meyer T, et al. Physical activity in patients with COPD. Eur Respir J. 2009;33:262–72.
171. Moy ML, Danilack VA, Garshick E. Daily step counts in a US cohort with COPD. Respir Med. 2012;106(7):962–9.
172. Garcia-Rio F, Lores V, Mediano O, et al. Daily physical activity in patients with chronic obstructive pulmonary disease is mainly associated with dynamic hyperinflation. Am J Respir Crit Care Med. 2009;180:506–12.
173. Moy ML, Teylan M, Weston NA, et al. Daily step count is associated with plasma C-reactive protein and IL-6 in a US Cohort with COPD. Chest. 2014;145(3):542–50.
174. Moy ML, Matthess K, Stolzman K, et al. Free-living physical activity in COPD: assessment with accelerometer and activity checklist. J Rehabil Res Dev. 2009;46(2):277–86.
175. Garcia-Rio F, Rojo B, Casitas R, et al. Prognostic value of the objective measurement of daily physical activity in patients with COPD. Chest. 2014;142(2):338–46.
176. Orsey AD, Wakefield DB, Cloutier MM. Physical activity (PA) and sleep among children and adolescents with cancer. Pediatr Blood Cancer. 2013;60(11):1908–13.
177. Rabinovich RA, Louvaris Z, Raste Y, et al. Validity of physical activity monitors during daily life in patients with COPD. Eur Respir J. 2013;42(5):1205–15.

178. Moy ML, Weston NA, Wilson EJ, et al. A pilot study of an Internet walking program and pedometer in COPD. Respir Med. 2012;106:1342–50.
179. DePew ZS, Novotny PJ, Benzo RP. How many steps are enough to avoid severe physical inactivity in patients with chronic obstructive pulmonary disease? Respirology. 2012;17:1026–7.
180. Katz ML, Ferketich AK, Broder-Aldach B, et al. Physical activity among Amish and non-Amish adults living in Ohio Appalachia. J Community Health. 2012;37(2):434–40.
181. Dallal CM, Brinton LA, Matthews CE, et al. Accelerometer-based measures of active and sedentary behavior in relation to breast cancer risk. Breast Cancer Res Treat. 2012;134 (3):1279–90.
182. Lynch BM, Friedenreich CM, Winkler EAH, et al. Associations of objectively assessed physical activity and sedentary time with biomarkers of breast cancer risk in postmenopausal women: findings from NHANES (2003–2006). Breast Cancer Res Treat. 2011;130 (1):183–94.
183. Loprinzi PD, Kohli M. Effect of physical activity and sedentary behavior on serum prostate-specific antigen concentrations: results from the National Health and Nutrition Examination Survey (NHANES), 2003–2006. Mayo Clin Proc. 2013;88(1):11–21.
184. Loprinzi PD, Lee H. Rationale for promoting physical activity among cancer survivors: literature review and epidemiologic examination. Oncol Nurs Forum. 2014;41(2):117–25.
185. Aznar S, Webster AL, SanJuan AF, et al. Physical activity during treatment in children with leukemia: a pilot study. Appl Physiol Nutr Metab. 2006;31(4):407–13.
186. Tillmann V, Darlington ASE, Elser C, et al. Male sex and low physical activity are associated with reduced spine bone mineral density in survivors of childhood acute lymphoblastic leukemia. J Bone Miner Res. 2002;17(6):1073–80.
187. Tan SY, Poh BK, Chong HX, et al. Physical activity of pediatric patients with acute leukemia undergoing induction or consolidation chemotherapy. Leuk Res. 2013;37(1):14–20.
188. Vermaete N, Wolter P, Verhoef G, et al. Physical activity and physical fitness in lymphoma patients before, during, and after chemotherapy: a prospective longitudinal study. Ann Hematol. 2014;93(3):411–24.
189. Lynch BM, Dunstan DW, Winkler E, et al. Objectively assessed physical activity, sedentary time and waist circumference among prostate cancer survivors: findings from the National Health and Nutrition Examination Survey (2003–2006). Eur J Cancer Care. 2011;20 (4):514–9.
190. Granger CL, McDonald CF, Irving L, et al. Low physical activity levels and functional decline in individuals with lung cancer. Lung Cancer. 2014;83(2):292–9.
191. Lynch BM, Dunstan DW, Healy GN, et al. Objectively measured physical activity and sedentary time of breast cancer survivors, and associations with adiposity: findings from NHANES (2003–2006). Cancer Causes Control. 2010;21(2):283–8.
192. Loprinzi PD, Lee H, Cardinal BJ. Objectively measured physical activity among US cancer survivors: considerations by weight status. J Cancer Surviv. 2013;7(3):493–9.
193. Broderick JM, Hussey J, Kennedy MJ, et al. Testing the 'teachable moment' premise: does physical activity increase in the early survivorship phase? Support Care Cancer. 2014;22 (4):989–97.
194. Ruiz-Casado A, Verdugo A, Soria S, et al. Objectively assessed physical activity levels in Spanish cancer survivors. Oncol Nurs Forum. 2014;41(1):E12–20.
195. Heath JA, Ramzy JM, Donath SM. Physical activity in survivors of childhood acute lymphoblastic leukaemia. J Paediatr Child Health. 2010;46(4):149–53.
196. Maddocks M, Byrne A, Johnson CD, et al. Physical activity level as an outcome measure for use in cancer cachexia trials: a feasibility study. Support Care Cancer. 2010;18(12):1539–44.
197. Ferriolli E, Skipworth RJE, Hendry P, et al. Physical activity monitoring: a responsive and meaningful patient-centered outcome for surgery, chemotherapy, or radiotherapy? J Pain Symptom Manage. 2012;43(6):1025–35.

198. Inoue Y, Kimura T, Noro H, et al. Is laparoscopic colorectal surgery less invasive than classical open surgery? Quantitation of physical activity using an accelerometer to assess postoperative convalescence. Surg Endosc. 2003;17(8):1269–73.
199. Takiguchi S, Fujiwara Y, Yamasaki M, et al. Laparoscopy-assisted distal gastrectomy versus open distal gastrectomy. A prospective randomized single-blind study. World J Surg. 2013;37(10):2379–86.
200. Novoa N, Varela G, Jimenz MF, et al. Influence of major pulmonary resection on postoperative daily ambulatory activity of the patients. Interact Cardiovasc Thorac Surg. 2009;9 (6):934–8.
201. Knols TH, deBruin ED, Aufdemkampe G, et al. Reliability of ambulatory walking activity in patients with hematologic malignancies. Arch Phys Med Rehabil. 2009;90(1):58–65.
202. Tonosaki A, Ishikawa M. Physical activity intensity and health status perception of breast cancer patients undergoing adjuvant chemotherapy. Eur J Oncol Nurs. 2014;18(2):132–9.
203. Vallance JKH, Courneya KS, Plotnikoff RC, et al. Analyzing theoretical mechanisms of physical activity behavior change in breast cancer survivors: results from the activity promotion (ACTION) trial. Ann Behav Med. 2008;35(2):150–8.
204. Rogers LQ, Shah P, Dunnington G, et al. Social cognitive theory and physical activity during breast cancer treatment. Oncol Nurs Forum. 2005;32(4):807–15.
205. Phillips SM, McAuley E. Social cognitive influences on physical activity participation in long-term breast cancer survivors. Psychooncology. 2013;22(4):783–91.
206. Vallance JKH, Courneya KS, Plotnikoff RC, et al. Randomized controlled trial of the effects of print materials and step pedometers on physical activity and quality of life in breast cancer survivors. J Clin Oncol. 2007;25(27):2352–9.
207. Vallance JKH, Courneya KS, Plotnikoff RC, et al. Maintenance of physical activity in breast cancer survivors after a randomized trial. Med Sci Sports Exerc. 2008;40(1):173–80.
208. Rogers LQ, Hopkins-Price V, Sandy M, et al. Physical activity and health outcomes three months after completing a physical activity behavior change intervention: persistent and delayed effects. Cancer Epidemiol Biomarkers Prev. 2009;18(5):1410–8.
209. Irwin ML, Cadmus L, Alvarez-Reeves M, et al. Recruiting and retaining breast cancer survivors into a randomized controlled exercise trial: the Yale Exercise and Survivorship Study. Cancer. 2008;112(11 Suppl):2593–606.
210. Matthews CE, Willcox S, Hanby CL, et al. Evaluation of a 12-week home-based walking intervention for breast cancer survivors. Support Care Cancer. 2007;15(2):203–11.
211. Nikander R, Sievanen H, Ojala K, et al. Effect of a vigorous aerobic regimen on physical performance in breast cancer patients—a randomized controlled pilot trial. Acta Oncol. 2007;46(2):181–6.
212. Frensham LJ, Zarnowiecki DM, Parfitt G, et al. The experiences of participants in an innovative online resource designed to increase regular walking among rural cancer survivors: a qualitative pilot feasibility study. Support Care Cancer. 2014;22(7):1923–9.
213. Blaaubroek R, Bourna MJ, Tuinier W, et al. The effect of exercise counselling with feedback from a pedometer on fatigue in adult survivors of childhood cancer: a pilot study. Support Care Cancer. 2009;17(8):1041–8.
214. Burton C, McKinstry B, Szentagotal T, et al. Activity monitoring in patients with depression: a systematic review. J Affect Disord. 2013;145(1):21–8.
215. Vallance JKH, Eurich D, Lavallee C, et al. Daily pedometer steps among older men: associations with health-related quality of life and psychosocial health. Am J Health Promot. 2013;27(5):294–8.
216. Drieling RI, Goldman R, Lisa M, et al. Community resource utilization, psychosocial health, and sociodemographic factors associated with diet and physical activity among low-income obese Latino immigrants. J Acad Nutr Diet. 2014;114(2):257–65.
217. Aronen ET, Simola P, Soininen M. Motor activity in depressed children. J Affect Disord. 2011;133(1–2):188–96.

218. McKercher C, Patton GC, Schmidt MD, et al. Physical activity and depression symptom profiles in young men and women with major depression. Psychosom Med. 2013;75 (4):366–74.
219. Yoshiuchi K, Nakabara R, Kumano H, et al. Yearlong physical activity and depressive symptoms in older Japanese adults: cross-sectional data from the Nakanojo study. Am J Geriatr Psychiatry. 2006;14(7):621–4.
220. Loprinzi PD. Objectively measured light and moderate-to-vigorous physical activity is associated with lower depression levels among older US adults. Aging Ment Health. 2014;17(7):801–5.
221. Song MR, Lee Y-S, Baek J-D, et al. Physical activity status in adults with depression in the National Health and Nutrition Examination Survey, 2005–2006. Public Health Nurs. 2012;29 (3):298–317.
222. Vallance JKH, Winkler E, Gardiner PA, et al. Associations of objectively-assessed physical activity and sedentary time with depression: NHANES (2005–2006). Prev Med. 2011;53 (4–5):284–8.
223. McKercher C, Schmidt MD, Sanderson KA, et al. Physical activity and depression in young adults. Am J Prev Med. 2009;36(2):161–4.
224. Ewald B, Attia J, McElduff P. How many steps are enough? Dose-response curves for pedometer steps and multiple health markers in a community-based sample of older Australians. J Phys Act Health. 2014;11(3):509–18.
225. Patel A, Keogh JW, Kolt GS, et al. The long-term effects of a primary care physical activity intervention on mental health in low-active, community-dwelling older adults. Aging Ment Health. 2013;17(6):766–72.
226. Prohaska TR, Wang P-P, Sarkisian CA. The longitudinal relationship between depression and walking behavior in older Latinos: the "Caminemos!" study. J Aging Health. 2013;25 (2):319–41.
227. Loprinzi PD, Fitzgerald EM, Cardinal BJ. Physical activity and depression symptoms among pregnant women from the National Health and Nutrition Examination Survey 2005–2006. J Obstet Gynecol Neonatal Nurs. 2012;41(2):227–35.
228. Alosco ML, Spitznagel MB, Miller L, et al. Depression is associated with reduced physical activity in persons with heart failure. Health Psychol. 2012;31(6):754–82.
229. Horne D, Kehler DD, Kaoukis G, et al. Impact of physical activity on depression after cardiac surgery. Can J Cardiol. 2013;29(12):1649–56.
230. Skotzko CE, Krichtern C, Zietowski G, et al. Depression is common and precludes accurate assessment of functional status in elderly patients with congestive heart failure. J Card Fail. 2000;6(4):300–5.
231. Piette JD, Richardson C, Himle J, et al. A randomized trial of telephonic counseling plus walking for depressed diabetes patients. Med Care. 2011;49(7):641–8.
232. Weinstock RS, Brooks G, Palmas W, et al. Lessened decline in physical activity and impairment of older adults with diabetes with telemedicine and pedometer use: results from the IDEATel study. Age Ageing. 2011;40(1):98–105.
233. White DK, Keysor JJ, Neogi T, et al. When it hurts, a positive attitude may help: association of positive affect with daily walking in knee osteoarthritis. Results from a multicenter longitudinal cohort study. Arthritis Care Res. 2012;64(9):1312–9.
234. Sabiston CM, Brunet J, Burke S. Pain, movement, and mind: does physical activity mediate the relationship between pain and mental health among survivors of breast cancer? Clin J Pain. 2012;28(6):489–95.
235. Suh Y, Moti RW, Mohr DC. Physical activity, disability, and mood in the early stage of multiple sclerosis. Disabil Health J. 2010;3(2):93–8.
236. Moti RW, McAuley E. Pathways between physical activity and quality of life in adults with multiple sclerosis. Health Psychol. 2009;28(6):682–9.
237. McLoughlin MJ, Colbert LH, Stegner AJ, et al. Are women with fibromyalgia less physically active than healthy women? Med Sci Sports Exerc. 2011;43(5):905–12.

238. Wichniak A, Skowerska A, Chojnacka-Wojtowicz J, et al. Actigraphic monitoring of activity and rest in schizophrenic patients treated with olanzapine or risperidone. J Psychiatr Res. 2011;45(10):1381–6.
239. Yasunaga A, Togo F, Watanabe E, et al. Yearlong physical activity and health-related quality of life in older Japanese adults: the Nakanojo study. J Aging Phys Act. 2006;14:288–301.
240. Yasunaga A, Togo F, Park H et al. (2008) Interactive effects of the intensity and volume of habitual physical activity on health-related quality of life: the Nakanojo study. Human Kinetics: Champaign, IL (AACC Archived articles)
241. Yoshiuchi K, Inada S, Nakahara R, et al. Stressful life events and habitual physical activity in older adults: 1-year accelerometry data from the Nakanonjo study. Mental Health Phys Act. 2010;3:23–5.
242. Anokye NK, Trueman P, Green C, et al. Physical activity and health related quality of life. BMC Public Health. 2012;12:624.
243. Withall J, Stahi A, Davis M, et al. Objective indicators of physical activity and sedentary time and associations with subjective well-being in adults aged 70 and over. Int J Environ Res Public Health. 2014;11(1):643–56.
244. Jepsen R, Aadland E, Andersen JR, et al. Associations between physical activity and quality of life outcomes in adults with severe obesity: a cross-sectional study prior to the beginning of a lifestyle intervention. Health Qual Life Outcomes. 2013;11:187.
245. Vallance JKH, Murray TC, Johnson ST, et al. Quality of life and psychosocial health in postmenopausal women achieving public health guidelines for physical activity. Menopause. 2010;17(1):64–71.
246. Kerr J, Sallis JF, Saelens BE, et al. Outdoor physical activity and self rated health in older adults living in two regions of the U.S. Int J Behav Nutr Phys Act. 2012;9:898.
247. Herman KM, Sabiston C, Tremblay A, et al. Self-rated health in children at risk for obesity: associations of physical activity, sedentary behaviour, and BMI. J Phys Act Health. 2014;11 (3):543–52.
248. Tsiros MD, Buckley JD, Howe PRC, et al. Day-to-day physical functioning and disability in obese 10- to 13-year-olds. Pediatr Obes. 2013;8(1):31–41.
249. Shoup JA, Gattshall M, Dandamudi P, et al. Physical activity, quality of life, and weight status in overweight children. Qual Life Res. 2008;17(3):407–12.
250. Freak-Poli R, Wolfe R, Wong E, et al. Change in well-being amongst participants in a four-month pedometer based workplace health program. BMC Public Health. 2014;14:953.
251. Mutrie N, Doolin O, Fitzsimons CT, et al. Increasing older adults' walking through primary care: results of a pilot randomized controlled trial. Fam Pract. 2012;29(6):633–42.
252. Fortuno G, Jesus R, Vinets J, et al. Medida de la capacidad funcional y la calidad de vida relacionada con la salud de un grupo de personas mayores que llevan a cabo un programa de caminatas: estudio piloto [Measurement of functional capacity and health related quality of life in an elderly group following a walking program: pilot study] [Spanish]. Rev Esp Geriatr Gerontol. 2011;46(3):147–50.
253. Hawkes AL, Patrao T, Green A, et al. CanPrevent: a telephone-delivered intervention to reduce multiple behavioural risk factors for colorectal cancer. BMC Cancer. 2012;12:560.
254. Long JE, Ring C, Bosch JA, et al. A life-style physical activity intervention and the antibody response to pneumococcal vaccination in women. Psychosom Med. 2013;75(8):774–82.
255. Morgan PJ, Callister R, Collins CE, et al. The SHED-IT community trial: a randomized controlled trial of internet- and paper-based weight loss programs tailored for overweight and obese men. Ann Behav Med. 2013;45(2):139–52.
256. Fitzsimons CF, Baker G, Gray SR, et al. Does physical activity counselling enhance the effects of a pedometer-based intervention over the long-term: 12-month findings from the Walking for Wellbeing in the West study. BMC Public Health. 2012;12:206.
257. Wallmann B, Frobiese I. Intervention Auswirkungen der 3000 Schritte mehr pro Tag [Intervention effects of 3000 steps more per day] [German]. Wien Klin Wochenschr. 2011;123 (11–12):369–77.

258. Harding J, Freak-Poli R, Backholer K, et al. Change in health-related quality of life amongst participants in a 4-month pedometer-based workplace health program. J Phys Act Health. 2013;10(4):533–43.
259. De Bock F, Genser B, Raat H, et al. A participatory physical activity intervention in preschools: a cluster randomized controlled trial. Am J Prev Med. 2013;45(1):64–74.
260. Wafa SW, Talib RA, Hamzaid NH, et al. Randomized controlled trial of a good practice approach to treatment of childhood obesity in Malaysia: Malaysian Childhood Obesity Treatment Trial (MASCOT). Int J Pediatr Obes. 2011;6(2–2):e62–9.
261. Müller J, Hess J, Hager A. Daily physical activity in adults with congenital heart disease is positively correlated with exercise capacity but not with quality of life. Clin Res Cardiol. 2011;101(1):55–61.
262. Muller JH, Hager A. Daily physical activity in adults with congenital heart disease is positively correlated with exercise capacity but not with quality of life. Clin Res Cardiol. 2012;101(1):55–61.
263. Izawa KP, Yamada S, Oka K, et al. Long-term exercise maintenance, physical activity, and health-relayed quality of life after cardiac rehabilitation. Am J Phys Med Rehabil. 2004;83:884–92.
264. Reid RD, Morrin LI, Beaton LJ, et al. Randomized trial of an internet-based computer-tailored expert system for physical activity in patients with heart disease. Eur J Prev Cardiol. 2012;19(6):1357–65.
265. Houle J, Doyon O, Vadeboncoeur N, et al. Effectiveness of a pedometer-based program using a socio-cognitive intervention on physical activity and quality of life in a setting of cardiac rehabilitation. Can J Cardiol. 2012;28(1):27–32.
266. Borland M, Rosenkvist A, Cider A. A group-based exercise program did not improve physical activity in patients with chronic heart failure and comorbidity: a randomized controlled trial. J Rehabil Med. 2014;46(5):461–7.
267. Gottlieb SS, Fisher ML, Freudenberger R, et al. Effects of exercise training on peak performance and quality of life in congestive heart failure patients. J Card Fail. 1999;5 (3):188–94.
268. Rand D, Eng JJ, Tang P-F, et al. Daily physical activity and its contribution to the health-related quality of life of ambulatory individuals with chronic stroke. Health Qual Life Outcomes. 2010;8:80.
269. Jehn M, Schindler C, Meyer A, et al. Daily walking intensity as a predictor of quality of life in patients with chronic obstructive pulmonary disease. Med Sci Sports Exerc. 2012;44 (7):1212–8.
270. HajGhanbari B, Garland SJ, Road JD, et al. Pain and physical performance in people with COPD. Respir Med. 2013;107(11):1692–9.
271. Mador MJ, Patel AN, Nadler J. Effects of pulmonary rehabilitation on activity levels in patients with chronic obstructive pulmonary disease. J Cardiopulm Rehabil Prev. 2011;31 (1):52–9.
272. Nguyen HQ, Gill DP, Wolpin S, et al. Pilot study of a cell phone-based exercise persistence intervention post-rehabilitation for COPD. Int J Chron Obstruct Pulmon Dis. 2009;4:301–13.
273. Hospes G, Bossenbroek L, Ten Hacken NHT, et al. Enhancement of daily physical activity increases physical fitness of outclinic COPD patients: results of an exercise counseling program. Patient Educ Couns. 2009;75(2):274–8.
274. de Blok BMJ, de Greef MHG, ten Hacken NHT, et al. The effects of a lifestyle physical activity counseling program with feedback of a pedometer during pulmonary rehabilitation in patients with COPD: a pilot study. Patient Educ Couns. 2006;61(1):48–55.
275. Mutrie N, Hillis S, MacIntyre PD. A randomized, controlled trial to study the effect of exercise consultation on the promotion of physical activity in people with type 2 diabetes: a pilot study. Diabet Med. 2002;18(11):877–82.

276. Vallance JKH, Boyle JK, Courneya TJ, et al. Associations of objectively assessed physical activity and sedentary time with health-related quality of life among colon cancer survivors. Cancer. 2014;120(18):2919–26.
277. Phillips SM, McAuley E. Physical activity and quality of life in breast cancer survivors: the role of self-efficacy and health status. Psychooncology. 2014;23(1):27–34.
278. Von Gruenigen VE, Frasure HE, Kavanagh MB, et al. Feasibility of a lifestyle intervention for ovarian cancer patients receiving adjuvant chemotherapy. Gynecol Oncol. 2011;122 (2):328–33.
279. Prioreschi A, Hodkinson B, Avidon I, et al. The clinical utility of accelerometry in patients with rheumatoid arthritis. Rheumatology. 2013;52(9):1721–7.
280. Harding P, Hollandd AE, Delany C, et al. Do activity levels increase after total hip and knee arthroplasty? Clin Orthop Relat Res. 2014;472(5):1502–11.
281. Brandes M, Ringling M, Winter C, et al. Changes in physical activity and health-related quality of life during the first year after total knee arthroplasty. Arthritis Care Res. 2011;63 (3):328–34.
282. Moti RW, McAuley E, Snook EM, et al. Physical activity and quality of life in multiple sclerosis: intermediary roles of disability, fatigue, mood, pain, self-efficacy and social support. Psychol Health Med. 2009;14(1):111–24.
283. Katayama A, Miyatake N, Nishi H, et al. Evaluation of physical activity and its relationship to health-related quality of life in patients on chronic hemodialysis. Environ Health Prev Med. 2014;19(3):220–5.
284. Shephard RJ. Aging, physical activity and health. Champaign, IL: Human Kinetics; 1997.
285. Paterson DH, Stathokostas L. Physical activity, fitness, and gender in relation to morbidity, survival, quality o life, and independence in older ages. In: Shephard RJ, editor. Gender, physical activity and aging. Boca Raton, FL: CRC Press; 2001. p. 99–120.
286. Aoyagi Y, Park H, Watanabe E, et al. Habitual physical activity and physical fitness in older Japanese adults: the Nakanojo study. Gerontology. 2009;55:523–31.
287. Park H, Park S, Shephard RJ, et al. Yearlong physical activity and sarcopenia in older adults: the Nakanojo study. Eur J Appl Physiol. 2010;109(5):953–61.
288. Cao Z-B, Miyatake N, Higuchi M, et al. Predicting VO2max with an objectively measured physical activity in Japanese women. Med Sci Sports Exerc. 2010;42(1):179–86.
289. Cao Z-B, Miyatake N, Higuchi M, et al. Prediction of VO2max with daily step counts for adult Japanese women. Eur J Appl Physiol. 2009;105:289–96.
290. Novoa NM, Varela G, Jimenez MF, et al. Value of the average basal daily walked distance measured using a pedometer to predict maximum oxygen consumption per minute in patients undergoing lung resection. Eur J Cardiothorac Surg. 2011;39(5):756–62.
291. Yasunaga A, Park H, Watanabe E, et al. Development and evaluation of the physical activity questionnaire for elderly Japanese: the Nakanojo study. J Aging Phys Act. 2007;15:398–411.
292. Orrell A, Doherty P, Miles J, et al. Development and validation of a very brief questionnaire measure of physical activity in adults with coronary heart disease. Eur J Cardiovasc Prev Rehabil. 2007;14:615–23.
293. Pilai V, Steenburg LA, Ciesla JA, et al. A seven day actigraphy-based study of rumination and sleep disturbance among young adults with depressive symptoms. J Psychosom Res. 2014;77(1):70–3.
294. Anderson KN, Katt M, Collerton J, et al. Assessment of sleep and circadian rhythm disorders in the very old: the Newcastle 85+ Cohort Study. Age Ageing. 2014;43(1):57–63.
295. Verwimp J, Amewye L, Bruyneel M. Correlation between sleep parameters, physical activity and quality of life in somnolent moderate to severe obstructive sleep apnea adult patients. Sleep Breath. 2013;17(3):903–4.

Chapter 9
Excessive Appetite vs. Inadequate Physical Activity in the Pathology of Obesity: Evidence from Objective Monitoring

Roy J. Shephard

Abstract Objective monitoring of physical activity confirms the impression formed from questionnaire responses that both adults and children who are overweight or obese take less physical activity than their peers who have a healthy body mass. Pedometer/accelerometer data provides relatively precise information on the magnitude of the deficit in physical activity, which amounts to around 2000 steps/day, or 15–20 minutes/day of moderate and/or vigorous physical activity. In some studies of those who are grossly obese, there is also evidence of an increase in sedentary time. The overall difference in daily energy expenditure between those of normal weight and those who are overweight or obese is quite small, underlining that the build-up of fat usually occurs over several years. Pedometers and accelerometers provide a useful initial stimulus to greater physical activity, although there remains a need to examine how to maximize the impact of instrumentation and to sustain its motivational effect. There is now good evidence that for at least a few months, the physical activity of an obese adult can be augmented by 2000–3000 steps/day, and that this initiates a slow but consistent loss of body fat (0.05–0.1 kg/week). Relative to dieting, the increase of physical activity also brings other health advantages, including increases of aerobic power and lean tissue, and a decrease of metabolic and cardiac risk factors.

9.1 Introduction

During the past two decades, the attention of many epidemiologists has been directed towards the obesity epidemic [1–4] (Chap. 2). Since matter can neither be created nor destroyed, it is plain that energy homeostasis has been disturbed in

Roy J. Shephard (✉)
Faculty of Kinesiology & Physical Education, University of Toronto, Toronto, ON, Canada
e-mail: royjshep@shaw.ca

R.J. Shephard, C. Tudor-Locke (eds.), *The Objective Monitoring of Physical Activity: Contributions of Accelerometry to Epidemiology, Exercise Science and Rehabilitation*, Springer Series on Epidemiology and Public Health,
DOI 10.1007/978-3-319-29577-0_9

the adults and children concerned. However, the cause of this disruption is less clear. Hypotheses that have been advanced include increased sales of dietary products with a high content of refined carbohydrates, particularly soft drinks [5], an ever-greater consumption of "fast food" [6], an increased portion size in many restaurants [7], and the decrease of habitual physical activity associated with automation of the workplace and personal transportation [8], prolonged television watching [9] and the use of personal computing devices [10].

However, there has been difficulty in deciding among these possibilities, in part because the process of weight gain is slow. Often obesity develops over many months and even years. One recent review set the likely maximal accumulation of fat at 6–10 % of body mass per year [11], or about 4–7 kg/year. Neither dietary nor physical activity questionnaires have had the sensitivity to detect a 6 % discrepancy between energy intake and expenditure, although some nutritionists (ignoring the slow nature of the process) have argued that an increase of physical activity was unlikely to correct an accumulation of body fat, because even a long bout of exercise burnt a relatively small quantity of fat. Often, these negative opinions were apparently supported by small changes in overall body mass after weeks of exercising; many nutritionists ignored the fact that an increase of daily physical activity could replace fat with lean tissue, with little net change in overall body mass. A further issue in some trials was that the additional energy expenditure developed in an out-patient exercise programme was countered, either by an increased intake of food or a reduction in the volume of activities performed outside of the rehabilitation centre. Twenty-four hour objective monitoring of study participants is important in ensuring that such compensatory reductions of physical activity do not occur.

The widespread availability of inexpensive objective monitoring equipment in recent years has offered the possibility of demonstrating both that the obese undertake insufficient physical activity for good health, and that their condition can be improved through an appropriately designed exercise programme. There have been quite a few papers examining relationships between careful estimates of habitual physical activity and body fat content. We will look first at a growing volume of cross-sectional epidemiological evidence showing differences in patterns of habitual physical activity between those with a healthy body mass and those who are overweight or obese. We will then comment on the potential motivational value of the pedometer as a means of stimulating an increase of physical activity among those who are initially obese, and we will note the uniform finding of a slow but consistent decrease of body mass among those who succeed in increasing their daily step count. After underlining additional advantages of an increase of physical activity, including increases of aerobic power, muscle mass and bone strength, and a reduction of cardiac and metabolic risk factors, we will finally point to some limitations of pedometers and accelerometers in the management of obesity.

9.2 Inadequate Habitual Physical Activity and Obesity

Many large-scale epidemiological studies have linked obesity with inadequate habitual physical activity, as demonstrated by objective monitoring of both adults and children.

9.2.1 Studies of Adults

Data for adults include the NHANES study in the U.S., and the Canadian Health Measures Survey in Canada. We link the obesity-related differences of activity with their likely impact upon the accumulation of body fat.

9.2.1.1 NHANES Study

Actigraph data were collected on a large and representative sample of U.S. adults aged >20 years during the 2005–2006 NHANES study [12]. The Actigraph readings were censored to eliminate individuals who did not provide at least one day of realistic data; information was accepted on 3522 individuals.

The Actigraph is recognized to be more sensitive to low accelerations than many commonly used pedometers; thus, to allow comparisons with other research, data were further censored to eliminate periods of the day when the counting rate for any participant dropped below 500 activity counts/minute (this reduced the overall reported activity by about 3000 steps/day). The censored values (Table 9.1) averaged 7190 steps/day for individuals with a healthy body mass. Those who were overweight took only slightly fewer steps (6879/day), but in those who were obese the count was reduced by 1406, to 5784 steps/day. The accelerometer records also showed that those with a normal body mass were meeting the minimal public health recommendation for aerobic activity (30 minutes/day), whereas the obese were falling some 10 minutes short of this target in terms of moderate and vigorous physical activity.

Table 9.1 Censored data from the 2005–2006 NHANES survey, showing Actigraph estimates of habitual physical activity[a] patterns for adults aged >20 years, classified as having a normal body mass, as overweight and as obese

Weight category	Total activity (steps/day)	Moderate activity (minute/day)	Vigorous activity (minute/day)
Normal	7190	25.7	7.3
Overweight	6879	25.3	5.3
Obese	5784	17.3	3.2

[a]Eliminating periods when the activity counting rate was <500/minute

Table 9.2 Associations of body mass index with patterns of physical activity and Actical accelerometer step count readings

Body mass index (kg/m^2)	Sedentary (minute/day)	Light activity (minute/day)	Moderate activity (minute/day)	Vigorous activity (minute/day)	Step count (steps/day)
Normal (<25)	575	252	29	5	10,577
Overweight (>25)	570	251	23	3	9,491
Obese (>30)	586	230	17	2	8,342

Based on data from the Canadian Health Measures Survey (2007–2009) [13]

9.2.1.2 Canadian Health Measures Survey

In Canada, the Canadian Health Measures Survey [13] collected similar objective information on habitual physical activity to that obtained during the NHANES survey, although the Canadian investigators used the Actical monitor. Each of the subjects included in the Canadian analysis had also yielded valid accelerometer records on 4 or more days, rather than a single day.

Differences in body mass index (BMI) among the Canadian participants were more closely associated with accelerometer step-count readings than with estimates of the intensity of physical activity (Table 9.2). The time spent on sedentary or light activities bore little relationship to BMI. However, as in the U.S. survey, there was a small obesity-related gradient in the time allocated to moderate and vigorous activity, so that those of normal BMI took 15 minutes more of such activity than those who were obese. As in the U.S., those of normal weight slightly exceeded the minimum recommended dose of activity, where the obese fell an average of 11 minutes short of this target. The BMI showed a closer relationship to the individual's daily step count, with those of normal body weight taking an average of 2235 more steps/day more than those who were obese.

9.2.1.3 Likely Impact of Differences in Habitual Physical Activity upon Body Fat

Taking the Canadian differential of daily step count (2235 steps/day higher in individuals with a healthy body mass), and making the debatable assumption of a consistent pace length of 0.7 m, this would equate to an additional walking distance of 1.56 km, or 18.7 minutes if covered at a speed of 5 km/hour. The net energy expenditure would be about 16 kJ/minute if a man of average body mass was walking purposefully at this pace, and with 18.7 minutes of such activity, energy usage would be increased by a total of 299 kJ/day or 2.09 MJ/week. Given that the metabolism of 1 g of fat yields approximately 29 kJ of energy, this equates to a fat consumption of about 72.2 g per week, or 3.75 kg/year. Assuming the sedentary individuals ate no less food than their more active peers, they would thus accumulate 3.75 kg of fat per year. This calculation demonstrates that although the difference in habitual physical activity is quite small, it could account for the accumulation of a substantial quantity of body fat over 20–30 years of adult life.

The minimum difference of body mass between a healthy individual (with a BMI of 25 kg/m^2 or less) and someone who is obese (BMI >30 kg/m^2) is about 15 kg. If 15 kg of fat was accumulated over a 10 year life span, this would amount to a weight gain of 1.5 kg per year. A grossly obese person (BMI 40 kg/m^2) would carry some 45 kg of additional body fat, corresponding to a gain of 4.5 kg/year for each of 10 years. These figures are plainly compatible with data from the Canadian Health Measures Survey.

9.2.2 Studies in Children and Adolescents

Associations between objective measurements of habitual activity and obesity have been studied much more frequently in preschool and school age children and adolescents than in adults. Leech and associates were able to review 18 studies looking at the clustering of low levels of physical activity, sedentary behaviour and obesity in children [14], and Parikh and Stratton collected 8 reports concerning the objectively measured intensity of effort and obesity [15]. As in the two studies of adults that we have discussed, the investigations in children present consistent evidence that those with an excessive body mass index (BMI) are engaging in less moderate and vigorous physical activity than those with a healthy BMI. The studies of children have often also examined the length and patterns of physical inactivity (Chap. 7), although relatively few reports have found associations between obesity and sedentary activities.

9.2.2.1 Preschool Children

A study of 357 U.S. preschool children demonstrated an association between obesity and accelerometer determinations of moderately vigorous physical activity in boys, but not in girls [16]. A second study by members of the same research group was based on a total of 418 preschool children. It found an association between accelerometer measurements of moderate to vigorous physical activity (MVPA) and BMI, but after allowing for effects of the MVPA, there was no independent effect attributable to sedentary time [17].

9.2.2.2 School-Age Children

Wittmeier and colleagues evaluated the odds of being overweight (>20 and >25 % body fat) in a sample of 251 Canadian children aged 8–10 years [18]. Regression equations suggested that the biggest effect upon both body fat content and BMI was exerted by vigorous physical activity, as indicated by triaxial accelerometer. Comparing children who undertook less than 5 minutes of vigorous activity with those who took more than 15 minutes, the first group were 4 times as likely to have >20 % body fat, and 5.2 times as likely to be classed as overweight. Comparing those who took less than 15 minutes of moderately vigorous activity with those who took >45 minutes, the risk of having >20 % fat was 4.2 greater for the former. These figures appear to support the current public health guidelines on the minimal physical activity needs of schoolchildren (at least 60 minutes of vigorous and moderately vigorous physical activity per day).

Steele and associates used 1-week Actigraph records to examine the relative impact of active and sedentary time upon the risk of obesity in a sample of 1862 10-year-old British children [19]. They found that the objectively measured sedentary time was associated with measures of obesity such as BMI and waist circumference, but that this association was greatly attenuated after co-varying for the time allocated to moderate to vigorous physical activity. They thus reasoned that the key to prevention of obesity was the promotion of moderately vigorous physical activity, rather than discouraging sedentary pursuits.

Ness, Leary and Riddoch (Fig. 9.1) recruited 5500 British children [20]. Actigraph measurements of habitual physical activity were examined in relation to dual photon x-ray estimates of body fat, with obesity arbitrarily defined as the top 10 % of the fat distribution in their sample. The data showed a graded negative relationship between fat mass and the daily volume of moderately vigorous physical activity that was undertaken, stronger in boys than in girls, and unaltered by adjustment of the data for the total daily volume of physical activity. Rowlands and colleagues [21] conducted a second study using dual photon x-rays

Fig. 9.1 Chris Riddoch has studied obesity and habitual physical activity in a large sample of British children

Fig. 9.2 Ian Janssen has carried out extensive research on the activity patterns of young children

estimates of body fat on a small sample of 38 girls and 38 boys aged 8–11 years. Fat levels were significantly correlated with accelerometer measurements of total physical activity and with vigorous physical activity; moderately vigorous physical activity was only related to body fat content in the girls.

In Dublin, Hussey and associates [22] evaluated physical activity in 224 children aged 7–10 years, using a triaxial accelerometer for periods of 4 days. The boys showed gradients of both body weight and waist circumference with the daily duration of vigorous physical activity, but no significant relationships were seen in the girls, perhaps because on average they performed only a half as much vigorous activity as the boys.

Mark and Janssen (Fig. 9.2) focussed upon inter-individual differences in patterns of physical activity. They obtained objective Actigraph data on 2498 youth aged 8–17 years from the U.S. NHANES studies of 2003–2004 and 2005–2006 [23]. After controlling for the total volume of moderately vigorous physical activity, they noted that the risk of being overweight was 0.38 for those who were in the top quartile in terms of the accumulated number of bouts of such activity. After controlling also for the times accumulated in short bouts of moderately vigorous physical activity, the risk ratio for those with frequent bouts of activity was still only 0.55. They thus concluded that multiple bouts of moderately vigorous activity protected against obesity; possibly, this finding could find a physiological explanation in terms of the persistent increase of oxygen consumption following a bout of vigorous activity.

A study of 10–12 year old children in Hungary and the Netherlands [24] divided the students into quartiles, based upon their sedentary time, both self-reported use of a television and personal computer and total sedentary time as measured by an accelerometer. In terms of BMI, the difference between subjects in the most sedentary and the least sedentary quartiles was greater for objective monitoring (21 vs. 19 kg/m^2) than for self-reports of participation in sedentary activities (20 vs. 19 kg/m^2); inter-group differences in waist circumferences supported this trend.

An extension of the study to 766 students in 5 European countries divided children of both sexes into 4 clusters, based on their sedentary times (ST) and duration of moderately vigorous physical activity (MVPA) [25]. In the girls, there was a substantial contrast of BMI (19.1 vs. 17.7 kg/m^2) and waist circumference (66.1 vs. 62.6 cm) between the sedentary (ST 571 minutes, MVPA 18 minutes) and the active (ST 453 minutes, MVPA 55 minutes) groups. In the boys, there were similar contrasts of BMI 20.5 vs. 17.6 kg/m^2) and waist circumference (70.5 vs. 63.7 cm) between the sedentary (ST 546 minutes, MVPA 26 minutes) and the active (ST 419 minutes, MVPA 70 minutes) groups, although in the boys the minutes of MVPA seemed more important than the time that they were sedentary.

9.2.2.3 Adolescents

Sanchez et al. [26] studied cross-sectional associations between self-reports, accelerometer measurements and BMI in 878 American youth aged 11–15 years. Failure to meet public health guidelines of 60 minutes of moderate to vigorous physical activity per week had a strong association with BMI values at or above the 85th percentile for participants in this investigation. Another report, apparently based on the same sample of 878 youth, classified subjects as falling above or below the 85th percentile of BMI [27]. In girls, the obese group undertook 3.5 minutes/day less moderate physical activity and 2.1 minutes/day less vigorous activity than those whose BMI fell below the 85th percentile; the heavier girls also devoted an additional 13.8 minutes/day to television watching, with a resulting decrease of 573 kJ in their daily energy expenditure. In boys, the corresponding differences were 6.6 minutes/day, 4.4 minutes/day, 33.1 minutes/day, and 1283 kJ/day.

Butte et al. [28] examined 897 Texan children and adolescents of Hispanic origin, ranging in age from 4 to 19 years. Objectively measured physical activity as determined by an Actiwatch accelerometer decreased progressively with age, and was consistently less in the girls than in the boys. Total activity counts were lower, and sedentary times were higher in obese than in non-obese students.

Lohman and associates [29] examined 1553 sixth grade U.S. female adolescents, noting an association between triceps skinfold estimates of body fat and Actigraph measurements of moderate and vigorous physical activity. Fat-free mass was also determined in this study, and it was positively correlated with habitual physical activity.

The HELENA study recruited 365 Spanish adolescents aged 12.5–17.5 years. Body fat was determined by several methods (the sum of six skinfolds, dual photon spectrophotometry and a pneumatic determination of body volume, the BodPod) [30]. After controlling for age, sex, and pubertal status, all measures of body fat were negatively correlated with accelerometer assessments of moderate, and especially with vigorous physical activity. Body fat was also negatively correlated with aerobic fitness and with the strength of the lower limbs [31]. The accelerometer data showed that muscular strength depended on engaging in either vigorous physical activity, or at least 60 minutes of moderate to vigorous physical activity [32].

Mitchell and associates made a prospective study of 789 children, following them from the age of 9–15 years [33]. Increases of BMI over the 6 years were negatively associated with accelerometer measurements of habitual physical activity for this period, and there were additional adverse effects from sedentary time, particularly in the heavier children.

9.2.2.4 Timing of Activity Bouts

Use of objective monitors with time-stamp devices allows the observer to identify those specific times of the day when physical activity is lower in the obese than in those who have a healthy weight; such information may be helpful in planning corrective tactics. An analysis of Canadian children aged 6–19 years demonstrated that boys (but not girls) who were obese took more sedentary time after 3 p.m. on weekdays (282 vs. 259 minutes). Each 60 additional minutes of sedentary time was associated with a gain in BMI of 1.4 kg/m^2 [34]. Further, the obese boys engaged in more prolonged bouts of sedentary activity.

An extension of this research demonstrated that frequent breaks in sedentary periods were associated with a favourable cardiometabolic risk profile [35], including lower values for BMI and waist circumference.

9.2.2.5 Conclusions Regarding Obesity in Children and Adolescents

General conclusions from the studies of children and adolescents are that associations between habitual physical activity and body fat are stronger for boys than for girls, that prevention of obesity is associated with the practice of moderate to vigorous and vigorous physical activity, and that any relationships with sedentary time are weaker than for those with vigorous physical activity.

9.3 Motivational Value of Pedometer

Given that there is sometimes a substantial reactive response to the wearing of a pedometer or an accelerometer (Chap. 1), one would anticipate that the use of such instrumentation would be helpful, at least initially, in motivating a person to an increase of habitual physical activity. However, given the rapid waning of any reactive response, it is less clear how long the boosting of motivation might persist, particularly in sedentary individuals without immediate manifestations of disease [36].

Much probably depends on how the pedometer is presented to the client, and details of the motivational plan. In some studies, participants have simply been told to aim for a daily count of 10,000 steps/minute, whereas in other investigations the physical activity target has been individually prescribed, and adjusted upwards as the physical condition of the client improved. There is some evidence that the provision of objective monitors is most effective in stimulating the physical activity of those who are initially overweight or obese [37].

9.4 Characteristics of Effective Pedometer-Based Walking Programmes

Several studies have suggested that over the first few months, the provision of a pedometer and a daily walking distance target can increase the step counts of initially sedentary individuals by 2500–3000 steps/day [38, 39]. Tudor-Jones and Lutes [40] have reflected upon the circumstances in which pedometers and accelerometers are successful in motivating greater physical activity. One striking weakness is that only about a third of studies introducing pedometers as motivational tools have to date given even token consideration to theories of behavioural change [41]. However, such a theoretical framework would probably have a substantial influence upon the response of the client.

We will comment briefly upon appropriate characteristics for the monitoring instrument, the setting of activity targets, and the continuing need for long-term studies of the motivational response.

9.4.1 Appropriate Instrument Characteristics

If the purpose of wearing a pedometer or accelerometer is to stimulate greater physical activity rather than to collect research data, then the instrument will be worn for a long period, and it needs to be unobtrusive. It must also be reasonably accurate but inexpensive, presenting a summary of the daily activity that has been performed in a manner that is easily understood by the average client. Rather than accumulating data to be down-loaded at the end of the day, week or month, the requirement is for an instantaneous feedback of the day's activity to date, allowing the wearer to plan further periods of physical activity if the prescribed activity target for the day has not been met.

9.4.2 Optimal Use of the Pedometer as a Motivational Tool

Motivation to an increase of physical activity is helped by the setting of clear goals. In some experimental studies, clients have simply been told to aim for a count of 10,000 steps/day. In other trials, the pedometer/accelerometer has been used as a part of a well-designed and regular, guided feedback process, with a focus upon the

Fig. 9.3 Mark Adams has investigated the impact of pedometer use upon the activity motivation of obese subjects

establishment of moderate and progressive goals of increased walking distance and the building of a sense of self-efficacy. Important variables influencing the effectiveness of activity monitors are (1) whether the daily activity target was set by the observer or the study participant, and (2) whether feedback was provided by a professional, a member of a peer group, or simply by a computer programme based on the recorded step count.

Engel and Lindner [42] compared the amount of walking undertaken by a group that received counselling with that observed in those who were also supplied with a pedometer. Somewhat to their surprise, the wearing of the pedometer did not increase the amount of walking relative to that of a group who received only thorough counselling. In contrast, Mark Adams (Fig. 9.3) and colleagues [43] compared responses between two small groups of women; both were issued with pedometers. One group was set a static goal of 10,000 steps/day, and the other was given a frequently adjusted daily step-count target. The latter group achieved a much larger increase in their daily step-count.

A further practical question is whether the wearing of a pedometer provides a unique stimulus to attaining the target volume of daily walking. Potentially, a similar target could be set by measuring a walking distance carefully with a car or a bicycle odometer, and then covering this distance on a regular daily basis. One obvious advantage of using a pedometer is that it allows flexibility in the choice of walking route, which should immediately encourage adherence.

9.4.3 *Need for Long-Term Studies*

In most motivational investigations, the pedometer or accelerometer has been used for only a short period, although the need is for a long-term change of lifestyle. Although the wearing of the monitor may stimulate an initial increase of physical activity, it is less clear how far such monitoring contributes to long-term programme adherence. Is there an optimal period of wear for the client to establish lifelong exercise habits?

9.5 Objective Monitoring in the Treatment of Obesity

Richardson and associates [39] made a meta-analysis of the response to pedometer-controlled walking programmes where no specific diet was imposed upon individuals who were initially obese. They found 9 trials involving at least 5 subjects; the investigations had continued for at least 4 weeks, with a total sample size of 307 participants. In general, the pedometer/accelerometer had been used as a motivational device, typically setting individual goals for an increase in the daily step count. However, the objective monitoring also served to ensure that study participants had adhered to the prescribed programme. Bravata et al. [38] listed a total of 8 controlled trials and 18 observational studies, only 4 of which were included in the review of Richardson et al. [39]. These various investigations are summarized in Table 9.3.

In many of the trials discussed by Richardson et al. [39] and Bravata et al. [38], the immediate increase in pedometer count averaged 3000 steps/day or more. This implies an increase in the daily walking distance of more than 2 km, perhaps 24 minutes of walking at a pace of 5 km/hour. The net increase of energy expenditure in a man of average body mass would be 384 kJ/day, and if we equate the metabolism of 1 g of fat with 29 kJ of energy, the fat loss would amount to 0.093 kg per week. The average length of the trial was 16 weeks, and the average decrease of body mass was 1.27 kg, or a loss of about 0.05 kg per week, of the same general order as the theoretical figure if there had been no change in the individual's food intake or resting metabolism.

One study evaluated possible changes of diet when healthy young adults increased their daily step count by 2667 steps/day [49]. No changes of energy intake were detected, although the authors admitted that a larger sample would be needed to be certain of this finding.

In the analysis of Bravata et al. [38], 18 of 26 reports noted changes of body mass and/or BMI. The BMI was decreased by an average of 0.38 kg/m^2 over a period averaging 18 weeks; assuming a height of 1.7 m, this would equate to a total fat loss of 1.1 kg, or 0.061 kg/week. Bravata et al. [38] commented that the response was larger in older individuals, was enhanced by goal setting, and was accompanied by an average 3.8 mmHg decrease in systolic blood pressure. However, these investigators observed no significant changes of blood lipids or glucose tolerance. Table 9.3 shows that the loss of body mass was very slow, but also remarkably consistent from one trial to another; only one trial observed a very small gain of body mass.

9.6 Associated Health Benefits

There are important secondary dividends if an obese individual increases their physical activity, and for the most part these gains are not realized if fat loss is attempted simply by a reduced intake of food. A modest exercise programme is

Table 9.3 Pedometer monitored changes in daily step count and associated weight loss (based in part on an analysis of Richardson et al. [39] and Bravata et al. [38])

Author	Sample	Initial step count	Post-intervention step count	Duration of trial (weeks)	Initial body mass (kg)	Change of body mass (kg, g/week)
Araiza et al. [44]	15 experimental patients with diabetes mellitus, 15 controls	7220	10,410	6	80.7*	–2.0*, –333*
Chan et al. [45]	106 sedentary workers	7029	10,480	12	79.4	–1.5, –125
Engel and Lindner [42]	57 overweight men and women, average age 62 year, overweight and with type 2 diabetes mellitus	Not recorded	7296	26	91.9	–2.1, –81
Hultquist et al. [46]	Previously inactive women	5603	10159	4	82.0	–0.5, –125
Jensen et al. [47]	18 obese women	4027	5883	13	100.4	–4.3, –331
Kilmer et al. [48]	20 patients with progressive neuromuscular disease	~4600	~5750	26	84.9	–1.4, –54
Koulouri et al. [49]	12 healthy young adults	9588	12,255	1	67.9	–0.3, –300
Miyatake et al. [50]	31 middle-aged structured obese Japanese men	7013	8840	52	82.3	–3.7, –71
Moreau et al. [51]	15 post-menopausal women	5400	9700	24	81.1	–1.3, –54
Pal et al. [52]	13 overweight obese women (13 controls)	6242	9703	12	77.8	–0.9, –75
Ransdell et al. [53]	28 grandmother/mother daughter triads (9 controls)	8422	11,517	30	60.4	–0.02, –0.6
Schneider et al. [54]	56 overweight adults	5123	9117	36	98.8	–2.4, –67
Sugiura [55]	14 post-menopausal women (13 controls)	6740	8,00 to 11,000	104	54.1	–0.9, –9
Swartz et al. [56]	18 obese women with family history of type 2 diabetes mellitus	4972	9213	12	94.0	+0.3, +25
Tudor-Locke et al. [57]	24 middle-aged overweight or obese adults (23 controls)	5753	9123	16	96.8	–0.7, –44
Van Wormer et al. [58]	22 patients in cardiac health programme	6152	8210	17	85.0*	–1.1*, –65*

(continued)

Table 9.3 (continued)

Author	Sample	Initial step count	Post-intervention step count	Duration of trial (weeks)	Initial body mass (kg)	Change of body mass (kg, g/week)
Wilson et al. [59]	African-American breast cancer survivors	4791	8297	14	86.7	–0.9, –64

Body mass values marked with an asterisk have been approximated from the BMI, assuming a height of 1.7 m

likely to increase aerobic fitness and muscle mass, strengthen bones strengthened, and reduce cardiac and metabolic risk factors. Moreover, the levels of activity required to induce such benefits seem of the order encountered in the fat loss programmes of Table 9.3.

9.6.1 Aerobic Fitness and Muscular Strength

The HELENA study noted that if the habitual activity of adolescents as measured by a uniaxial Actigraph was sufficient to control obesity, the students concerned also showed a higher level of aerobic fitness and greater muscular strength in the lower limbs [31]. In order to see beneficial effects upon muscular strength, it was necessary for individuals to engage either in vigorous activity, or to engage in at least 60 minutes per day of moderate to vigorous physical activity.

Likewise, when Miyatake et al. [50] provided 31 obese Japanese men with a Seiko pedometer, instructing them to make a modest increase in their daily physical activity for 1 year, aerobic performance and leg strength were increased, and visceral adipose tissue was decreased—the last change was most closely correlated with a decreased insulin resistance.

We found that among seniors living in Nakanojo, Japan [60], the preferred and maximal walking speeds and the peak knee extension torque were all correlated with pedometer/accelerometer measurements of habitual physical activity, whether reported as steps/day or as minutes of moderate to vigorous physical activity. When subjects were divided into quartiles, significant advantage in these measures of fitness was seen among men taking an average of >8218 steps/day or >22.0 minutes/day of moderate to vigorous physical activity, while for the women, the corresponding thresholds were a count >7030 steps/day, or >19.1 minutes/day of moderate to vigorous physical activity.

Further observations on the Nakanajo population, carried out by Hyuntae Park (Fig. 9.4) and associates, used double absorption spectrophotometry to estimate the lean tissue mass in the limbs [61]. The muscle mass in the legs, and to a lesser extent

Fig. 9.4 Hyuntae Park has worked with Yukitoshi Aoyagi and myself on many studies of seniors in Nakanajo

in the arms, increased progressively with measures of habitual physical activity. When subjects were divided into quartiles, muscle mass was significantly greater in men taking an average of >7972 steps/day or >21.6 minutes of moderate to vigorous physical activity, and in women >6942 steps/day or >18.5 minutes/day of moderate to vigorous physical activity.

9.6.2 Bone Strength

If an individual undertakes a programme of vigorous physical activity to reduce body fat content, a further by-product is likely to be an increase of bone health. This issue does not seem to have been evaluated in young adults, but we have observed such an effect in both cross-sectional and longitudinal studies of seniors living in Nakanojo, Japan [62]. The calcaneal osteosonic index was significantly greater among subjects in the upper two quartiles of the habitual physical activity distribution (for the men, an average count >7697 steps/day, >21.4 minutes/day of moderate to vigorous activity, and for the women, a count >6356 steps/day, >19.6 minutes/day of moderate to vigorous physical activity).

9.6.3 Cardiac and Metabolic Risk Factors

In the study of Miyatake et al. [50] a modest increase in the daily physical activity of 31 obese Japanese men was accompanied by a decreased insulin resistance.

Swartz et al. [56] engaged 18 obese middle-aged women with a family history of type 2 diabetes mellitus in an 8-week walking programme that was intended to bring their step counts progressively to a target level of 10,000 steps/day. This programme induced favourable changes in both blood pressure and glucose tolerance.

Schneider and colleagues [54] followed 56 overweight adults for 36 weeks; 38 of the group maintained a step count >9500 steps/day between weeks 4 and 36, with associated decreases in body mass and fat mass, and also increases of HDL cholesterol. In contrast, the 18 subjects who were non-compliant showed little change in any of these variables.

Iwane and associates [63] recruited an initial sample of 730 male manufacturing workers with an average age of 47 years; 83 of this 730 were persuaded to walk at least 10,000 steps/day, as monitored by a pedometer. Thirty of those individuals who were willing to exercise were initially hypertensive, but after walking an average of 13,510 steps/ day for 12 weeks, the systolic blood pressure fell from 149 to 139 mmHg, and the diastolic pressure from 103 to 93 mmHg, although (possibly because the subjects were not initially obese) there were no changes of BMI or serum lipids over this period. In terms of the blood pressure changes, a ceiling of benefit was reached at a count of 15,000 steps/day. Blood pressures were unchanged in those who remained sedentary, whether these individuals were initially hypertensive or normotensive.

Moreau and associates [51] used a pedometer to monitor the totality of imposed changes of physical activity in experimental subjects, at the same time confirming that the activity of control subjects remained unchanged. In their small sample of postmenopausal women, the experimental group increased their step count from 5000 steps/day to an average of 10,000 steps/day. However, the intensity of activity was not recorded. The resting systolic blood pressure of the experimental group was reduced by 6 mmHg at 12 weeks, and by 11 mmHg at 24 weeks, but the diastolic pressure remained unchanged. The body mass of the experimental subjects had also decreased by 1.3 kg at 24 weeks, but there were no significant change in the fasting plasma insulin.

As in the study of Moreau et al. [51], Yamanouchi et al. [64] used a pedometer to confirm the existence of inter-group differences of physical activity over a 6–8 week programme. Their comparison was between an exercise + diet group (19,200 steps/day, 4–6 MJ/day negative energy balance) and a diet group (4500 steps/day, 4–6 MJ/day negative energy balance). The exercised group showed a larger decrease of body mass (7.8 vs. 4.2 kg) and an increase of insulin sensitivity that was not seen in those who only dieted.

We studied the health of 220 seniors living in Nakanojo, Japan [65], relating objectively measured year-long habitual physical activity to the presence of five markers of the metabolic syndrome (a BMI >25 kg/m^2; a fasting serum triglyceride >1.7 mmol/L, a fasting HDL cholesterol <1.0 mmol/L in men or 1.3 mmol/L in women, a systolic blood pressure >130 mmHg or a diastolic pressure >85 mmHg, and a fasting plasma glucose >6.1 mmol/L or a haemoglobin A_{1c} >5.5 %). A person showing two or less of these markers was recognized as being in good metabolic health. The odds ratio for showing three or more markers of the metabolic syndrome was strongly related to habitual physical activity (Table 9.4), but much of the benefit was associated with the most active quartile, where subjects

Table 9.4 Relationship between quartiles of habitual physical activity and risk of presence of three or more markers of the metabolic syndrome

Habitual physical activity	Odds ratio	Range
Stepping rate (counts/day)		
Quartile 1 (average 3427)	4.55	1.81–11.41
Quartile 2 (average 5581)	3.10	1.22–7.88
Quartile 3 (average 7420)	2.63	1.02–6.75
Quartile 4 (average 10,129)	1	1
Minutes of activity >3 METs		
Quartile 1 (average 4.4 minutes/day)	3.67	1.50–8.97
Quartile 2 (average 12.1 minutes/day)	2.29	0.92–5.71
Quartile 3 (average 19.4 minutes/day	2.1	0.83–5.27
Quartile 4 (average 33.5 minutes/day)	1	1

Based on data from the Nakanojo study [65]

were averaging over 10,000 steps/day, and taking an average of 33.5 minutes/day of moderate physical activity at an intensity of >3 METs. These findings suggest that in studies where no benefit was associated with physical activity, the intensity and/or the volume of activity was inadequate.

Interestingly, in our study, the main associations of physical activity were with blood pressure and with the concentration of HBA_{1c}. The BMI was similar for the four quartiles. Possibly, the BMI is an inappropriate index for Japanese; there may have been a confounding of a high BMI due to muscle rather than fat, or possibly too few of the group was sufficiently obese to be influenced by an increase of habitual physical activity.

Critique of Objective Monitoring One important advantage of exercise relative to dieting in the treatment of obesity is that lean tissue tends to be conserved; this in turns limits the reduction of resting energy expenditure that negates the effect of dieting alone. Resistance training should thus be an important part of the therapeutic regimen for an obese person, particularly if the excess weight has led to osteoarthritis of the knees. Unfortunately, the usual form of objective monitor (a pedometer/accelerometer) makes little or no response to resisted exercise, so that if a rehabilitation programme is to be based on resistance exercise, this type of objective instrumentation is of little help in either motivation or monitoring adherence.

One potential alternative method of objective monitoring total energy expenditure is the doubly-labelled water technique. It is at present a relatively costly procedure, but several studies of small subject samples (Ferro-Luzzi and Martino, Schoeller et al., Klaas Westerterp (Fig. 9.5) and Goran [66–68], using this approach, have shown a negative relationship between energy expenditures levels averaged over a 2-week period and body mass. Unfortunately, the doubly-labelled water technique does not allow us to determine the intensity or the duration of physical activity that is associated with maintenance of a healthy body mass.

Fig. 9.5 Klaas Westerterp has evaluated many techniques for the objective monitoring of physical activity, including the ingestion of doubly-labelled water

9.7 Conclusions

Objective monitoring of physical activity confirms the impression, formed with less certainty from questionnaire responses, that both adults and children who are overweight or obese take less physical activity than their peers who have a healthy body weight. Accelerometer data also provides relatively precise information on the magnitude of the difference in daily activity between the 2 groups, which amounts to around 2000 steps/day, or 15–20 minutes/day of moderate or vigorous physical activity. In those who are grossly obese, some studies also provide evidence of an increase in sedentary time. The overall difference in daily energy expenditure between those of healthy weight and the obese is quite small, but assuming there is no difference of food intake between the two groups, then the latter could accumulate 3.5–4.0 kg of fat per year. This underlines the point that in many obese people the build-up of fat has occurred gradually over several years, and that measures to correct the fat accumulation are also likely to be slow.

The wearing of a pedometer or an accelerometer provides a useful initial stimulus to greater daily physical activity, although there remains a need to examine how this motivational effect can be maximized and how long it is likely to persist. Multiple studies of obese adults show that over the first few months, physical activity can be augmented by 2000–3000 steps/day, and this consistently initiates a slow loss of body fat (0.05–0.1 kg/week). The required increase of physical activity carries other health advantages, including increases of aerobic power, lean tissue and skeletal health, plus a decrease of metabolic and cardiac risk factors. If motivation can be sustained until greater habitual activity becomes a part of normal lifestyle, greater physical activity thus seems a preferable approach to crash dieting. The latter can produce dramatic results in the short term, but fat lost by dieting is usually regained within a few months.

References

1. Dollman J, Olds T, Norton K, et al. The evolution of fitness and fatness in 10–11-year-old Australian schoolchildren: changes in distributional characteristics between 1985 and 1997. Pediatr Exerc Sci. 1999;11(1):108–21.
2. Flegal KMT. The obesity epidemic in children and adults: current evidence and research issues. Med Sci Sports Exerc. 1999;33 Suppl 11:S509–14.
3. Tremblay MS, Willms JD. Secular trends in the body mass index of Canadian children. Can Med Assoc J. 2000;163:1429–33.
4. World Health Organisation. Obesity: preventing and managing the global epidemic. Report of a WHO consultation (WHO Technical Report Series 894). Geneva, Switzerland: World Health Organisation; 2000.
5. Malik VS, Schulze MB, Hu FB. Intake of sugar-sweetened beverages and weight gain: a systematic review. Am J Clin Nutr. 2014;84:274–88.
6. DeVogli R, Kouvonen A, Gimeno D. The influence of market deregulation on fast food consumption and body mass index: a cross-national time series analysis. Bull World Health Org. 2014;92:99–107A.
7. Young LR, Nestle M. The contribution of expanding portion sizes to the U.S. obesity epidemic. Am J Public Health. 2002;92(2):246–9.
8. Korbonits M. Obesity and metabolism. Basel, Switzerland: Karger; 2008.
9. Robinson TN. Television viewing and childhood obesity. Pediatr Clin North Am. 2001;48 (4):1017–25.
10. Gõmez-Arbeáez D, Camacho PA, Cohen DD, et al. Higher household incomes and the availability of electronic devices and transport at home are associated with higher waist circumferences in Coloombian children: the ACFIES study. Int J Environ Res Public Health. 2014;11(2):1834–43.
11. Puhl RM, Heuer CA. Obesity stigma: important consideration for public health. Am J Public Health. 2010;100(6):1019–28.
12. Tudor-Locke C, Brashear MM, Johnson WD, et al. Accelerometer profiles of physical activity and inactivity in normal weight, overweight, and obese US men and women. Int J Behav Nutr Phys Act. 2010;7:60.
13. Colley RC, Garriguet D, Janssen I, et al. Physical activity of Canadian adults: accelerometer results from the 2007 to 2009 Canadian Health Measures Survey. Ottawa, ON: Statistics Canada; 2014.
14. Leech RM, McNaughton SA, Timperio A. The clustering of diet, physical activity and sedentary behavior in children and adolescents: a review. Int J Behav Nutr Phys Act. 2014;11:4.
15. Parikh T, Stratton G. Influence of intensity of physical activity on adiposity and cardiorespiratory fitness in 5–18 year olds. Sports Med. 2011;41(6):477–88.
16. España-Romero V, Mitchell JA, Dowda M, et al. Objectively measured sedentary time, physical activity and markers of body fat in preschool children. Pediatr Exerc Sci. 2013;25 (1):154–63.
17. Byun W, Liu J, Pate RR. Association between objectively measured sedentary behavior and body mass index in preschool children. Int J Obes. 2013;37:961–5.
18. Wittmeier KDM, Mollard RC, Kriellaars DJ. Overweight and adiposity in children. Obesity. 2008;16:416–20.
19. Steele RM, van Sluijs EMF, Cassidy A, et al. Targeting sedentary time or moderate- and vigorous-intensity activity: independent relations with adiposity in a population-based sample of 10-yr-old British children. Am J Clin Nutr. 2009;90:1185–92.
20. Ness AR, Leary SD, Riddoch C. Objectively measured physical activity and fat mass in a large cohort of children. PLoS One. 2007;4(3), e97.
21. Rowlands AV, Eston R, Powell SM. Total physical activity, activity intensity and body fat in 8–11-year-old boys and girls. J Exerc Sci Fit. 2006;4(2):96–102.

22. Hussey J, Bell C, Bennett K, et al. Relationship between the intensity of physical activity, inactivity, cardiorespiratory fitness and body composition in 7–10-year-old Dublin children. Br J Sports Med. 2007;41:311–6.
23. Mark AE, Janssen I. Influence of bouts of physical activity in overweight youth. Am J Prev Med. 2009;36(5):416–21.
24. Chipinaw MJM, Yidirim M, Altenburg TM, et al. Objective and self-rated sedentary time and indicators of metabolic health in Dutch and Hungarian 10–12 year olds: the ENERGY Project. PLoS One. 2012;7(5), e36657.
25. DeBourdeaudhuij I, Verloigne M, Maes L, et al. Associations of physical activity and sedentary time with weight and weight status among 10- to 12-year-old boys and girls in Europe: a cluster analysis within the ENERGY project. Pediatr Obes. 2013;8(5):367–75.
26. Sanchez A, Norman GJ, Patrick K. Patterns and correlates of physical activity and nutrition behaviors in adolescents. Am J Prev Med. 2007;32(2):124–30.
27. Patrick K, Norman GJ, Calfas KJ, et al. Diet, physical activity and sedentary behaviors as risk factors for overweight in adolescence. Arch Pediatr Adolesc Med. 2004;158(4):385–90.
28. Butte NC, Puyau MR, Adolph AL, et al. Physical activity in non-overweight and overweight Hispanic children and adolescents. Med Sci Sports Exerc. 2007;39(8):1257–66.
29. Lohman TG, Ring K, Schmitz KH, et al. Associations of body size and composition with physical activity in adolescent girls. Med Sci Sports Exerc. 2006;38(6):1175–81.
30. Moliner-Urdiales D, Ruiz JR, Ortega FB, et al. Association of objectively assessed physical activity with total and central body fat in Spanish adolescents: the HELENA study. Int J Obes. 2009;33:1126–35.
31. Moliner-Urdiales D, Ruiz JR, Vicente-Ropdriguez G, et al. Associations of muscular and cardiorespiratory fitness with total and central body fat in Spanish adolescents: the HELENA study. Br J Sports Med. 2014;45:101–8.
32. Moliner-Urdiales D, Ortega FB, Vicente-Rodriguez G, et al. Association of physical activity with muscular strength and fat-free mass in adolescents: the HELENA study. Eur J Appl Physiol. 2010;109:1119–27.
33. Mitchell JA, Pate RR, Beets MW, et al. Time spent in sedentary behavior and changes in childhood BMI: a longitudinal study from ages 9 to 15 years. Int J Obes. 2013;37:54–60.
34. Colley RC, Garriguet D, Janssen I, et al. The association between accelerometer-measured patterns of sedentary time and health risk in children and youth: results from the Canadian Health Measures Survey. BMC Public Health. 2013;13:200.
35. Saunders TJ, Tremblay MS, Mathieu M-È, et al. Associations of sedentary behavior, sedentary bouts and breaks in sedentary time with cardiometabolic risk in children with a family history of obesity. PLoS One. 2013;8(11), e79143.
36. Burton NW, Walsh A, Brown WJ. It just doesn't speak to me: mid-aged men's reactions to '10,000 Steps a Day'. Health Promot J Austr. 2008;19(1):52–9.
37. Tudor-Locke C, Chan CB. An exploratory analysis of adherence patterns and program completion of a pedometer based physical activity intervention. J Phys Act Health. 2006;3 (2):210–20.
38. Bravata DM, Smith-Spangler C, Sundaram V, et al. Using pedometers to increase physical activity and improve health. JAMA. 2007;298(19):2296–304.
39. Richardson CR, Newton TL, Abraham JJ, et al. A meta-analysis of pedometer-based walking interventions and weight loss. Ann Fam Med. 2008;6:69–77.
40. Tudor-Locke C, Lutes L. Why do pedometers work? A reflection upon the factors related to successfully increasing physical activity. Sports Med. 2009;39(12):981–93.
41. Painter JE, Borba CP, Hynes M, et al. The use of theory in health behavior research from 2000 to 2005: a systematic review. Ann Behav Med. 2008;35(3):358–62.
42. Engel L, Lindner H. Impact of using a pedometer on time spent walking in older adults with type 2 diabetes. Diabetes Educ. 2006;32(1):98–107.
43. Adams MA, Sallis JF, Norman GJ, et al. An adaptive physical activity intervention for overweight adults: a randomized controlled trial. PLoS One. 2013;8(12), e82901.

44. Araiza P, Hewes H, Gashetawa C, et al. Efficacy of a pedometer-based physical activity programme of diabetes control in type 2 diabetes mellitus. Metabolism. 2006;55(10):1382–7.
45. Chan CB, Ryan DAJ, Tudor-Locke C. Health benefits of a pedometer-based physical activity intervention in sedentary workers. Prev Med. 2004;39(6):1215–22.
46. Hultquist C, Albright C, Thompson DL. Comparison of walking recommendations in previously inactive women. Med Sci Sports Exerc. 2005;37(4):676–83.
47. Jensen GL, Roy M-A, Buchanan AE, et al. Weight loss intervention for obese older women: improvements in performance and function. Obes Res. 2004;12:1814–20.
48. Kilmer DD, Wright NC, Aitkens S. Impact of a home-based activity and dietary intervention in people with slowly progressive neuromuscular diseases. Arch Phys Med Rehabil. 2005;86:2150–6.
49. Koulouri AA, Tigbe WW, Lean ME. The effect of advice to walk 2000 extra steps daily on food intake. J Hum Nutr Diet. 2006;19(4):263–6.
50. Miyatake N, Nishikawa H, Morishita A, et al. Daily walking reduces visceral adipose tissue areas and improves insulin resistance in Japanese obese subjects. Diabetes Res Clin Pract. 2002;58(2):101–7.
51. Moreau KL, Degarmo R, Langley J. Increasing daily walking lowers blood pressure in postmenopausal women. Med Sci Sports Exerc. 2001;33(11):1825–31.
52. Pal S, Cheng C, Donovan R. Using pedometers to increase physical activity in overweight and obese women: a pilot study. BMC Public Health. 2009;9:309.
53. Randell LB, Robertson L, Ornes L, et al. Generations exercising together to improve fitness (GET FIT): a pilot study designed to increase physical activity and improve health-related fitness in three generations of women. Women Health. 2004;40(3):77–94.
54. Schneider PL, Bassett DR, Thompson DL, et al. Effects of a 10,000 steps per day goal in overweight adults. Health Promot. 2006;21(2):85–9.
55. Sugiura H, Kajima K, Mirbod SM, et al. Effects of long-term moderate exercise and increase in number of daily steps on serum lipids in women: randomised controlled trial. BMC Womens Health. 2002;2(1):3.
56. Swartz AM, Strath SJ, Bassett DR. Increasing daily walking improves glucose tolerance in overweight women. Prev Med. 2003;37(4):356–62.
57. Tudor-Locke C, Bell RC, Myers AM, et al. Controlled outcome evaluation of the First Step Program: a daily physical activity intervention for individuals with type II diabetes. Int J Obes Relat Metab Disord. 2004;28(1):113–9.
58. VanWormer JJ, Boucher JL, Pronk NP, et al. Lifestyle behavior change and coronary artery disease: effectiveness of a telephone-based counseling program. J Nutr Educ Behav. 2004;36 (6):333–4.
59. Wilson DB, Porter JS, Parker G, et al. Anthropometric changes using a walking intervention in African American breast cancer survivors: a pilot study. Prev Chronic Dis. 2005;2(2):A16.
60. Aoyagi Y, Park H, Watanabe E, et al. Habitual physical activity and physical fitness in older Japanese adults: the Nakanojo study. Gerontology. 2009;55(5):523–31.
61. Park H, Park S, Shephard RJ, et al. Yearlong physical activity and sarcopenia in older adults: the Nakanojo study. Eur J Appl Physiol. 2010;109(5):953–61.
62. Park H, Togo F, Watanabe E, et al. Relationship of bone health to yearlong physical activity in older Japanese adults: cross-sectional data from the Nakanojo study. Osteoporos Int. 2007;18 (3):285–93.
63. Iwane M, Arita M, Tomimoto S, et al. Walking 10,000 steps/day or more reduced blood pressure and sympathetic nerve activity in mild essential hypertension. Hypertens Res. 2000;23:573–80.
64. Yamanouchi K, Shinozaki T, Chikada K, et al. Daily walking combined with diet therapy is a useful means for obese NIDDM patients not only to reduce body weight but also to improve insulin sensitivity. Diabetes Care. 1995;18:775–8.

65. Park S, Park H, Togo F, et al. Yearlong physical activity and metabolic syndrome in older Japanese adults: cross-sectional data from the Nakanojo study. J Gerontol A Biol Sci Med Sci. 2008;63(10):1119–23.
66. Ferro-Luzzi A, Martino L. Obesity and physical activity. In: Chadwick DJ, Cardew G, editors. The origins and consequences of obesity Ciba Foundation Symposium 201. Chichester: Wiley; 1996.
67. Schoeller DA, Shay K, Kushner RF. How much physical activity is needed to minimize weight gain in previously obese women? Am J Clin Nutr. 1997;66:551–6.
68. Westerterp KR, Goran MI. Relationship between physical activity related energy expenditure and body composition: a gender difference. Int J Obes. 1997;21:184–8.

Chapter 10
Objective Monitoring and the Challenge of Defining Dose/Response Relationships for the Prevention of Chronic Disease

Roy J. Shephard

Abstract The form of the dose/response relationship linking habitual physical activity to major health outcomes is of practical importance for those who are formulating public health recommendations on minimum daily physical activity needs. However, available information on the form of this relationship is conflicting. One might anticipate that the optimal physical activity pattern would mimic that of our hunter-gatherer ancestors. Data gathered on traditional Inuit hunters point to a large daily energy expenditure in many types of hunting; this is accumulated mainly through long days of moderate physical activity. Studies based upon occupational classifications also suggest health benefit from vigorous physical exercise at work, and this view is supported by early questionnaire analyses of leisure behaviour. However, a systematic survey of questionnaire-based reports revealed no clear picture, because of differing methods of measuring and classifying physical activity, substantial inaccuracies in self-reports, and examination of a wide variety of health outcomes. Other studies, using simple measures of fitness as surrogate indicators of accumulated physical activity, have pointed to the largest health benefits being gained from a moderate increment of fitness. Objective physical activity monitors allow a much finer gradation of physical activity patterns, and thus have the potential to clarify dose/response relationships. However, this potential has not as yet been realized. The samples tested have been small, and the health outcomes examined have been physiological changes rather than clinical events. Confidence intervals for the benefits realized at different levels of physical activity have thus been as large or larger than those found in questionnaire studies. Future developments should permit the use of objective monitors on a much larger scale; a 100-fold increase in subject numbers seems likely to overcome current problems.

Roy J. Shephard (✉)
Faculty of Kinesiology & Physical Education, University of Toronto, Toronto, ON, Canada
e-mail: royjshep@shaw.ca

R.J. Shephard, C. Tudor-Locke (eds.), *The Objective Monitoring of Physical Activity: Contributions of Accelerometry to Epidemiology, Exercise Science and Rehabilitation*, Springer Series on Epidemiology and Public Health,
DOI 10.1007/978-3-319-29577-0_10

10.1 Introduction

The issue of the shape of the dose/response relationship that links habitual physical activity with various markers of clinical health has been mentioned briefly at several points throughout this text. It is a question that has considerable practical importance for those who are formulating public health recommendations about optimal daily levels of physical activity. Should the advocates of greater physical activity be content with the small gains that are fairly easily achieved, or should they insist that the public progress to quite vigorous effort? Unfortunately, the nature of the dose/response relationship is still far from clear. We will here note anthropological expectations based upon the lifestyle of our ancestors, evidence drawn from occupational classifications and the dilemma posed by conflicting reports based upon questionnaire, diary and physical fitness-based assessments of habitual physical activity. Finally, we will explore the current contribution of objective monitoring techniques, and will speculate how far the issue can be resolved by further developments in objective monitoring technology.

10.2 Anthropological Expectations

It can be argued that the ideal pattern of habitual physical activity can be sought in our anthropological history. Over the course of many centuries, it is reasonable to suppose that evolutionary forces adapted the human body to the activity patterns associated with a hunter-forager lifestyle, and that the lifespan of the current generation is likely to be maximized by following the activity patterns inherent to success in the neolithic economy [1].

A painstaking Kofranyi-Michaelis respirometer survey of the Canadian Inuit population as they engaged in various forms of hunting across the high arctic established that the intensity of effort demanded throughout much of the day was no more than moderate, but since the activity continued for long periods, the overall energy expenditures demanded by traditional forms of hunting averaged 10.6–18.6 MJ/day (Table 10.1), levels that would be considered as very heavy industrial work. Moreover, substantial bouts of resistance activity were interspersed with aerobic activity as animal carcasses were hauled and carried.

Such observations might suggest that human health would be maximized by a physical activity plan that involved a large volume of moderate intensity physical activity, supplemented by bouts of resistance activity.

Table 10.1 Field estimates of the daily energy expenditures involved in eight traditional forms of Inuit hunting, based on Kofranyi-Michaelis respirometer data and the timing of physical activity patterns [2]

Type of hunt	Estimated energy cost (MJ/day)
Summer fishing (kayak or umiak)	18.6
Ice-fishing	16.9
Summer overland caribou hunt	16.2
Winter caribou hunt (dog sled)	16.1
Walrus hunt (kayak or umiak)	15.5
Seal hunt (ice-holes)	14.5
Seal hunt (kayak)	14.4
Seal hunt (at floe edge)	10.6
Average (across all 8 hunts)	15.4

10.3 Evidence from Occupational

Three of the earliest investigators of the physical activity/health relationship were Ralph Paffenbarger, Jeremy Morris and Henry Taylor (Chap. 2). Their initial differentiation of individual daily energy expenditures was based upon the occupational categorization of their subjects, rather than the completion of physical activity questionnaires.

Morris first compared sedentary bus drivers with the physically more active conductors who worked on double-deck vehicles. They found that the respective annual coronary arterial disease incidence rates were 2.7/1000 for the drivers and 1.9/1000 for the conductors [3]. Even after allowance for self-selection of employment, this study pointed to the value of vigorous daily physical activity, but as there were only two job categories, it could not clarify the form of the relationship. Morris next examined British post-office employees. The lowest rates of cardiovascular disease were seen in postal carriers who cycled or walked to deliver the mail. Rates were higher in physically less active employees (counter-hands, postal supervisors, and higher grade postal staff) and were highest in totally sedentary workers (telephone operators, clerks and executives). Respective annual rates for the incidence of coronary arterial disease were 1.8/1000, 2.0/1000, and 2.4/1000, and annual case fatality rates were 0.6/1000, 0.9/1000 and 1.2/1000 in the three groups of workers. A gradient was thus demonstrated, but in the absence of precise data for energy expenditures in the three categories of employee, the form of this relationship could not be specified.

Henry Taylor classified railroad employees into three categories: clerks, switchmen and section workers [4]. The clerks clearly had sedentary employment, and the other two groups of workers were physically more active, but again interpretation of the data was limited because no precise estimates of occupational energy expenditures were available. Age-adjusted death rates for the three categories of worker were, respectively, 11.8/1000, 10.3/100 and 7.6/1000 for all-cause mortality, and 5.7/1000, 3.9/1000, and 2.8/1000 for deaths ascribed to arteriosclerotic heart disease.

Paffenbarger and his associates made a binary classification of male Californian longshore workers in terms of their presumed energy expenditure per shift. "Heavy" workers expended an estimated 4 MJ more than their peers, and this was associated with a lower coronary artery disease death rate (59 vs. 80 incidents per 10,000 person-years of work) [5]. A second analysis of data for the same population distinguished three categories of employee, with estimated average energy expenditures of 7.8, 6.1 and 3.6 MJ/shift. The respective age-adjusted coronary arterial disease death rates for the three groups were 26.9 vs. 46.3 vs. 49.0 deaths per 10,000 person-years of work, and sudden death rates were 5.6 vs. 19.9 vs. 15.7 deaths per 10,000 person-years) [6]. This study thus pointed to the health value of a high rather than a moderate or light volume of occupational energy expenditure; the expenditure of 7.8 MJ/shift would bring the dockworkers on a par with some of the levels of physical activity seen during Inuit hunting expeditions (Table 10.1).

Although some progress was being made in defining dose/response relationships from these occupational comparisons, there were necessarily problems in the initial self-selection of employment, and in changes of occupational categorization if an employee developed symptoms of cardiac disease such as angina. The precise energy cost of most tasks was also poorly defined, and no account was taken of inter-occupational differences in leisure activity. Further exploitation of this approach was hampered by automation, which progressively reduced the energy demands of what had previously been hard physical work.

10.4 Conflicting Conclusions from Physical Activity Questionnaire, Diary and Physical Fitness Data

Morris and Paffenbarger later turned their attention to patterns of leisure activity, using questionnaires and diaries of varying complexity (Chap. 2). As in their occupational research, they argued that there was little health benefit from participation in moderate physical activity, and that vigorous physical activity was needed to enhance cardiovascular health. In contrast, many subsequent observers concluded that the largest health benefit was realized from small increments of physical activity and physical fitness, at the lower end of the population distribution for these variables [7]. The issue is as yet unresolved.

10.4.1 Proponents of Vigorous Physical Activity

Morris and his associates [8] had 16,682 male executive-class civil servants recollect their leisure activities on the preceding Friday and Saturday. Observers then noted periods >5 minutes/day that the subjects had ascribed to active recreation, "keeping fit," or "vigorous getting about" (activities demanding near maximal effort), as well as bouts of heavy work lasting >15 minutes (gardening, building or moving heavy objects) and stair climbing, all regarded as demanding peak energy outputs >31 kJ/minute. The first 214 subjects who sustained a clinical attack of ischaemic heart disease were less likely to have recorded such vigorous activity than their peers. Vigorous physical activity also protected against fatal heart attacks, but no protection was seen from moderate activity. Those engaging in much vigorous exercise seemed to gain more benefit (risk ratio 0.18) than those doing some vigorous exercise (risk ratio 0.42–0.55), although it could be argued that participation in vigorous physical activity was serving as a marker of some other favorable personal characteristic.

A second study of male executive civil servants [9] found 9 % of employees reporting that they had often participated in vigorous sports, undertaken considerable amounts of cycling, and/or rated their walking pace as >6.4 km/hour over the preceding 4 weeks. This sub-group of employees experienced less than a half as many non-fatal and fatal heart attacks as their peers, over a 9-year follow-up. However, no protection was observed unless the sport or exercise was reported as vigorous.

Paffenbarger and his associates [10] questioned a large sample of male Harvard alumni on the number of city blocks walked, the number of stairs climbed, and the intensity and duration of any sport involvement during a typical week. Based on this information, they estimated gross weekly leisure energy expenditures. Setting the risk of cardiovascular disease in the least active group as 1.00, benefit was seen from walking (<5 vs. >15 km/week, relative risk 0.67), stair-climbing ((<20 vs. >55 flights/week, 0.75), sports (none vs. moderate sports play >4.5 METs, 0.63), and a large total energy expenditure (<2 to 8 vs. 8 to >14 MJ/week, 0.70). The data were interpreted as showing a need to spend more than 8 MJ/week in order to enhance cardiovascular prognosis, although risk ratios did not differ greatly between participants with estimated expenditures of 2–4 MJ/week (0.63) and 12–14 MJ/week (0.68). An 8-year follow-up of the same population examined health outcomes in relation to changes in physical activity. The baseline risk was set as that found in subjects who failed to meet the 8 MJ/week standard in either survey. There was an insignificant increase of cardiovascular risk if physical activity had diminished over the 8 years, but in those whose physical activity had increased, the cardiac risk diminished to a similar level to that seen in subjects who had remained active over the survey (risk ratios for moderate sports, 0.71 vs. 0.69, and for weekly energy expenditures >8 MJ, 0.74 vs. 0.79).

Thus, the studies of Morris pointed the need for periods of activity with an intensity >31 kJ/minute, and Paffenbarger's interpretation of his observations suggested the need to accumulate a gross energy expenditure >8 MJ/week in order to enhance cardiovascular prognosis.

10.4.2 Other Questionnaire Studies

Blair and colleagues [7] conducted an exhaustive review of published articles relating to physical activity and/or fitness and health outcomes (morbidity from coronary heart disease, stroke, cardiovascular disease and cancer and/or mortality from cardiovascular disease, cancer, or all causes); 67 papers published to August 2000 were given detailed review. Forty-nine articles based upon self-reports divided subjects into at least three activity categories, and nine studies categorized subjects in terms of physical fitness as assessed during a treadmill or cycle ergometer test.

In terms of the questionnaire data, Blair and his associates [7] concluded that with a few exceptions that included a study of mortality in very old men [11] and papers on the risks of breast [12] and testicular [13] cancer, most studies showed an inverse relationship between health risk and estimates of habitual physical activity. Nevertheless, because of the differing methods of measuring and categorizing physical activity and differing health outcomes, it was difficult to combine information from the various studies and specify an overall dose-response relationship. The apparent shape of the dose/response curve varied widely between studies, and was particularly inconsistent in women.

10.4.3 Critique of Questionnaire Data

The use of questionnaire data to determine dose/response relationships is limited by differing methodologies, differing health outcomes, and above all by the limited accuracy of questionnaire assessments of physical activity.

10.4.3.1 Differing Methodologies

Inter-study differences in the methods of measuring and categorizing levels of physical activity make it impossible to combine data from the various published sources in order to gain an overall impression of dose/response relationships. Some investigators have estimated gross energy expenditures per week, dividing their sample of subjects into three to six categories of habitual physical activity, albeit with differing cut-points between categories. Others have used questionnaire responses to separate subjects into three or four arbitrary physical activity categories, with different authors using differing methods of classification. Some have considered both leisure and occupational activities, and some investigators have focussed their attention upon the reported frequency rather than the intensity of physical activity.

10.4.3.2 Differing Health Outcomes

Many observers have focussed upon all-cause and/or cardiovascular mortality, but others have examined such outcomes as cancer and stroke. It seems unlikely that the shape of the dose/response curve will be identical for differing health outcomes, although there is at present little information on this question. For some outcomes such as rectal cancer, there is little relationship between habitual physical activity and disease incidence or mortality. In the case of fat accumulation and the metabolic syndrome, one might anticipate a relatively linear dose-response relationship, with fat loss proportional to any increase in the level of daily energy expenditure. In the case of anxiety and depression, a very modest increase of physical activity might be sufficient to cause arousal and thus an enhancement of mood state, but in order to counter sudden cardiac death it might be necessary to condition the body to the high levels of energy expenditure sometimes required in an emergency.

10.4.4 Lack of Absolute Accuracy of Questionnaires

The most important factor limiting the interpretation of questionnaire data is probably the limited absolute accuracy of the estimates of habitual physical activity. Although it is generally accepted that questionnaires can make a useful three or four level categorization of physical activity patterns, it is also widely acknowledged that the absolute estimates of energy expenditure may have a three-fold error relative to "gold standard" measurements. Moreover, the magnitude of these gross errors varies between individuals, and it is likely to differ systematically between those reporting high and low levels of physical activity. Partly as a consequence of such errors, and partly as a consequence of the limited number of cardiac incidents when quite a large sample of subjects are followed for 5–10 years, the confidence limits to the estimated relative risks associated with various levels of physical activity are very broad, and this makes it difficult to determine the shape of the dose/response relationship. This is well illustrated by the data of Haapanen and colleagues [14, 15], who examined the risks of coronary heart disease over a 10-year follow-up of 1340 men and 1500 women aged 35–63 years (Table 10.2).

10.4.5 Aerobic Fitness Dose/Response Data

Blair et al. [7] identified nine studies that related simple indices of aerobic fitness (treadmill endurance times or heart rates at specific exercise intensities) to health

Table 10.2 Relative risk of coronary heart disease over 10-year follow-up of middle-aged adults, showing broad confidence intervals [15]

Activity measure	Relative risk	Confidence intervals
Men		
Total energy expenditure		
High	1.00	
Moderate	1.33	0.78–2.27
Low	1.98	1.22–3.23
Intensity		
Vigorous >1/week	1.00	
Vigorous <1/week	1.42	0.92–2.17
Women		
Total energy expenditure		
High	1.00	
Moderate	0.73	0.38–1.39
Low	1.25	0.72–2.15
Intensity		
Vigorous >1/week	1.00	
Vigorous <1/week	1.13	0.62–2.07

outcomes; five of the nine reports were drawn from the Aerobics Center in Dallas, TX, where he was then working. All of these studies showed cardiovascular risk and/or all-cause mortality decreasing with increasing aerobic fitness, commonly as much as three to four fold, [16]. The overall gradient was steeper for fitness than for questionnaire data, probably reflecting the greater accuracy in classification of the subjects when using aerobic fitness data.

Most analyses divided subjects into fitness tertiles, quartiles or quintiles. The trend seemed to demonstrate the largest health benefit on moving from the least fit to the next fitness quartile, but despite very large sample sizes, the confidence limits to estimates of risk ratios and thus the shape of the dose/response relationship were again very broad, as exemplified by the studies of Ekelund et al. on 3106 men [16] and Blair et al. on 10,224 men and 3120 women [17] (Table 10.3).

10.5 New Dose/Response Information from Objective Monitoring

Objective monitoring in principle provides a much finer gradation of physical activity patterns than occupational classifications, physical activity questionnaires or even simple measures of physical fitness. Does this imply the likelihood of obtaining further detail on the shape of dose/response relationships between physical activity and health outcomes? Critical issues are sample size, range of physical activity within the chosen sample, ability to measure the most relevant components of overall activity, the nature of the outcome measures, and overall confidence levels.

Table 10.3 Dose/response relationships between estimates of aerobic fitness quartile and cardiovascular [16] or all-cause mortality [17]

Fitness measure	Risk ratio	Confidence limits
Heart rate in submaximal treadmill test (all men)		
Q1 (least fit)	2.21	1.65–3.25
Q2	1.56	0.68–2.44
Q3	1.30	0.49–2.11
Q4 (most fit)	0.26	0.00–0.62
Treadmill endurance time		
Men		
Q1 (least fit)	3.44	2.05–5.77
Q2	1.37	0.76–2.50
Q3	1.46	0.81–2.63
Q4	1.17	0.63–2.17
Q5 (most fit)	1.00	
Women		
Q1 (least fit)	4.65	2.22–9.75
Q2	2.42	1.09–5.37
Q3	1.43	0.60–3.44
Q4	0.76	0.27–2.11
Q5 (most fit)	1.00	

10.5.1 Sample Size and Range of Physical Activity

Because of the tedious work involved in checking pedometer and accelerometer records for the presence of various types of artifact, the number of subjects recruited for objective studies of physical activity and health has to date been relatively small, particularly when compared with the occupational questionnaire and fitness-based surveys noted above. Even with a 5–10 year period of observation, the number of cardiac or all-cause deaths has thus been too few to permit the calculation of accurate statistics, and it has been necessary to consider the physical activity data in relation to surrogate measures of health such as arterial stiffness or the ultrasonic index of bones.

A related issue, applicable to all methods of classifying physical activity, is to ensure that there are an adequate number of active individuals within the sample. This is becoming progressively more difficult in sedentary urban populations. In a study of 57 eighty-year-old women, Gerdhem and associates [18] found no significant correlations between accelerometry measurements of physical activity and balance, muscle strength or bone density. However, there was little chance that a significant relationship would be seen, since only 8 of the 57 subjects were engaging in moderate or vigorous physical activity. This negative conclusion was quickly reversed by recruiting a larger population of 152 men and 206 women aged 50–80 who were followed for 10 years; in the second study, annual bone loss was 0.6 % less in those who were classified as active relative to those who were inactive [19]. The larger group also showed benefits in terms of balance, although habitual physical activity still showed no relationship with muscle strength or gait velocity.

The limitations of a small sample size are likely to be overcome in the next few years, as effective computer programmes are developed to upload and scan accelerometer records.

10.5.2 Measured Components of Physical Activity

Although objective monitors give a finely-graded report on an individual's physical activity, they under-estimate some activities such as cycling and swimming, resistance exercise and the work performed in climbing a hill. These may be some of the more vigorous activities in a person's weekly repertoire, and it is thus necessary to assume that either (1) they represent only a small fraction of then total daily energy expenditure, or (2) that their volume is closely correlated with the volume of activities that are measured.

In an elderly person, a large fraction of the significant daily activity takes the form of deliberate level walking, but this assumption is much less certain in an adolescent or a young adult. The development of multiphasic monitors may in future overcome this limitation, but for the present it may be best to restrict objective studies of the physical activity/health dose/response relationship to older individuals where one can be certain that the instrumentation has recorded most of the significant activity that has been undertaken.

10.5.3 Choice of Outcome Measures

To date, objective studies of health outcomes have tended to use surrogate measures such as arterial stiffness, because a statistically adequate number of clinical incidents such as cardiac or all-cause death were unlikely to be seen over the course of a 5–10-year follow-up. However, dose/response relationships may differ for the surrogate and for the clinical outcome of interest.

This is particularly likely when arterial pulse wave velocities are substituted for a heart attack. Changes in the arterial wall are in the first instance due to an accumulation of lipid that can potentially be modified by a steady daily energy use. Lipid accumulation in the coronary arteries certainly augments the risk of a heart attack, but a second factor in precipitating an attack during exercise is the intensity of effort relative to the individual's peak aerobic power, and this is likely to be enhanced mainly by engaging in vigorous physical activity [20], as originally advocated by Morris and Paffenberger.

The problem of using inappropriate surrogates of health outcome should again be overcome as technical developments allow objective studies based upon a much larger pool of subjects.

Fig. 10.1 Fumiahara Togo has used objective monitors to study dose/response relationships for a number of health outcomes, including bone health

10.5.4 Empirical Data

For many beneficial health outcomes, including greater bone health as inferred from an ultrasonic index [21] (Fig. 10.1), improved cardiovascular health as indicated by lesser arterial stiffness [22], and a lesser risk of sarcopaenia as shown by a greater lean tissue mass [23], objective monitoring seems to confirm the impression gained from many questionnaire studies in showing that the biggest improvement of health status occurs on moving from a completely sedentary status to a modest level of physical activity. However, this is not true of all health outcomes; in particular, the risk of showing manifestations of the metabolic syndrome [24] decreases across each of the four quartiles of habitual physical activity.

10.5.5 Confidence Limits of Estimates of Health Benefit

One practical assessment of the current value of objective tests is the magnitude of the confidence limits within which dose/response relationships can be established. At present, with studies that are based upon samples of 160–200 subjects, the published confidence limits of the estimated odds ratios are as broad and in some instances broader than those achieved with questionnaire research (Table 10.4). A part of the answer to improving this situation would be to increase subject numbers—if this could be boosted by two orders, to a level seen in some of the major questionnaire surveys, the confidence limits would drop to a tenth of their current level, and this would give a much clearer indication of the shape of dose/relationships. Emerging technology for the loading and interpretation of data from electronic monitors should soon permit objective studies of this magnitude. This may also permit analyses to be based on clinical events such as myocardial infarction and cardiac death, rather than their surrogates. Finally, greater standardization of measuring techniques should allow the pooling of information from several large scale studies.

Table 10.4 Confidence limits in objective assessments of multifactor-adjusted dose/response relationships, arranged by quartiles

	Odds ratio	Confidence limits	Odds ratio	Confidence limits
Activity level	Steps/day		Activity >3 METs (minute/day)	
Metabolic syndrome [24]				
Q1	4.55	1.51–11.41	3.67	1.50–8.97
Q2	3.10	1.22–7.88	2.29	0.92–5.71
Q3	2.61	1.02–6.75	2.10	0.83–5.27
Q4 (most active)	1.00		1.00	
Bone health [21]				
Q1	1.20	0.23–3.96	2.23	1.36–9.47
Q2	1.31	0.48–5.59	1.10	0.20–6.33
Q3	0.82	0.19–2.93	0.53	0.33–3.75
Q4 (most active)	1.00			
Sarcopaenia [23]				
Men				
Q1	2.00	1.01–5.03	3.39	1.21–7.10
Q2	1.20	0.20–3.22	2.03	1.00–4.31
Q3	0.79	0.19–1.96	1.05	0.28–3.14
Q4 (most active)	1.00		1.00	
Women				
Q1	2.66	1.21–4.99	4.55	1.12–7.12
Q2	1.57	0.96–4.04	3.15	1.02–4.91
Q3	1.02	0.31–2.36	1.23	0.29–3.25
Q4 (most active)	1.00		1.00	
Pulse wave velocity [22][a]				
Q1	1.176	1.093–1.259	1.121	1.061–1.181
Q2	1.068	1.002–1.134	1.080	1.002–1.162
Q3	1.005	0.950–1.050	1.008	0.955–1.062
Q4	1.000		1.000	

[a]2 SD range approximated from graph

10.6 Conclusions

Although objective monitoring devices have the potential to analyze physical activity/disease relationships more precisely than other approaches that have been used by epidemiologists, their potential has not yet been realized, in part because sample sizes have been relatively small, and in part because surrogate markers of health have been substituted for clinical events. Developments in equipment and methods of data analysis may soon permit much larger scale objective studies, overcoming some of these difficulties.

References

1. Shephard RJ. An illustrated history of health and fitness: from prehistory to our postmodern world. New York, NY: Springer; 2015.
2. Godin G, Shephard RJ. Activity patterns of the Canadian eskimo. In: Edholm O, Gunderson G, editors. Human polar biology. Cambridge: Heinemann; 1973.
3. Morris JN, Heady J, Raffle PA, et al. Coronary heart disease and physical activity of work. Lancet. 1953;262:1053–7. 1111–20.
4. Taylor HL, Klepetar E, Keys A, et al. Death rates among physically active and sedentary employees of the railroad industry. Am J Publ Health. 1962;52:1697–707.
5. Paffenbarger RS, Laughlin ME, Gima AS, et al. Work activity of longshoremen as related to death from coronary heart disease and stroke. N Engl J Med. 1970;20:1109–14.
6. Paffenbarger R, Hale WE. Work activity and coronary heart mortality. N Engl J Med. 1975;292:545–50.
7. Blair SN, Cheng Y, Holder JS. Is physical activity or physical fitness more important in defining health benefits? Med Sci Sports Exerc. 2001;33 Suppl 6:S379–99.
8. Morris JN, Chave SP, Adam C, et al. Vigorous exercise in leisure time and the incidence of coronary heart disease. Lancet. 1973;1(7799):333–9.
9. Morris JN, Clayton DG, Everitt MG, et al. Exercise in leisure time: coronary attack and death rates. Br Heart J. 1990;63:325–34.
10. Paffenbarger R, Hyde RT, Wing AL, et al. Some interrelations of physical activity, physiological fitness, health and longevity. In: Bouchard C, Shephard RJ, Stephens T, editors. Physical activity, fitness and health. Champaign, IL: Human Kinetics; 1994. p. 119–33.
11. Linsted KD, Tonstad S, Kuzma JW. Self-report of physical activity and patterns of mortality in Seventh-day Adventist men. J Clin Epidemiol. 1991;44:355–64.
12. Dorgan JF, Brown C, Barrett M. Physical activity and risk of breast cancer in the Framingham heart study. Am J Epidemiol. 1994;139:662–9.
13. Srivastava A, Kreiger N. Relation of physical activity to risk of testicular cancer. Am J Epidemiol. 2000;151:78–87.
14. Haapanen N, Miilunpalo SI, Vuori I, et al. The characteristics of leisure time physical activity associated with decreased risk of premature all-cause and CVD-mortality in middle-aged men. Am J Epidemiol. 1996;143:870–80.
15. Haapanen N, Miilunpalo S, Vuori I, et al. Association of leisure time physical activity with the risk of coronary heart disease, hypertension and diabetes in middle aged men and women. Int J Epidemiol. 1998;26:739–47.
16. Ekelund LG, Haskell WL, Johnson JL, et al. Physical fitness as a predictor of cardiovascular mortality in asymptomatic North American men: the Lipid Research Clinics follow-up study. N Engl J Med. 1988;319:1379–84.
17. Blair SN, Kohl HW, Paffenbarger RS, et al. Physical fitness and all-cause mortality: a prospective study of healthy men and women. JAMA. 1989;262:2395–401.
18. Gerdhem P, Dencker M, Ringsberg K, et al. Accelerometer measured daily physical activity among octogenarians: results and associations to other indices of physical performance and bone density. Eur J Appl Physiol. 2008;102:173–80.
19. Daly RM, Ahlborg HG, Ringsberg K, et al. Association between changes in habitual physical activity and changes in bone density, muscle strength, and functional performance in elderly men and women. J Am Geriatr Soc. 2008;56(12):2252–60.
20. Shephard RJ. Intensity, duration and frequency of exercise as determinants of the response to a training regime. Int Z Angew Physiol. 1968;26:272–8.
21. Park H, Togo F, Watanabe E, et al. Relationship of bone health to yearlong physical activity in older Japanese adults: cross-sectional data from the Nakanojo Study. Osteoporosis Int. 2007;18:285–93.
22. Aoyagi Y, Park H, Kakiyama T, et al. Year-long physical activity and arterial stiffness in older adults: the Nakanojo Study. Eur J Appl Physiol. 2010;109(3):455–64.

23. Park H, Park S, Shephard RJ, et al. Year-long physical activity and sarcopenia in older adults: the Nakanojo Study. Eur J Appl Physiol. 2010;109(5):953–61.
24. Park S, Park H, Togo F, et al. Yearlong physical activity and metabolic syndrome in older Japanese adults: cross-sectional data from the Nakanojo Study. J Gerontol (Biol Sci Med Sci). 2008;63:1119–23.

Chapter 11
The Economic Benefits of Increased Physical Activity as Seen Through an Objective Lens

Roy J. Shephard

Abstract Over a lifetime, the annual per capita direct and indirect costs associated with cardiovascular disease amount to around 1 year's salary for the average wage-earner. Additional costs arise from other forms of chronic disease where physical activity could potentially play a preventive role. Questionnaires and other methods of categorizing an individual's habitual physical activity suggest that per capita health-care costs may be half as great in those members of the community who are classed as physically active, and some authors have thus argued that overall costs could be halved if everyone were to become active. Objective monitoring is now offering the potential for a more precise gradation of the costs attributable to individual diseases in relation to levels of habitual physical activity. Rather than assuming a generic beneficial effect of "activity," it has become possible to quantitate the magnitude of the economic benefits likely from the small increases of activity that can be achieved in sedentary populations with respect to each of a range of chronic diseases. The application of objective monitoring demonstrates that the greatest economic benefit is likely from changing the behaviour of the most sedentary individuals in a given population. It also identifies specific clinical conditions where an increase of habitual physical activity should yield large financial dividends. If applied on a large scale, objective monitoring offers the potential for prospective monitoring of the effects of defined increases in exercise behaviour upon immediate charges to the health care system, as well as an objective assessment of the costs of motivating defined changes in physical activity patterns. To date, objective monitors have only been applied to two economic analyses in elderly people (where the primary activity of walking is readily monitored, but also where a substantial fraction of population health care costs are incurred). Extension of these analyses to younger adults is desirable, but it will require the development of a second generation of objective monitors that can respond accurately to the full range of sports and pastimes pursued by the younger generation.

Roy J. Shephard (✉)
Faculty of Kinesiology & Physical Education, University of Toronto, Toronto, ON, Canada
e-mail: royjshep@shaw.ca

R.J. Shephard, C. Tudor-Locke (eds.), *The Objective Monitoring of Physical Activity: Contributions of Accelerometry to Epidemiology, Exercise Science and Rehabilitation*, Springer Series on Epidemiology and Public Health,
DOI 10.1007/978-3-319-29577-0_11

11.1 Introduction

Health economics is a relatively recent discipline within the broad field of economics. It developed following World War II, as many governments around the world began to provide tax-payer supported systems for the delivery of health services to their populations [1], and it has grown substantially in complexity and sophistication over the past 20 years [2]. A review by the World Bank noted that over 33,000 papers in Health Economics had been published between 1969 and 2001 [3]. Moreover, material on this topic is still appearing. For example, one reviewer found that 34 articles estimating the economic costs of cardiovascular disease in developed countries had been published between 1998 and 2006 [4].

The main concerns of the health economist are the overall cost of state-sponsored health services, commonly expressed as a percentage of the country's gross national product, and the efficiency and effectiveness of such services in enhancing population health and improving the overall quality of life. An article by Kenneth Arrow [5] noted a number of distinctive features of health economics, including extensive government involvement in both the provision of funding and the financing of individual services, many intangible components of enhanced health, an asymmetry of knowledge between the physician-provider and the patient-consumer, and the ability of the physician-provider to determine the extent of the services that are deemed necessary, while usually being insulated from the costs of his or her decisions [6].

Although some health economists have focussed upon the overall burden of ill-health, others have examined the costs associated with individual diseases and conditions, particularly the chronic conditions that have become widely prevalent since World War II. The earliest of the modern epidemics was cardiovascular disease (Chap. 2), and disease-specific analyses quickly looked at the fiscal consequences of this epidemic. Subsequently, attention turned to other recent epidemics, for example, obesity, diabetes mellitus, and cancer. Organizations seeking support for research and treatment of each of these conditions were anxious to impress governments by maximizing the estimated costs associated with "their" disease. Often, there was an exaggeration and/or an attribution of the same costs to several conditions (for example, obesity, diabetes mellitus and cardiovascular disease). Such exaggeration is particularly prevalent in analyses based upon risk factors, since a given risk factor may increase a person's risk of several different chronic conditions [7]. Exercise scientists were also eager to impress governments by calculating the fiscal savings that would result if an increase in population activity was effective in reducing the risk of these various illnesses, although as noted in the final section of this chapter there are as yet many limitations to such predictions.

We will consider firstly how the costs of disease are calculated, taking diseases of the cardiovascular system as a working example. We will next look at the costs associated with other major diseases, particularly those whose prevalence is likely to be reduced by an increase of habitual physical activity. We will then consider how questionnaire data and other epidemiological findings have been used to estimate the likely economic benefits of increased physical activity, and finally we will discuss potential new approaches offered by the objective monitoring of habitual physical activity, noting the limitations both of currently available objective activity monitors and of overall cost-benefit analyses for an increase of physical activity.

11.2 Economic Costs Associated with Cardiovascular Disease

The economic costs associated with any disease depend upon its prevalence in the community, and the direct and indirect costs incurred by each individual who develops that particular disease at each point in his or her life course. The calculations contain many uncertainties, and estimates of potential benefits differ widely between investigators.

11.2.1 Disease Prevalence

It is often difficult to ascertain the prevalence of a disease in a community. A proportion of those who are affected by the disease may be unaware that they are unwell; if they do sense ill-health, they may be misinformed as to its cause, or (for various reasons) they may fail to report it. One investigator estimated that of 1000 participants in a household survey, 150 would have some form of heart disease, but only 60 of these individuals would report their problem [8]. Estimates of the occurrence rate would be substantially larger if a one-visit medical examination was applied to the chosen population, but unless extensive tests were undertaken at the medical visit and care was taken to include the elderly and institutionalized patients in the survey, the prevalence of disease was still likely to be underestimated.

Mortality statistics might seem a more accurate source of information, although in the case of coronary heart disease, prevalence of the condition may be overestimated by mortality data; by the time of death, almost everyone has some coronary atheroma, and in the absence of an obvious alternative cause of decease, death is often attributed to a heart attack. Moreover, mortality data necessarily cannot assess the extent of the morbidity from coronary vascular disease over the earlier part of the individual's life course.

11.2.2 Direct Costs

The direct costs of an illness include all expenditures on personal services and supplies. Items contributing to the total are the imputed costs of intensive-, acute- and chronic-care beds in hospitals and institutions, the services of physicians, nurses, physiotherapists and other paramedical professionals, payments for diagnostic procedures and medications, and a variety of non-personal items such as research, the training of health care professionals, the construction of medical facilities, and the administration of health insurance programmes.

A U.S. analysis completed in 1962 set the direct costs of cardiovascular disease at $3.07 billion, with $2.58 billion attributed to personal services (particularly hospital care), and $0.49 billion attributed to non-personal items [9]. Such estimates inevitably vary with regional and secular differences in the pattern of treatment provided, particularly the length of hospital stay associated with any given cardiac episode and the extent to which out-patient follow-up is provided.

In Klarman's analysis [9], direct costs accounted for 10 % of the total expenditures upon cardiovascular disease. Other more recent analyses from Canada, the U.S. and Europe have suggested that direct counts are a larger fraction of the total expenditure, accounting for 37–71 % of the total outlay [4]; this is in part because more recent analyses have not attempted to assess the third-party component of indirect costs.

11.2.3 Indirect Costs

The indirect costs of physical inactivity include loss of production from illness and premature death, intangible items such as the need to care for a sick relative, and the impact of pain, suffering and orphanhood upon family members.

The loss of production due to illness is perhaps the least controversial of the indirect items. For example, the average work-time lost from a cardiac episode is generally known, and this can then multiplied by the average sex-specific personal income, adjusted downwards to allow for the less than complete participation of the working age population in the national labour force. Nevertheless, the average income is less than fully representative; earnings vary over the life course, and many episodes of cardiac disease develop after retirement from the labour force.

Estimates of the economic loss from premature death must postulate an age of retirement; once between 60 and 65 years, is progressively shifting to an older age as life expectancy increases. Calculations must also make assumptions about future changes in per-capita productivity, and must then apply a by no means fixed discount rate to the value of these future earnings. Most estimates of the effects of premature death have also neglected to consider the goods and services that a person would have consumed if he or she had not died prematurely. Applying life insurance data to citizens dying of cardiovascular disease in middle age, Klarman estimated the economic impact of premature death at $19.4 billion/year, and the additional losses from non-fatal cardiac episodes at a further $3.0 billion/year [9].

The estimated allowance for intangible items is particularly controversial. In 1964, Klarman estimated that cardiovascular disease was likely to make 3 % of U.S. children orphans before they reached the age of 18 years, and he set the cost of intangible losses from cardiovascular disease in the U.S. as high as $5.2 billion/year [9]. The influence of sorrow, anxiety, and orphanhood upon productivity is certainly greater than zero, but because of uncertainty as to its magnitude, most investigators have omitted this item from their calculations.

11.2.4 Total Costs of Cardiovascular Disease

Klarman set the total cost of cardiovascular disease to the U.S. economy at $30.7 billion/year, or with a 1964 population of 192 million, about $160 per capita, and with a 58.5 % labour force participation rate in that era [10], $274 per worker. Over a career of 50 years, cardiovascular disease was thus robbing each employee of $13,700, measured in 1962 U.S. dollars.

John Armstrong made similar calculations for Canada in 1965 [11]. Probably because at that time Canada had a somewhat younger population than the U.S., he estimated the prevalence of cardiovascular disease at 2.5 million cases in a population of 20 million. The cost of personal services was set at $120 million, or $6 per head, substantially less than the $13.4 per capita figure in the U.S.; this reflected lower hospital and medical costs in Canada, and possibly the choice of less expensive forms of treatment. The estimated per-capita costs from losses of output, at $80 million, were also about a quarter of the U.S. figure. Armstrong did not attempt to calculate charges due to the use of non-personal services, premature death or intangible factors, but setting these at 60 % of U.S. estimates (to allow for the younger Canadian population and differences in earning power), the annual cost to the Canadian economy would have amounted to around $1.7 billion, $85 per capita, or $170 per wage-earner. The differences between the two sets of calculations underlines the point that although the health economist can estimate the general order of costs associated with chronic disease, the precise figure is open to considerable interpretation.

11.2.5 Factors Limiting Comparisons Between Investigators

Comparisons of disease costs between investigators are limited by differing data sources (population surveys, Medicaid data, the experience of health maintenance organizations and employer health plans), and assumptions regarding intangible items, the discount rate, the value of future earnings, and gains in future productivity. Further, adjustments for differences in the year of the analysis and the effects of inflation are commonly treated using the overall inflation rate for a given country; however, the medical inflation rate is generally much steeper than the overall inflation rate, in part because recent years have seen the introduction of expensive tests and forms of treatment.

11.2.6 *Specific Estimates for Cardiovascular Disease*

Recent U.S. data for the economic costs attributable to all forms of cardiovascular disease (ischaemic heart disease, chronic heart failure, stroke, hypertension, and peripheral vascular disease) have shown greater consistency, perhaps because investigators have worked from similar base statistics. Costs are still substantially greater in the U.S. than in Canada and the European Union (Table 11.1), as might be expected from differences in the proportion of the gross national product committed to health care (figures for 2012, were 17.9 % in the U.S., 10.9 % in Canada, and about 10 % in the European Union). Notice that the annual per capita costs, when

Table 11.1 Published estimates of the costs of cardiovascular disease, expressed in 2004 Canadian dollars/capita

Country	Data source	Year of analysis	Direct annual costs, per capita	Total annual costs, per capita
United States	Population surveys [9]	1962	$129	$1281
Canada	Population data [11]	1965		$505[b]
United States	Medicaid data base [12]	1995	$866	No estimate
United States	Population surveys [13]	1996	$245	No estimate
Canada	Health Canada [14]	1998	$234	$632
United States (women)	Employer data base [15]	2000	$3966	$5586
United States (women)	Health claims in excess of control subjects [16]	2002	$1673[a]	No estimate
European Union	Cost of illness [17]	2003	$221	$356
United States	Am. Heart Assoc. estimate [18]	2006	$1071	$1673
United States	Population data [19]	2006	$1057	$1654
United States	Medical expenditure panel survey, 2001–2005 [20]	2010	$1019	$1661

[a]Lifetime cost, divided by U.S. female life-expectancy in 2002 (79.9 years)
[b]Author's estimate adjusted as described in text

accumulated over a working career of 40 or 50 years, remain more than the average person's annual salary.

11.3 Specific Costs Associated with Other Chronic Diseases

Health economists have estimated the direct and indirect costs of a variety of other chronic conditions, using a similar approach to that adopted for cardiovascular disease. Figures for Canada in the year 2000 are summarized in Table 11.2. The distribution of costs between the various diseases is controversial; in this analysis, the costs of diabetes mellitus are ranked as relatively small, and no specific cost is attributed to obesity. The costs for cardiovascular disease are somewhat greater than in other analyses of Canadian data; thus the per capita figure of $791 2004 dollars for the year 2002 (Table 11.1) may be compared with the $632 of Health Canada [14] and the $601 derived from the 1998 EBIC study [21]. An independent analysis conducted by the Alberta Institute of Health Economics [22] set the 2004 Canadian per-capita costs for six of the main health conditions at: neuropsychiatric disorders $1078, cardiovascular disease $734, musculoskeletal disorders, $734, malignant neoplasms, $593, diabetes $344, and respiratory diseases $339, a total of $3822 per capita; the corresponding entries in Table 11.2 totaled $3161, with diabetes and respiratory diseases being rated as much less costly in the Federal Government report than in the Alberta analysis. The Alberta study attributed an annual per capita cost of $337 2004 dollars to physical inactivity, and $385 to obesity, without discussing possible overlap between the two totals.

In the U.S., the Milliken Foundation [23] set the combined costs of five important chronic health conditions (cardiovascular disease, mental disorders, respiratory disease, cancers and diabetes) at $369 billion for treatment, and $1396 billion due to losses of productivity, a total cost of $1765 billion 2004 Canadian dollars, or $6086 dollars per capita, more than twice the estimated $2363 per capita cost of these five conditions in Canada, Table 11.1). The total per capita costs attributed to four major conditions in the U.S. were: heart disease and stroke $561 billion/year [19], diabetes $226 billion/year [24], lung diseases $200 billion/year [25] and Alzheimer's Disease $192 billion/year [26], or more than $3800/year, all of these figures being expressed in Canadian 2004 dollars.

Plainly, overall costs differ from one country to another, and estimates of the relative economic importance of individual conditions also show a wide variation from one study to another, depending on the precise grouping of conditions. Interestingly, there appear to be inflation-adjusted changes in costs with time; thus, the total annual Canadian costs due to cardiovascular disease were estimated at $23.4 billion in 1986, $21.0 billion in 1993, and $18.5 billion in 1998, despite a 16 % growth in population over this interval [21], reflecting a substantial decrease in the prevalence of cardiac disease over this period (Chap. 2). Nevertheless, the overall annual per capita costs of ill health seem in the range $3000–6000 2004 Canadian dollars for both Canada and the U.S., and many of the illnesses

Table 11.2 Estimated direct and indirect costs of 20 diagnostic categories for Canada in the year 2000 (as approximated from published graph)

Condition	Direct cost (billion)	Indirect cost (billion)	Total cost (billion)	Annual per capita cost[b]
Musculo-skeletal diseases	$3.9	$18.4	$22.3	$798
Cardiovascular diseases	$7.4	$14.7	$22.1	$791
Neuro-psychiatric conditions	$9.5	$11.1	$20.6	$737
Malignant neoplasms	$2.4	$14.9	$17.3	$619
Injuries	$4.3	$10.3	$14.6	$522
Digestive diseases	$3.9	$3.0	$6.9	$269
Respiratory diseases	$2.9	$3.2	$6.1	$215
Sense organ diseases	$1.3	$2.9	$4.2	$150
Respiratory infections	$2.1	$1.6	$3.7	$132
Genito-urinary diseases	$2.6	$0.7	$3.3	$118
Infectious and parasitic diseases	$1.1	$1.6	$2.7	$96
Diabetes mellitus	$0.6	$1.8	$2.4	$85
Endocrine disorders	$1.2	$0.7	$1.9	$68
Perinatal conditions	$0.8	$1.2	$2.0	$71
Maternal conditions	$1.5	$0.1	$1.6	$57
Skin diseases	$1.1	$0.2	$1.3	$46
Congenital anomalies	$1.1	$0	$1.1	$39
Other neoplasms	$0.2	$0.4	$0.6	$22
Nutritional deficiencies	$0.1	$0.1	$0.2	$8
Oral conditions	$0.05	$0.05	$0.1	$3
Other	$6.1	$2.1	$8.2	$293
Total	$54.2	$89.1	$143.3[a]	$5128

Source: http://www.phac-aspc.gc.ca/cd-mc/cvd-mcv/cvd_ebic-mcv_femc-eng.php; accessed 13 October, 2014). Short-term disability costs are included for six conditions (cardiovascular diseases, musculo-skeletal diseases, neuro-psychiatric conditions, digestive diseases, respiratory diseases, and respiratory infections

[a]Because of measurement and rounding errors, the total cost is slightly less than the figure of $147.9 billion reported by the Public Health Agency of Canada

[b]Costs are adjusted to 2004 Canadian dollars to allow comparison with Table 11.1

contributing to this total could potentially be addressed by preventive measures, including an increase in habitual physical activity.

11.4 Potential Benefits of Habitual Physical Activity

Earliest analyses of the potential financial benefits of physical activity were directed specifically at cardiovascular disease. It was noted that a review of coronary vascular disease epidemiology, based upon occupational studies and questionnaire data) indicated that mortality was about 50 % lower in active than in sedentary individuals [27]. Thus, it was suggested that the costs of cardiovascular disease could potentially be lowered by as much as 50 % through an increase in the habitual physical activity of the general population [28]. It was less clear that physical activity offered an equal measure of protection against hypertension and strokes, but it certainly was associated with some decrease in the risk of these conditions. Thus, it was thought reasonable to set the overall dividend from an active lifestyle at a 26–50 % reduction in the costs associated with cardiovascular disease.

Katzmarzyk (Fig. 11.1) and Janssen extended the earlier analysis of Katzmarzyk et al. [29], making a systematic analysis of differences in the relative risks of several chronic conditions between active and inactive, and obese versus normal weight people (Tables 11.3 and 11.4).

An exhaustive meta-analysis of the literature [30] examined all publications that provided point estimates of the relative risk of physical inactivity and its 95 % confidence limits. A 2-way classification of activity status was generally made, using questionnaire data on leisure activity (although occasionally low vs. high tertiles, quartiles or even quintiles were compared), and in the case of cancer, information from occupational activity questionnaires was sometimes included. The relative risk of inactivity ranged from 1.30 for hypertension to 1.60 for stroke. The population attributable risk (PAR%) was calculated by multiplying the relative risk for a given condition by the percentage of the population affected by that condition; the PAR% ranged from 13.8 % for hypertension to 24.1 % for stroke. The direct costs of various chronic conditions influenced by habitual physical activity

Fig. 11.1 Peter Katzmarzyk has made several analyses of the health-care costs of physical inactivity in Canada

Table 11.3 Relative risk of developing chronic conditions in inactive segment of population, relative to active segment (based on data from a 2001 meta-analysis completed by Katzmarzyk and Janssen [30] and findings of a repeat analysis in 2009)

Condition	Number of comparisons	Relative risk in inactive individuals	Range of risk	Population attributable risk (%)[a]	
				2001	2009
Coronary artery disease	26	1.45	1.38–1.54	19.4	18.2
Stroke	17	1.60	1.42–1.80	24.1	22.8
Hypertension	6	1.30	1.16–1.46	13.8	12.9
Colon cancer	16	1.41	1.31–1.53	18.0	16.8
Breast cancer	17	1.31	1.23–1.38	14.2	13.3
Type 2 diabetes mellitus	11	1.50	1.37–1.63	21.1	19.8
Osteoporosis	13	1.59	1.40–1.80	24.0	22.5

[a]The PAR% reflects the relative risk of a given condition due to physical inactivity, multiplied by the prevalence of the risk factor in the community

Table 11.4 Estimates of direct costs of physical inactivity in Canada in 1999, and estimated direct costs for Ontario, allowing for differences in population and converted to 2009 dollars (based on the data of Katzmarzyk et al., [29, 30])

Condition	Direct costs (billion)	PAR% of direct costs (billion)	Estimated direct costs for Ontario, in 2009 dollars (billion)
Coronary disease	$2.49	$0.891	$0.490
Stroke	$1.73	$0.345	$0.190
Hypertension	$1.58	$0.314	$0.173
Colon cancer	$0.332	$0.066	$0.036
Breast cancer	$0.282	$0.031	$0.017
Type 2 diabetes mellitus	$0.620	$0.123	$0.068
Osteoporosis	$1.30	$0.352	$0.194
Total	$8.33	$2.12	$1.17

were derived from reports on the economic burden of illness in Canada for 1993 [31] and 1998 [14, 21]. The published costs for many of the diseases of interest were global figures, and efforts were thus made to estimate the distribution of costs within categories, for instance the division of cardiovascular costs between coronary disease, hypertension and stroke, of cancer costs between colon and breast cancer and other forms of neoplasm, of endocrine disorders between diabetes and other forms of endocrine disease, of digestive disorders between gall bladder disease and other digestive conditions, and of musculo-skeletal disorders between osteoarthritis, osteoporosis and other musculo-skeletal complaints (Table 11.4). The final column in this table provides comparable estimates for Ontario, measured in 2009 dollars, to allow comparisons with Table 11.5.

Table 11.5 Estimated direct and indirect costs of seven conditions associated with physical inactivity in the Province of Ontario, expressed in 2009 Canadian dollars

Disease	Total direct cost (billion)	Total indirect cost (billion)	Total PAR% direct cost (billion)	Total PAR% indirect cost (billion)	Total PAR % cost (billion)
Coronary disease	$1.63	$4.24	$0.297	$0.770	$1.07
Stroke	$1.14	$0.98	$0.259	$0.224	$0.483
Hypertension	$1.03	$1.00	$0.133	$0.128	$0.261
Colon cancer	$0.188	$0.900	$0.032	$0.151	$0.183
Breast cancer	$0.302	$1.44	$0.040	$0.191	$0.231
Type 2 diabetes mellitus	$0.534	$0.396	$0.107	$0.078	$0.185
Osteoporosis	$0.682	$3.54	$0.154	$0.797	$0.951
Total			$1.02	$2.34	$3.36

Based on data of Katzmarzyk [32]

A more recent analysis by Katzmarzyk [32] covered both direct costs (as in the earlier report) and also the indirect costs from short and long-term disability and premature death for the Province of Ontario. This study made 2-way analyses of both habitual physical activity ("active" vs, "inactive") and "obese" vs. "non-obese," using data from the 2009 Canadian Community Health Survey (CCHS). The original CCHS had classed people as "inactive," "moderately active" (daily energy expenditure >6.2 kJ/kg of body mass) and "active" (daily expenditure > 12.4 kJ/kg), but Katzmarzyk defined as "active" all those reporting a daily energy expenditure >6.2 kJ/kg. Some 50.7 % of Canadian adults met this criterion. The prevalence of obesity (based on reported heights and weights) was 17.2 %, but this figure was adjusted substantially upwards to 25.4 % in order to allow for an anticipated under-reporting of body mass. The findings were linked to data from the earlier meta-analysis [30]; the 1998 costs of chronic illnesses were inflated to 2009 values, using an index based upon the increase in health care expenditures over this interval. The PAR% values were slightly lower in 2009 than in 1998, due to a minor decrease in those classed as "inactive" (49.3 % vs. 53.9 %), although this change was more than offset by increases in medical costs and indirect costs over the same period.

The final step in the analysis for Ontario was thus to apply the PAR% values to estimates of the direct and indirect costs of the seven conditions of interest (Table 11.5). Given the similar data sources, the direct per capita costs in Ontario in 2009 were similar to those calculated for Canada as a whole in 1999, after adjusting for medical inflation during the interim (Table 11.4).

Estimates that other investigators have made for Nova Scotia and British Columbia (Table 11.6) are of the same general order as those for Ontario. When expressed on a per capita basis, all of these recent figures are lower than some previous values (Table 11.1), particularly those for the U.S.; in part, this is because

Table 11.6 Estimates of the direct and indirect costs arising from seven conditions linked to physical inactivity in Canada in recent years

Author	Direct costs (billion)	Indirect costs (billion)	Total and per capita costs (billion)
Katzmarzyk: Ontario 2001 [30]	$1.1[a]	$2.5[a]	$3.6[a] ($288)
Katzmarzyk Ontario 2009 [32]	$1.02	$2.34	$3.4 ($272)
Colman [33] Nova Scotia 1998	$0.107	$0.245	$0.354 ($378)
Colman and Walker [34] British Columbia 1998	$0.21	$0.36	$0.57 ($143)

[a]Values adjusted to 2009 Canadian dollars

Table 11.7 Influence of habitual physical activity in reducing health care charges relative to sedentary subjects, as seen in 42,520 retirees in relation to body mass index Based on data of Wang et al. [37]

	Annual savings relative to inactive subjects		
Habitual physical activity	Normal BMI	Overweight	Obese
2–3 times/week	$1456	$1731	$1177
4+ times/week	$1823	$581	$1379

of a more accurate definition of the economic costs associated with various illnesses. However, the more recent calculations are also necessarily conservative since: (1) Table 5 does not include all of the diseases listed in Table 11.2, (2) the threshold of daily physical activity required for the assumption of cost avoidance (6.2 kJ/kg) is less than that associated with evidence of benefit in some of the estimates included in Table 11.3, and (3) the true prevalence of physical inactivity is probably greater than suggested by questionnaire estimates [35, 36] (Chap. 6).

11.5 The Potential Overlap Between Costs Attributed to Inactivity and Obesity

One questionnaire-based study [37] addressed the potential overlap between inactivity and obesity-related costs, making three-way classifications of both habitual physical activity and of body mass index in 42,520 individuals who had retired from a U.S. manufacturer (Table 11.7). In this study, the estimated financial savings associated with physical activity seemed to be relatively independent of the person's body mass index.

11.6 New Information Obtained from Objective Monitoring of Physical Activity

To the author's knowledge, there have as yet been only two studies relating health-care costs to the objective monitoring of habitual physical activity [38, 39]. Both have been limited to expenditures in elderly subjects, and there is plainly a need to apply this approach to younger adults.

Aoyagi and Shephard [39] based their analysis upon seniors living in their own homes in the city of Nakanojo (Japan). Data were obtained for the actual amounts paid to service providers (doctors, nurses, and hospital out-patient departments) by the nursing care and national health insurance schemes of the Nakanojo Health and Welfare Plan for seniors, and the Gumma Prefecture National Health Insurance Group. Information was collected for a somewhat longer list of medical conditions than those included in most of the earlier subjective analyses: dependency, depression, osteoporosis, fractures, hypertension, diabetes mellitus, hyperlipidaemia, ischaemic or coronary heart diseases (angina pectoris and myocardial infarction), stroke or cerebrovascular diseases (cerebral infarction, cerebral haemorrhage and subarachnoid haemorrhage), dementia (vascular dementia and Alzheimer's disease) and cancer (colon/ rectum, lung, breast and endometrial cancers). Pedometer/accelerometer readings were collected for an entire year, and study participants were grouped into five categories: (dependent, $n = 800$, and 4 other activity quartiles, $n = 1200$ each, corresponding to activity increments of 2000 steps/day, and 5–10 minutes/day spent at an intensity > 3 METs). The prevalence of the various clinical conditions was estimated for each of the five activity categories (Table 11.8) using mainly observations from the Nakanojo study, but supplemented where necessary with questionnaire data that was arbitrarily stepped by reported activity increments of 10 MET-hours/week [40]. Multiplication of the disease-specific cost by the prevalence of that disease in each physical activity category allowed calculation of the average annual per capita direct health-care costs (Table 11.9). Because all subjects were in the age range 65–85 years, total per capita expenditures were relatively high, averaging $4016 per year. Nevertheless, a large part of this cost was attributable to those who were dependent, and figures showed a very steep downward gradient with increasing habitual physical activity; the range was from $18,948 in those classed as dependent to only $332 in the most active category. Because of the steep downward gradient in costs, Aoyagi and Shephard [38] emphasized the substantial fiscal benefits that could be realized, not only by a total population adoption of vigorous physical activity, but simply be encouraging a small percentage of the population to shift their physical activity upward by a single category. Such gains would be particularly large if a proportion of the population were persuaded to move out of the dependent category.

The study of Simmonds et al. [39] also focussed upon the relationship between health system usage and objectively monitored physical activity in the elderly. A group of 240 adults with an initial mean age of 77 years were followed for 4–5 years.

Table 11.8 Per capita disease costs, and prevalence of selected disease conditions categorized by physical activity level

Medical condition	Disease cost (per person-year)	Disease prevalence				
		Dependent (%)	Activity quartile 1 (%)	Activity quartile 2 (%)	Activity quartile 3 (%)	Activity quartile 4 (%)
Dependency	$20,513	73	0	0	0	0
Depression	$2497	6	4	2	1	0
Osteoporosis	$1991	18	13	8	3	1
Fractures	$503	15	11	7	3	1
Hypertension	$2115	48	36	24	12	8
Diabetes mellitus	$3894	16	12	8	4	2
Hyperlipidaemia	$1919	19	14	9	4	2
Ischaemic heart disease	$3613	12	7	2	1	1
Cerebrovascular disease	$2694	15	9	3	1	1

Based on data from the Nakanojo study for 2009, expressed in 2009 Canadian dollars (Aoyagi, 2011 [39])

Table 11.9 Average annual per capita direct costs of disease, by activity category (2009 Canadian dollars, based on the data of Aoyagi and Shephard [38] for senior citizens living in the community of Nakanojo, Japan

Medical Condition	Annual per capita costs associated with specific medical conditions				
	Dependent	Activity quartile 1	Activity quartile 2	Activity quartile 3	Activity quartile 4
Dependency	$14,974	$0	$0	$0	$0
Depression	$150	$100	$50	$25	$0
Osteoporosis	$358	$259	$159	$60	$20
Fractures	$75	$55	$35	$15	$5
Hypertension	$1015	$761	$508	$254	$127
Diabetes mellitus	$623	$467	$312	$156	$78
Hyperlipidaemia	$365	$269	$173	$77	$39
Ischaemic heart disease	$434	$253	$72	$36	$36
Cerebrovascular disease	$404	$242	$81	$27	$27
TOTAL	$18,398	$2406	$1390	$650	$332

Habitual physical activity was determined by accelerometry, with individual recordings continuing for > 10 hours on 5 or more days. Participants were classified into three activity categories, based upon steps/day and minutes of moderate or vigorous physical activity (Table 11.10), with the two methods of categorization yielding very similar results. Health system usage over the follow-up period was assessed in terms of new diagnoses, the number of medical consultations and

Table 11.10 Usage of health care system, usage based on a 3-level accelerometer classification of habitual physical activity

Habitual physical activity	Usage of health-care system				
	New diagnoses	Consultations	Prescriptions	Hospital admission days	Secondary care referrals
Steps per day					
Low	68.8	17.5	12.8	1.4	1.4
Moderate	62.7	13.8	7.5	0.7	1.2
High	51.4	14.4	7.0	0.7	1.1
Moderate/vigorous physical activity					
Low	69.8	17.5	12.0	1.4	1.4
Moderate	58.6	14.9	7.8	0.8	1.3
High	55.1	13.2	7.1	0.6	1.1

Based on the data of Simmonds et al. for senior citizens in the U.K. [39]

prescriptions written, the number of unplanned hospital admissions and the number of secondary referrals. On all of these measures, there was a gradient favouring the more active individuals, although a part of this was attributable to the fact that the less active individuals had a larger number of diseases at entry to the study. No attempt was made to ascribe specific costs to the health services that were utilized.

A systematic review of health-care utilization and physical activity, with particular reference to the experience of older adults [41], noted that 11 reports covered physical activity interventions, and the remaining reports that were cited all used questionnaires rather than objective monitoring to assess physical activity (for example, [37, 42–47]).

11.7 Potential New Information from Objective Monitors

As noted above, the traditional epidemiological approach to studying the economic consequences of physical inactivity has been to use some form of questionnaire to divide a given population into active and inactive groups, and then to apply this simple two-way classification of the prevalence of inactivity to a combination of risk ratios and the overall health-care costs of the main conditions thought susceptible to benefit from an increase of physical activity.

However, objective monitoring has demonstrated that a simple active/inactive categorization is inappropriate for this calculation; the dose of exercise associated with enhanced health differs between the various medical conditions [48]. Because objective monitoring allows a finer gradation of physical activity patterns, it becomes possible to evaluate within any given study which components of the overall cost are likely to be reduced by any given increase in physical activity (Table 11.9). This is an important advance, because the likely increase of population activity in response to an exercise promotional campaign is unlikely to be as large as that accepted in many of the questionnaire-based calculations of the relative risk of inactivity. For some conditions, a small increase of habitual physical activity may be insufficient to reduce the risk of disease, and if only a small increase of physical activity is achieved, it may be inappropriate to claim any economic benefit.

Objective monitors potentially allow a detailed analysis of the relationship between increases in physical activity and the likely reduction in health care costs for each of a wide range of medical conditions. This not only yields a more accurate assessment of the total costs of inactivity, but it also permits a more cost-effective deployment of motivational efforts. For instance, overall cost-savings are likely to be much larger if a totally sedentary or dependent individual is encouraged to participate in a little physical activity than if a person who is already moderately active is encouraged to become highly active (Table 11.9). Moreover, measures to encourage greater habitual physical activity should have a greater impact upon health care costs when they are applied to individuals who are dependent, those with hypertension, and those with type-2 diabetes mellitus than would occur if, for instance, an increase of habitual activity was induced among those who are liable to fractures.

Objective monitoring should further allow accurate assessments of the costs of efforts to achieve specific increments of physical activity, which is another important element in health economics equations. Questionnaires are unreliable as a means of assessing the success of exercise motivation, because many of the people who are questioned are anxious to report the changes that they have been asked to make, whether these have actually occurred or not.

Finally, objective monitoring is well-suited to prospective analyses where individual health costs are recorded concurrently with the individual's physical activity profile over a period of several years, as in the study of Simmonds and associates [39].

The one current major challenge in objective monitoring is the difficulty in translating measurements between instruments, between studies, and between the research laboratory and the hospital clinic. This issue is exacerbated by the secret nature of the equations used in many commercial instruments, shifting technologies and the obsolescence of much previously used equipment.

11.8 Current Limitations to the Objective Monitoring of Physical Activity

Currently, one important limitation of objective monitoring is that it is labour intensive, at least if accurate information is to be obtained. The details of each day's record must be inspected by the investigator, to determine possible periods when artifacts occurred, or the instrument failed to record adequately. The need to check pedometer/accelerometer output inevitably limits sample size, and this in turn shifts indicators of disease from hard end-points such as a heart attack or sudden death to surrogate measures; thus, pulse wave velocity has been used as a measure of arteriosclerosis [49], and the osteosonic index has served as a measure of osteoporosis and the risk of fractures [50].

The cost of most physical activity monitors has now decreased to the point where such instrumentation could be used on a large scale; it remains for effective computer programmes to be developed that will avoid the need for human scanning of each day's record. Possibly, devices and software such as the recently developed FitBit, which uploads data, will enable investigators to track larger groups of subjects successfully.

The application of objective monitoring has to date been limited to two economic analyses [38, 39], and it is significant that in both of these investigations the subjects were old people. This is perhaps not a major criticism, as a large part of overall health care costs arises in the treatment of the elderly. Walking is the main significant form of physical activity for the elderly, and walking is recorded quite accurately by the current generation of monitors. However, application of the same type of approach to younger adults will require documentation of how much of their activity is taken in the form of moderately-paced walking. Possibly, there may be a need for a second generation of objective monitors that can respond accurately not only to walking, but also to a much wider range of daily sports and pastimes.

11.9 Overall Limitations of Health Cost Analyses

There are many limitations to calculations of both the direct and the indirect costs of physical inactivity, even when the most precise objective monitoring of an individual's movement patterns is available [51]. Some health economists such as Louise Russell [52, 53] (Fig. 11.2), although accepting the health benefits of regular exercise, have thus vigorously criticized the conclusion that the promotion of physical activity is likely to yield large fiscal savings to health-care providers.

In general terms, health care expenditures are currently capped by national and/or provincial governments, and if the incidence of one medical condition is reduced, then expenditures are likely to be correspondingly increased in some other area of health care. Neither individual physicians nor hospital administrators are likely to relish a decrease in their annual income, so that the projected savings are unlikely to be realized. If the net effect is the provision of additional services in some other needy area of health care, this is a net benefit to society, and it is certainly not a strong argument against seeking savings in those areas of health care expenditure that are likely to be reduced by an increase of physical activity.

Much of the controversy concerning direct costs centres around the fact that a large fraction of these expenditures are incurred in the final year of a person's life. Critics argue that even if increased physical activity reduces the risk of one type of illness, the individual will eventually succumb to some alternative medical condition, likely making equal terminal demands upon the health care system. Fries [54] has argued that this is not the case; whereas the sedentary individual often suffers

Fig. 11.2 When working at the Brookings Institute, Louise Russell vigorously criticized the view that the promotion of greater physical activity would reduce the costs of health care

from an expensive chronic illness over the final 10 years of life, the active individual is likely to show a compression of morbidity, remaining in good health and making few medical demands until shortly before death. This hypothesis should be verified by longitudinal studies of the type already initiated by Simmonds [39], with an objective recording of physical activity patterns and actual medical costs in the years leading up to death.

Indirect costs are a major element in the total expense of ill-health, and some indirect costs are difficult to quantify. Furthermore, critics have argued that although better health may contribute savings from reduced time loss during a person's working years, the active individual negates any accrued benefits by requiring larger pension and benefit payments over an extended lifespan. There are several counter-arguments to this criticism: any extension of lifespan due to a vigorous lifestyle is relatively small, and the active person is likely to eliminate this charge by accepting a later retirement than someone who is sedentary. Further, the active senior is likely to make substantial unpaid contributions to society during his or her years of retirement, something that is no longer possible for a sedentary individual who is now suffering from chronic disability.

Finally, although there is good evidence that physical activity has a role in preventing ill-health, the effectiveness of an increase of physical activity in reversing chronic disease is less clearly established. Perhaps the most important gain from a rehabilitation programme is a deferral of dependency [55], and as noted above this is a major cost in an elderly population. The recent Life Study demonstrated that a moderate aerobic, resistance and flexibility intervention reduced the incidence of disability (demonstrated by an inability to walk 400 m) over a 2.6 year period among subjects aged 70–89 years from 35.5 % in controls to 30.1 % in the experimental group [56]. If an exercise programme enhances other areas of health in some people, the fraction of the population who will benefit remains to be clarified, and it is unclear how long a lag period will be involved. Any delay in realizing benefits in turn leads to a discussion about an appropriate discount rate to apply to the postulated savings [51].

Debate continues as to whether the promotion of physical activity could avert the looming fiscal crisis associated with ever-increasing medical expenditures. But as Woolf and associates point out, this may not be the most appropriate question to argue [57]. It is better to ask whether value is being obtained from the dollars that are expended on national health services. Much of the current cost of end-of-life care could be replaced by preventive and/or palliative treatment, with major fiscal savings, and probably an enhanced overall quality of life. If greater physical activity can be promoted for a moderate cost, this will add greatly to quality-adjusted life expectancy, and it is an appropriate use of public funds.

References

1. Rebelo LP. The origins and the evolution of health economics: a discipline by itself? Led by economists, practitioners or politics? Porto, Portugal: Universita Católica Portuguesa; 2007.
2. Blumenschein K, Johannesson M. Economic evaluation in healthcare. A brief history and future directions. Pharmacoecon. 1996;10(2):114–22.
3. Wagstaff A, Cuyler AJ. Four decades of health economics through a bibliometric lens. Washington, DC: World Bank e-Library; 2011.
4. Tarride J-E, Lim M. Desmeules Mea. A review of the cost of cardiovascular disease. Can J Cardiol. 2009;25(6):e195–202.
5. Arrow K. Uncertainty and the welfare economics of medical care. Am Econ Rev. 1963;53 (5):941–73.
6. Phelps CE. Health economics. Boston, MA: Addison Wesley; 2003.
7. Taylor B, Rehm J. When risk factors combine: the interaction between alcohol and smoking for aerodigestive cancer, coronary heart disease, and traffic and fire injury. Addict Behav. 2006;31(9):1522–35.
8. White KL, Ibrahim MA. The distribution of cardiovascular disease in the community. Ann Int Med. 1963;58:627–36.
9. Klarman HE. Socio-economic impact of heart disease. In: Andrus EC, editor. The heart and circulation- Second National Conference on Cardiovascular Disease, vol. 2. Washington, DC: U.S. Public Health Service; 1964. p. 693–707.
10. Juhn C, Potter S. Changes in labor force participation in the United States. J Econ Perspect. 2006;20(3):27–46.
11. Armstrong JB. Costs of cardiovascular disease in Canada. Notes on Public Health & Preventive Medicine. Dept. of Public Health, University of Toronto, Toronto, ON; 1965.
12. Garis RI, Farmer KC. Examining costs of chronic conditions in a Medicaid population. Manag Care. 2002;11:43–50.
13. Wang G, Pratt M, Macera CA, et al. Physical activity, cardiovascular disease, and medical expenditures in U.S. adults. Ann Behav Med. 2004;28:88–94.
14. Health Canada. Economic burden of illness in Canada. Health Canada, Ottawa, ON, http://www.phac-aspc.gc.ca/publicat/ebic-femc98/ (2002).
15. Sasser AC, Rousculp MD, Birnbaum H, et al. Economic burden of osteoporosis, breast cancer, and cardiovascular disease among postmenopausal women in an employed population. Womens Health Issues. 2005;15:97–108.
16. Birnbaum H, Leong S, Kabra A. Lifetime medical costs for women: cardiovascular disease, diabetes, and stress urinary incontinence. Womens Health Issues. 2003;13:204–13.
17. Leal J, Luengo-Fernandez R, Gray A, et al. Economic burden of cardiovascular diseases in the enlarged European Union. Eur Heart J. 2006;27:1610–9.

18. American Heart Association. Heart disease and stroke statistics —2006 update: a report from the American Heart Association Statistics Committee and Stroke Statistics Subcommittee. Circulation. 2006;113:e85–151.
19. Mensah GA, Brown DW. An overview of the cardiovascular disease burden in the United States. Health Aff. 2007;26(1):38–48.
20. Heidenrich PA, Trogdon JC, Khavjou OA, et al. Forecasting the future of cardiovascular disease in the United States: a policy statement from the American Heart Association. Circulation. 2011;123:933–44.
21. Health Canada. Economic burden of illness in Canada, 1998. Ottawa, ON: Health Canada; 1998.
22. Patra J, Popova S, Rehm J, et al. Economic cost of chronic disease in Canada 1995-2003. Edmonton, AL: Alberta Institute of Health Economics; 2007. http://www.cdpac.ca/media.php?mid=260.
23. deVol R, Bedroussian A. An unhealthy America: the economic burden of chronic disease-charting a new course to save lives and increase productivity and economic growth. Santa Monica, CA: Milken Institute; 2007.
24. American Diabetes Association. Direct and indirect costs of diabetes in the United States. Alexandria, VA: American Diabetes Association; 2007. http://www.diabetes.org/diabetes-statistics/cost-of-diabetes-in-us.jsp.
25. National Heart Lung and Blood Institute. Morbidity and mortality 2004; 2004 chart book on cardiovascular, lung and blood diseases. Bethesda, MD: National Institutes of Health; 2004.
26. Alzheimer's Association. Alzheimer's disease facts and figures. http://www.alz.org/national/documents/Report_2007FactsAndFigures.pdf (2007).
27. Fox S, Skinner JS. Physical activity and cardiovascular health. Am J Cardiol. 1964;14:731–46.
28. Shephard RJ. Endurance fitness. Toronto, ON: University of Toronto Press; 1969.
29. Katzmarzyk P, Gledhill N, Shephard RJ. The economic burden of physical inactivity in Canada. Can Med Assoc J. 2000;163(11):1435–40.
30. Katzmarzyk P, Janssen I. The economic costs associated with physical inactivity and obesity in Canada. Can J Appl Physiol. 2004;29(1):90–115.
31. Moore R, Mao Y, Zhang J, Clarke K. Economic burden of illness in Canada Health Canada. Ottawa, ON: Canadian Public Health Association; 1997.
32. Katzmarzyk P. The economic costs associated with physical inactivity and obesity in Ontario. Health Fitness J Canada. 2011;4(4):31–40.
33. Colman R. The cost of physical inactivity in Nova Scotia. Halifax, NS: GPI Atlantic for Recreation Nova Scotia and Sport Nova Scotia; 2002.
34. Colman R, Walker S. The cost of physical inactivity in British Columbia. Victoria, BC: GPI Atlantic for the BC Ministry of Health; 2004.
35. Dyrstad SM, Hansen BH, Holme IM, et al. Comparison of self-reported versus accelerometer-measured physical activity. Med Sci Sports Exerc. 2014;46(1):99–106.
36. Prince SA, Adamo KB, Hanmel ME, et al. A comparison of direct versus self- report measures for assessing physical activity in adults: a systematic review. Int J Behav Nutr Phys Act. 2008;5:56.
37. Wang F, McDonald T, Reffitt B, et al. BMI, physical activity, and health care utilization/costs among Medicare retirees. Obes Res. 2005;13:1450–7.
38. Aoyagi Y, Shephard RJ. A model to estimate the potential for a physical activity-induced reduction in healthcare costs for the elderly, based on pedometer/accelerometer data from the Nakanojo study. Sports Med. 2011;41(9):1–14.
39. Simmonds B, Fox K, Davis M, et al. Objectively assessed physical activity and subsequent health service use of UK adults aged 70 and over: a four to five year follow up study. PLoS One. 2014;9(5), e97676.
40. Woolcott JC, Ashe MC, Miller WC, et al. Does physical activity reduce seniors' need for healthcare? A study of 24,281 Canadians. Br J Sports Med. 2010;44(12):902–4.

41. Sari N. Exercise, physical activity and healthcare utilization: a review of literature for older adults. Maturitas. 2011;70:285–9.
42. Colditz GA. Economic costs of obesity and inactivity. Med Sci Sports Exerc. 1999;31 (11 Suppl):663–7.
43. Garrett NA, Brasure M, Schmitz KH, et al. Physical inactivity. Direct cost to a health plan. Am J Prev Med. 2004;27(4):304–9.
44. Haapanen-Niemi N, Miilunpalo S, Vuori I, et al. The impact of smoking, alcohol consumption, and physical activity on use of hospital services. Am J Publ Health. 1999;89(5):691–8.
45. Keeler EB, Manning WG, Newhouse JP, et al. The external costs of a sedentary lifestyle. Am J Public Health. 1989;79:975–81.
46. Martin MY, Powell MP, Peel C, et al. Leisure-time physical activity and health- care utilization in older adults. J Aging Phys Act. 2006;14(4):392–410.
47. Sari N. Physical inactivity and its impact on healthcare utilization. Health Econ. 2009;18:885–901.
48. Aoyagi Y, Shephard RJ. Steps per day: the road to senior health? Sports Med. 2009;39:423–38.
49. Aoyagi Y, Park H, Kakiyama T, et al. Yearlong physical activity and regional stiffness of arteries in older adults: the Nakanojo Study. Eur J Appl Physiol. 2010;109:455–64.
50. Park H, Togo F, Watanabe E, et al. Relationship of bone health to yearlong physical activity in older Japanese adults: cross-sectional data from the Nakanojo Study. Osteoporosis Int. 2007;18(3):285–93.
51. Shephard RJ. The economic benefits of enhanced fitness. Champaign, IL: Human Kinetics; 1986.
52. Russell LB. Is prevention better than cure? Washington, DC: Brookings Institution; 1986.
53. Russell LB. Preventing chronic disease: an important investment, but don't count on cost savings. Health Aff. 2009;28:42–5.
54. Fries JF. Aging, natural death and the compression of morbidity. N Engl J Med. 1980;303:130–6.
55. Shephard RJ. Maximal oxygen intake and independence in old age. Br J Sports Med. 2009;43 (5):342–6.
56. Pahor M, Guralnik KJM, Ambrose WT, et al. Effect of structured physical activity on prevention of major mobility disability in older adults. The LIFE S tudy randomized clinical trial. JAMA. 2014;311(23):2388–96.
57. Woolf SH, Husten CG, Lewin LS, et al. The economic argument for disease prevention: distinguishing between value and savings. Washington, DC: Partnership for Prevention; 2009.

Chapter 12
Limitations of Current Objective Monitors and Opportunities to Overcome These Problems

Catrine Tudor-Locke

Abstract The myriad of objective monitors currently available, their differing measurement features and capacity, and the tactics of commercial enterprise, when mixed together with the multiple purposes of scientific studies and the inability of investigators to reach an academic consensus on standardized protocols for the measurement of habitual physical activity are all grounds for concern as research in this area seeks to move forwards. On the academic front, collaboration, sharing, standardization, and systematic knowledge creation seem to offer the optimum path if the promise of objective monitoring is to be realized and data are to become accessible equally to researchers, practitioners and the general public.

12.1 Introduction

The objective monitoring of physical activity and sedentary time has become a mainstay scientific measurement tool for exercise epidemiologists and clinical physiologists; this approach has transformed the ability of investigators to conduct surveillance, screen research participants, characterize dose-response relationships precisely, communicate behavioural prescriptions effectively, motivate behavioural change and track adherence, and evaluate physical activity programmes and other initiatives. However, opportunities that must be considered thoughtfully in order to select, use, and interpret the data derived from objective physical activity monitors effectively and/or improve their functionality in the face of their current limitations.

C. Tudor-Locke (✉)
Department of Kinesiology, University of Massachusetts Amherst, Amherst, MA, USA
e-mail: catrine.tudor.locke@gmail.com

R.J. Shephard, C. Tudor-Locke (eds.), *The Objective Monitoring of Physical Activity: Contributions of Accelerometry to Epidemiology, Exercise Science and Rehabilitation*, Springer Series on Epidemiology and Public Health,
DOI 10.1007/978-3-319-29577-0_12

12.2 Well Known Limitations of Objective Physical Activity Monitors

Matthews et al. [1] pointed out three clear weakness of objective physical activity monitors. Firstly, many currently available monitors are, by design, most sensitive to ambulatory movements and lack accuracy in detecting other movements, for example, those performed by the upper body and/or not performed in an upright position. Such movements occur during a number of familiar activities, including cycling, resistance training and other types of conditioning exercise, load carrying, and when climbing stairs or hills. Moreover, non-waterproof devices cannot be used to assess swimming or other water-based activities (and indeed, no one has yet confronted the issue of what to do with the accelerometer signals potentially emerging from such monitored activities). Although these are indeed familiar types of physical activity, this does not necessarily imply that they are commonly performed across a population, on a daily basis and for long periods of time [2]. Considerable recent effort has been put into detecting such "fringe" behaviours [1], but this effort may be disproportionate relative not only to their impact upon total energy expenditure and/or health, but also relative to their importance as a component of behavioural interventions.

Secondly, objective monitors typically do not collect contextual information about the location and purpose of the detected movement/non-movement signals. Matthews et al. [1] acknowledges that this weakness may be a limiting factor for some specific research questions (most obviously those that include the need for contextual information), but it is not a universal issue. Furthermore, this weakness can be addressed by combining objective physical activity monitoring with data obtained from other information sources such as a GPS monitor.

Finally, the infrastructure required to support automated and standardized data processing of a large number of "increasingly complex and bulky data files" [1] collected in the course of population-based studies has not been fully developed at the present time. The practical implication is that sufficient resources and expertise must be budgeted to accommodate the time and financial burdens associated with deliberate and systematic post-processing and quality control checks by investigators, leading to locked data sets

12.3 Advancing Technology

Whether you choose to call it the Information Age, the Computer Age, the Digital Age, or the Age of Technology, we currently live in a fast-paced and knowledge-based period of human history. And whether you consider it a limitation or an opportunity perhaps depends upon where you stand on the playing field at any given moment. An increasingly high-tech global economy is the engine that drives ever-continuing efforts to simplify daily decisions and provide goods and services in more efficient and convenient ways. The rapid evolution of objective physical activity monitoring that we have witnessed over recent decades is just a single example of this overall trend. On the scientific front, simple mechanical pedometers gave way to electronic pedometers, which opened the field to piezoelectric cantilever sensor-based accelerometers and inclinometers that are now being replaced with monolithic differential capacitance sensors with dual axial and triaxial measurement capabilities [3]. Analog displays transitioned to digital displays. Displayed data were first recorded by hand and dropped off, mailed, or faxed to study centers. Data were also once captured in the limited internal memories of on-board microprocessors and/or transmitted to electronic spreadsheets by wire. Now wireless, internet, and cloud-based technologies enable more readily accessible remote monitoring. At the same time, microprocessor memory, computational power, and battery life have expanded immensely. Firmware and software have evolved to process data signals and configure them into manageable outputs that end users can more easily consume. Data transfer and storage barriers are crumbling. Such technological improvements have enabled longer-term monitoring at increasingly greater resolution.

Despite these exciting advances in measurement technology, a number of limitations are still apparent, and in some cases have even emerged as a result of this technological explosion. Perhaps foremost, comparability between objective monitors produced by different companies, and even between successive generations of the same brands, remains an important measurement issue. For example, activity counts, and any related cut-points, are well known to be instrument-specific. As a single example, across model generations of the popular ActiGraph accelerometer, the CSA was shown to be more sensitive compared to accepted research-grade pedometers [4, 5], the 7164 was also shown to be more sensitive [6], but the GT1M had a similar sensitivity [7, 8], the GT3X was less sensitive [9], and the GT3X+ was either more or less sensitive, depending on the filter selected for data processing [10]. Comparability of output data is perhaps the most important concern threatening our ability to conduct surveillance programmes, compare study results, and provide direction to public health guidelines in the form of precise objectively monitored metrics. It is important to underline that some available objective physical activity monitors, notably research-grade pedometers, have resisted change throughout this evolution; they apparently offer the same technology as they did when they were originally introduced. The advantage of this apparent technological inertia is that data comparability is assured both over time and between studies.

Another important wave of technological advances has been the surging availability of personal health monitoring devices [11]. Although generally targeted to consumers, scientists are already starting to embrace these relatively low cost devices for purposes of both measurement [12] and intervention [13]. Some devices offer a wireless interface with other personal technologies, including mobile phones, computers, tablets, and music players. Although there are undeniable privacy and ethical issues to navigate, in the future some personal health monitoring companies may allow scientists access to their stored consumer-generated data in a form appropriate for ethical human research.

The ubiquitous nature of mobile phones represents a particularly tantalizing opportunity to "leverage consumer investment" [11] and take advantage of a major opportunity to sense, store, process and network data. Since consumers are already conditioned to re-charge such devices regularly, it may be possible to tap into this behaviour to extend the battery life of objective physical activity monitors [11]. Challenges that will need to be addressed include non-continuous wear, technological differences (including operating systems) between phones, competing demands for personal technologies (thus reducing researcher control over a specific device) [11], determining how to track and compensate research participants for any data plan costs, etc.

12.3.1 Data Processing

Manufacturers of objective physical activity monitors typically provide summary output metrics (e.g., activity counts, "active time", etc.) but the methods used in treating, filtering and ultimate distillation of these summary values is often not disclosed. Further post-processing by investigators in attempts to infer meaning from activity counts in terms of the time allocated to differing intensities of physical activity quickly revealed that activity counts were arbitrary unitless digits ascribed to raw accelerometer signals in a unique and proprietary manner by the device manufacturer. Attempts to calibrate activity count-based data related to direct measures of energy expenditure were heralded at first [14], but were subsequently all but abandoned as multiple calibration studies disagreed on absolute cut-point values [15]. Recent best practice recommendations have called for the discontinued development and use of activity count-based calibrations [16]. As a result, the current trend in accelerometer technology is to process raw data. Although manufacturers may continue to provide proprietary summary metrics (such as activity counts), investigators have suggested that providing access to raw accelerometer data and standardizing accelerometer units as gravitational constants (g, m·second^{-2}) or time integrated units (e.g., m·second^{-1}) will help to increase comparability between instruments [3, 17]. Since 2008, ActiGraph has released software options, firstly for their GT1M, and then for their GT3X and GT3X+ accelerometers, that allow users to collect raw, unfiltered data [3]. The GENEActiv device also provides access to the raw accelerometer signal [18]. Intille et al. [11] have suggested the prudent collection and storage of accelerometer data in raw format, whenever possible, thus providing the potential for more advanced data processing in the future should such opportunity arise. Optimists continue to archive data in preparation for analytical breakthroughs anticipated over a period of years or decades [19]. Access to raw data is still not available for all objective physical activity monitors, however, so researchers who desire this feature find themselves severely restricted in terms of their choice of accelerometer.

Access to raw data would theoretically permit application of non-proprietary "universal" algorithms (yet to be produced, validated, and accepted) that could be applied to summarize signals in terms of energy expenditure and/or activity type [11]. For example, Finnish researchers [20] have offered an open source method for processing raw accelerometer data and classifying the intensity of physical activity. A "universal step" algorithm would be immensely useful as well. Creative applications of statistical pattern recognition, machine learning techniques, and other mathematical modeling approaches may further propel an investigator's ability to identify the types of activity that are performed and to refine the precision of energy expenditure estimates [16, 17]. Such efforts are challenging to validate under free-living conditions [11], outside of the laboratory, although such validation is not impossible [21]. We can anticipate that future generations of scientists and statisticians will continue to create yet more elaborate data processing algorithms, and that these will then need to be validated and optimized. The utility of such algorithms will be assured if they are also built into easy-to-use software and/or open source shareware platforms [19]. As a final note concerning anticipated changes in data processing, the way that we conduct analyses may also change from time-consuming and scheduled batch processing on desktop computers to cloud-based computing platforms that facilitate faster and more readily accessible rolling data analyses [19].

Such likely advances in data processing are exciting. However, they are also increasingly complex and require more and more sophisticated technological and statistical expertise; such understanding is presently shared by only a relatively small (but growing) cadre of physical activity-focused investigators worldwide. The opportunities and limitations of nascent raw signal based analyses have yet to be fully explored. It is also clear that the "activity count," despite its acknowledged limitations [22] and calls to discontinue its use [16, 23], is not going to disappear quickly. Many accelerometer-based devices continue to provide this summary output (with or without also providing access to raw data signals), and historic data captured using this metric still have a potential to contribute to future analyses. It would thus be foolish to discard such information or to prevent its publication. Further, there are still longitudinal studies that must continue to use the same technology (if it remains available) in their future data collection cycles, both for economic reasons and to ensure the comparability of data.

12.3.2 Data Collection Protocols

As advances in technology have generated smaller and more waterproof devices, and with extended memory and battery life, data collection protocols have shifted more and more to collecting 24-hour data. This obviates the concern about prematurely curtailed wearing times and also offers the opportunity to collect information regarding sleep-related movement and non-movement behaviours. Although there are now open source algorithms available to extract the nocturnal sleep episode time from the 24-hour period [24], much more research is needed to classify the meaning of movement and non-movement signals detected during this time, and to modify and validate these algorithms in different populations.

Remote monitoring is a current reality that allows almost real-time data acquisition and immediate intervention as required. For example, evidence indicative of non-wear can trigger immediate contact with a subject to encourage compliance. Future advances may include sensors that can detect whether or not a device is being worn as indicated, thus improving compliance to wearing regimens. Interactive and automated compliance aids have also been suggested [1, 11]; however, related concerns about measurement reactivity remain to be explored [11]. Behaviour modification programmes, for example, those focused on increasing physical activity and/or decreasing sedentary time, can also be made more effective by facilitating timely intervention as needed.

12.4 The Commercial Enterprise

12.4.1 Intellectual Property

Objective physical activity monitors are generally commercial items, developed to meet some market need and to generate profit for the manufacturer. The design and configuration of the new technology is thus held closely as proprietary intellectual property and is legally protected by a range of patents, copyrights, trademarks, trade secrets, and non-competition agreements. Manufacturers may highlight novel features in an effort to distinguish their product from that of their rivals, but in general it is not advantageous for them to share the underlying algorithms that are used to detect and/or process data. This exposes manufacturers to both criticism and competition. But it is also frustrating for scientists who are confronted with a "black box" approach to measurement. Their naturally inquisitive nature propels them to "peek under the hood" [25] and ultimately to conduct studies that pick apart algorithms [26] in an attempt to "reverse engineer" the device and provide transparent [27] and/or universal [20] algorithms. Investigators have repeatedly called for manufacturers to be more transparent in releasing relevant information [1, 3, 11, 19]. The reality, however, is that protection of intellectual property and the need to differentiate a product/service from that of competitors is elemental to the commercial enterprise if the manufacturer is to remain profitable, stay in business, and provide a product that will continue to be available.

12.4.2 Obsolescence: Planned and Unplanned

The same economic forces that have compelled manufacturers to create increasingly more sophisticated sensors and microprocessor technology to support objective monitoring have also caused the manufacture of some devices to be discontinued. An example of this is the Caltrac, which was one of the first wearable accelerometers to be used in large-scale research studies [28]. More recently, the multi-sensor IDEEA accelerometer [29] no longer appears to be available. Since the SenseWear armband [30] was recently acquired by the Jawbone company, it may be only a matter of time before the former technology is pushed into otherwise unplanned obsolescence. This is a source of concern for investigators in any measurement-based science. Current and future comparability is compromised when monitoring devices continually enter and exit the market.

Planned obsolescence is a business strategy intended to push the consumer to refresh their purchases frequently under the guise that all changes in the latest format of a monitor are "improvements." Small changes in fasteners, cable connections, battery requirements, etc. can render an existing device obsolete if adequate replacement parts can no longer be sourced or are relatively expensive to repair versus replacing the entire instrument. This is not only a problem for the end-consumer; Troiano et al. [19] noted that small-capacity on-board memory chips and piezoelectric sensors are now expensive and difficult for device manufacturers to find. Continual firmware and software changes also make technologies incompatible, even within the same brand line [31]. Although investigators could demand that manufacturers provide comparability data between predecessor and successor objective monitors within the same brand line [3], independent calibration would still always be required, involving a great deal of repeated and misplaced effort. Unfortunately, there is a real possibility that a device and its supporting software could evolve even before validation results painstakingly collected on the original product can be published. As Pedisic and Bauman [32] lamented, it appears that the pace at which "improved" accelerometers are marketed outstrips investigators' ability to evaluate and share findings about predecessor models thoroughly.

12.4.3 Pricing

A number of review articles have compiled prices of different objective physical activity monitors, including a publication by McClain et al. in 2009 [33]. Unit prices for objective monitors with published validity/reliability evidence ranged from US$17 for the Yamax brand pedometer to US$450 for the Actical accelerometer. Despite the diminishing costs of accelerometer sensor technology and the widespread availability of consumer-grade objective monitors, the cost of a research quality accelerometer remains relatively high (around US$200 [32]). The continuing relatively higher unit costs of accelerometer-based research devices is likely due to the fact that the collective buying power of the scientific market is relatively small when compared to that of the very large consumer market for simpler wearable devices. Unit pricing aside, accelerometry-based objective monitoring data is also expensive to collect and manage. At a minimum, investigators must budget for instrument purchase (including planning for loss/damage, peripherals, software, and batteries), and the personnel expertise to prepare/initialize and distribute/collect devices, monitor compliance, and download, transfer, store, and process the data. Researchers need to distribute/collect accelerometers, instruct participants, recharge/replace batteries, replace broken devices, initialize/calibrate devices, and store and process the data. Pedisic and Bauman [32] pointed out that, although objective physical activity monitoring does indeed provide important insights in small scale studies, in the face of limited resources, associated costs should be considered seriously before governments invest in large-scale accelerometer-based surveillance efforts.

12.4.4 The Lack of Industry Standards

In Japan, the Ministry of Industry and Trading regulations proposed Japanese Industrial Standards that set a maximal miscounting rate of 3 % for step detection [34]. Outside of Japan, the manufacture of objective monitors is an unregulated commercial enterprise. There are no industry standards, and there is no watchdog to evaluate manufacturer claims or to command quality control. Outside of Japan, this remains a "buyer-beware" situation; investigators must educate themselves and perform their own validation and quality control studies as part of continual due diligence. Investigators may demand lower cost and higher quality objective monitors for research, but the impact even of their collective voices may not resound against the cry of the larger potential market of consumers, who are motivated to engage in the conspicuous consumption of slick and trendy gadgetry, regardless of its accuracy. Without industry standards, it is unlikely that the current state of non-comparability between objective monitors will change any time soon. Japan serves as a model that the rest of the world should emulate if comparability of instrumentation, and ultimately of data, is truly valued.

12.5 The Way Forward

It appears that the way forward will be marked by continued technological innovation, with increasingly complex methods of data processing and sophisticated statistical treatments. But does this mean that the commercial enterprise should set the pace? It is difficult not to get caught up in the churning excitement created by dispatching one device/method in pursuit of the next. Academics who do not readily jump on this bandwagon risk being labeled technological dinosaurs and being treated as a "research underclass" [35]. To an uninformed observer, however, it might appear that there has been a lot of research but the amount of true progress is much more questionable. Pedisic and Bauman [32] recently called the "Emperor's clothing" into question by publishing a carefully considered inventory of accelerometry issues; they question the current utility of the accelerometer, at least in large scale surveillance applications. This article offers a timely and refreshingly balanced examination of the state of the science of physical activity measurement. It also opens up an opportunity for scientists to work with manufacturers to re-think and purposively plan and shape the way forward, rather than simply reacting to the commercial release of new devices, features, and firmware.

The greatest threat to objective monitoring is non-comparability at all levels. Exciting efforts to address this issue include collaboration and open-source sharing of data and data processing approaches [19]. However, objective monitors, metrics, and protocols for data collection, management, processing and reporting are likely to continue to differ between studies in major and minor ways, including also the type and purpose of the investigation, the research question that is being asked and the available resources, both human and financial. Clearly, science is a creative vocation, and within our critical thinking we must evolve a greater tolerance for some choices, as long as they are thoughtfully considered, reasonably defended, and ultimately supported by evidence. Consensus is perhaps not always realistic, but continued efforts to establish *best practice standards* are useful for guiding and evaluating future efforts.

So, what are examples of a pragmatic, charted way forward to establish best practice standards?

- Pursue technological innovation in tandem with pragmatic research that is focused on building a sustainable knowledge foundation
- Continue to compile and publish expected values for different devices, metrics, protocols, and populations
- Catalogue the various rules, variables, and definitions applied to objective monitoring.
- Plug gaps in methodological knowledge. For example, we need more measurement reliability research focused on behavioural stability in order to have a better understanding of factors that influence "how many days are enough?"
- Create a "tool box" of checklists, flow charts, decision trees, templates, open source software, etc. to assist investigators in choosing objective monitors, designing protocols, and analyzing, interpreting, and reporting data
- Report results in terms of multiple metrics simultaneously, instead of a choice few, as a way to illuminate which metrics are the most important to track for various objectives
- Evaluate the possibility of standardized reporting of a "simple common metric" (e.g., steps/day) across all types and costs of objective monitors and study designs, regardless of purpose, in an effort to bridge science, clinical practice, and the real world

- Lobby for industry standards similar to those in Japan for an acceptable minimum accuracy of step counting
- Conduct larger scale studies of device comparability across the lifespan
- Evaluate the possibility of correction factors to allow a comparison of outputs between different objective monitors
- Evaluate successor objective monitors against their predecessors when a new version is released
- Publish protocols, including detailed manuals of operation
- Resist the misguided temptation to equate self-reported behaviour with objectively-determined metrics and purposefully collect more objectively monitored data related to different health outcomes, building the future for public health guidelines based on objectively determined physical activity and sedentary time metrics that the public can track with their own affordable technologies

12.6 Conclusions

The emergence and development of objective physical activity monitoring has been a game changer for those studying and promoting physical activity. However, researchers have struggled to keep pace with technological developments as commercial enterprises have churned out a broad diversity of instruments, each with proprietary measurement properties. The promise of objective monitoring is threatened by non-comparability between different classes and brands of objective monitors, and even across successive generations of the same brand. Researchers are now increasingly taking the reins in a collaborative push to share more compatible and open-source technologies, to establish best practice standards, and to plug methodological gaps systematically by extending the underlying knowledge necessary to establish objective monitoring as a scientifically grounded and sustainable methodology.

References

1. Matthews CE, Hagstromer M, Pober DM, et al. Best practices for using physical activity monitors in population-based research. Med Sci Sports Exerc. 2012;44(1 Suppl 1):S68–76.
2. Tudor-Locke C, Johnson JD, Katzmarzyk PT. Frequently reported activities by intensity for U.S. adults: the American time use survey. Am J Prev Med. 2010;39(4):e13–20.
3. Chen KY, Janz KF, Zhu W, et al. Redefining the roles of sensors in objective physical activity monitoring. Med Sci Sports Exerc. 2012;44(1 Suppl 1):S13–23.
4. Le Masurier GC, Tudor-Locke C. Comparison of pedometer and accelerometer accuracy under controlled conditions. Med Sci Sports Exerc. 2003;35(5):867–71.
5. Tudor-Locke C, Ainsworth BE, Thompson RW, et al. Comparison of pedometer and accelerometer measures of free-living physical activity. Med Sci Sports Exerc. 2002;34(12):2045–51.
6. Feito Y, Bassett DR, Thompson DL. Evaluation of activity monitors in controlled and free-living environments. Med Sci Sports Exerc. 2012;44(4):733–41.

7. Clemes SA, O'Connell S, Rogan LM, et al. Evaluation of a commercially available pedometer used to promote physical activity as part of a national programme. Br J Sports Med. 2009;44 (16):1178–83.
8. Ramirez-Marrero FA, Rivera-Brown AM, Nazario CM, et al. Self-reported physical activity in Hispanic adults living with HIV: comparison with accelerometer and pedometer. J Assoc Nurses AIDS Care. 2008;19(4):283–94.
9. Connolly CP, Coe DP, Kendrick JM, et al. Accuracy of physical activity monitors in pregnant women. Med Sci Sports Exerc. 2011;43(6):1100–5.
10. Barreira TV, Brouillette RM, Foil HC, et al. Comparison of older adults' steps/day using NL-1000 pedometer and two GT3X+ accelerometer filters. J Aging Phys Act. 2012;21 (4):402–16.
11. Intille SS, Lester J, Sallis JF, et al. New horizons in sensor development. Med Sci Sports Exerc. 2012;44(1 Suppl 1):S24–31.
12. Tully MA, McBride C, Heron L, et al. The validation of Fibit Zip physical activity monitor as a measure of free-living physical activity. BMC Res Notes. 2014;7:952.
13. Lyons EJ, Lewis ZH, Mayrsohn BG, et al. Behavior change techniques implemented in electronic lifestyle activity monitors: a systematic content analysis. J Med Internet Res. 2014;16(8), e192.
14. Freedson PS, Melanson E, Sirard J. Calibration of the Computer Science and Applications, Inc. accelerometer. Med Sci Sports Exerc. 1998;30(5):777–81.
15. Loprinzi PD, Lee H, Cardinal BJ, et al. The relationship of actigraph accelerometer cut-points for estimating physical activity with selected health outcomes: results from NHANES 2003-06. Res Q Exerc Sport. 2012;83(3):422–30.
16. Strath SJ, Pfeiffer KA, Whitt-Glover MC. Accelerometer use with children, older adults, and adults with functional limitations. Med Sci Sports Exerc. 2012;44(1 Suppl 1):S77–85.
17. Butte NF, Ekelund U, Westerterp KR. Assessing physical activity using wearable monitors: measures of physical activity. Med Sci Sports Exerc. 2012;44(1 Suppl 1):S5–12.
18. Rowlands AV, Rennie K, Kozarski R, et al. Children's physical activity assessed with wrist- and hip-worn accelerometers. Med Sci Sports Exerc. 2014;46(12):2308–16.
19. Troiano RP, McClain JJ, Brychta RJ, et al. Evolution of accelerometer methods for physical activity research. Br J Sports Med. 2014;48(13):1019–23.
20. Vaha-Ypya H, Vasankari T, Husu P, et al. A universal, accurate intensity-based classification of different physical activities using raw data of accelerometer. Clin Physiol Funct Imaging. 2015;35(1):64–70.
21. Lyden K, Keadle SK, Staudenmayer J, et al. A method to estimate free-living active and sedentary behavior from an accelerometer. Med Sci Sports Exerc. 2014;46(2):386–97.
22. Staudenmayer J, Zhu W, Catellier DJ. Statistical considerations in the analysis of accelerometry-based activity monitor data. Med Sci Sports Exerc. 2012;44(1 Suppl 1):S61–7.
23. Freedson P, Bowles HR, Troiano R, et al. Assessment of physical activity using wearable monitors: recommendations for monitor calibration and use in the field. Med Sci Sports Exerc. 2012;44(1 Suppl 1):S1–4.
24. Barreira TV, Schuna JM Jr, Mire EF, et al. Identifying children's nocturnal sleep using 24-hour waist accelerometry. Med Sci Sports Exerc 2014; Sept 8, epub ahead of print.
25. John D, Freedson P. ActiGraph and Actical physical activity monitors: a peek under the hood. Med Sci Sports Exerc. 2012;44(1 Suppl 1):S86–9.
26. Van Hoye K, Boen F, Lefevre J. Validation of the SenseWear armband in different ambient temperatures. J Sports Sci. 2015;33(10):1007–18.
27. Tudor-Locke C, Martin CK, Brashear MM, et al. Predicting doubly labeled water energy expenditure from ambulatory activity. Appl Physiol Nutr Metab. 2012;37(6):1091–100.
28. Tudor-Locke C, Ainsworth BE, Adair LS, et al. Physical activity in Filipino youth: the Cebu Longitudinal Health and Nutrition Survey. Int J Obes Relat Metab Disord. 2003;27(2):181–90.
29. Zhang K, Werner P, Sun M, Pi-Sunyer FX, Boozer CN. Measurement of human daily physical activity. Obes Res. 2003;11(1):33–40.

30. Johannsen DL, Calabro MA, Stewart J, et al. Accuracy of armband monitors for measuring daily energy expenditure in healthy adults. Med Sci Sports Exerc. 2010;42(11):2134–40.
31. John D, Sasaki J, Hickey A, et al. ActiGraph activity monitors: "The Firmware Effect". Med Sci Sports Exerc. 2014;46(4):834–9.
32. Pedisic Z, Bauman A. Accelerometer-based measures in physical activity surveillance: current practices and issues. Br J Sports Med. 2015;49(4):219–23.
33. McClain JJ, Tudor-Locke C. Objective monitoring of physical activity in children: considerations for instrument selection. J Sci Med Sport. 2008;12(5):526–33.
34. Hatano Y. Use of the pedometer for promoting daily walking exercise. J ICHPER. 1993; 1–4.
35. Rowe DA. Back to the future? Algorithms and equipment vs. simplicity and common sense in physical activity measurement. Int J Hum Movement Sci. 2011;5(2):25–45.

Chapter 13
Objective Measurement in Physical Activity Surveillance: Present Role and Future Potential

Adrian Bauman, Željko Pedišić, and Kevin Bragg

Abstract Measuring physical activity at the population level is the central component of a physical activity surveillance system. These systems are used to monitor trends at the national level, identify sub-groups at risk of low activity, and to assess the long-term impact of policies and public health programs targeting physical activity. Surveillance measures assess physical activity in large samples to provide estimates generalizable to whole populations, and must be used in identical ways over many years. They need to be convenient, affordable and feasible to implement in population-representative samples, with good participants' adherence to the measurement tasks required. Thus, surveillance measures pose special challenges for objective assessment of physical activity. Objective assessment of physical activity in populations started in the 1980s, with efforts to measure cardiorespiratory fitness and energy expenditure. These measures were costly, placed a substantial burden on participants, and selection effects limited their generalizability. Since around 2000, population objective assessment has been carried out in several countries using motion-sensing accelerometers. These can measure the intensity and duration of most ambulatory activities, walking and running. They produce population prevalence estimates of meeting physical activity guidelines of around 5 % for adults, rates that are much lower than those assessed by self-reports. In addition, simpler pedometers that record step-counts are also used in population studies. When compared to accelerometers, which are

A. Bauman (✉) • K. Bragg
Prevention Research Collaboration, School of Public Health, Sydney University, Sydney, Australia
e-mail: adrian.bauman@sydney.edu.au

Ž. Pedišić
Prevention Research Collaboration, School of Public Health, Sydney University, Sydney, Australia

Institute of Sport, Exercise and Active Living, Victoria University, Footscray, Melbourne, Australia

Faculty of Kinesiology, University of Zagreb, Zagreb, Croatia

R.J. Shephard, C. Tudor-Locke (eds.), *The Objective Monitoring of Physical Activity: Contributions of Accelerometry to Epidemiology, Exercise Science and Rehabilitation*, Springer Series on Epidemiology and Public Health,
DOI 10.1007/978-3-319-29577-0_13

influenced by regular changes in device models and cut points, pedometers generally provide more comparable estimates over time. However, their comprehensiveness is limited, as they only assess step counts. Most recently, 'wearable technology' has burgeoned, with accelerating growth in the sales of devices such as Fitbits and Jawbone since around 2012. These devices, alongside smart phones and smart watches (that also have accelerometers and other motion sensors built in), provide the potential to reach millions of people. However, the reliability and validity of these devices is not well understood. Further, the models are updated very frequently, which does not influence their usefulness for individuals, but does compromise their potential in population surveillance. The challenge will be to identify affordable technology that is exactly comparable over many years to monitor population physical activity levels.

13.1 Introduction

To tackle the high rates of insufficient physical activity, in 2004 the World Health Organization (WHO) developed the Global Strategy on Diet, Physical Activity and Health [1]. The more recent WHO Global Monitoring Framework set the 10 % reduction in physical inactivity rates by 2025 as one of the nine main targets in preventing non-communicable disease [2]. Long-term and continuous surveillance plays a key role in assessing and monitoring the effects of national and global physical activity strategies. To maximize the utility of population surveillance efforts, we need to use valid physical activity indicators and standardized data-collection procedures.

This chapter focuses on the potential use of objective measures in large-scale surveillance systems to measure and monitor physical activity at the population level. We start with a description of the elements that comprise a good physical activity surveillance system, and indicate the main properties required of population assessment tools. Typically, large-scale surveys and surveillance systems include self-report physical activity measures, which have well-known strengths and limitations in assessing population trends over time. In this chapter, we consider cardio-respiratory fitness measures, accelerometers and pedometers, appraising how these tools conform to the principles of physical activity surveillance. We then present examples of population representative surveys that have used objective measures; although they are not always part of a comprehensive surveillance system, much useful information can be gleaned from them.

Finally, we examine the potential future of surveillance by noting the recent and rapid growth in the field of wearable technology, much of which is capable of monitoring physical activity. We discuss the potential of wearable technology to be scaled up to achieve population-level assessment, and its potential place in population surveillance. We conclude by appraising the suitability of objective measures for physical activity surveillance by comparing their properties with traditionally used self-reports.

13.2 Physical Activity Surveillance

13.2.1 Definition and Examples of Physical Activity Surveillance

Public health surveillance is a process that systematically collects, analyses, interprets, and disseminates population health data. Such surveillance can be performed at the, regional, national or international level, and is usually supported by government. Health surveillance systems initially monitored communicable and vaccine preventable diseases, such as malaria, smallpox and measles. However, in recent decades the burden of non-communicable diseases (NCDs) has become the leading global health problem [3] and the scope of health surveillance was expanded to include tracking of NCDs and their behavioural determinants, such as smoking, dietary habits and physical activity. Nowadays, population data on physical activity is collected within: (a) comprehensive NCD surveillance systems, that cover chronic diseases and their risk factors; (b) behavioural risk factor surveillance systems, focused on a range of health behaviours; (c) specialized physical activity surveillance systems, aiming for a more thorough examination of physical activity behaviour and its antecedents, determinants and correlates; and (d) surveillance systems using representative population data from 'non-health' sectors, such as population censuses and school, sport, recreation, transport and time-use surveys.

Physical activity is assessed within many international- and national-level surveys. For example, by 2015, nearly half of the World's countries have implemented the *WHO STEPwise Approach to Surveillance (STEPS),* a standardized survey that includes data on physical activity and other NCD risk factors [4]. In addition, nearly as many countries are included in the WHO's *Global School-based Student Health Survey (GSHS)* that collected physical activity data among 13–17 years old students [5]. In 43 (mainly European) countries, physical activity of schoolchildren has been monitored over three decades using the *WHO Health Behaviour in School-aged Children (HBSC)* survey [6]. One attribute of national surveys is their diversity, with a recent review finding more than 30 national-level systems that assess physical activity in the European Union countries [7]. Sometimes, more than one survey exists, which may provide different estimates. For example, in the UK only, more than ten national-level health systems monitor physical activity levels [8–13]. Examples of multiple surveys that monitor physical activity can be found even in small states. For instance, among the population of ~65,000 inhabitants of Bermuda, physical activity has been repeatedly monitored within at least five independent national surveys across different age groups [14, 15]. This demonstrates the substantial, but mostly uncoordinated efforts to understand and assess physical activity levels, and identifies a plethora of potentially contrasting estimates and trends. This may confuse policymakers, in the way that no single estimate reflects 'national progress towards physical activity goals', and adding objective measures makes this situation even more complicated.

Fig. 13.1 Michael Pratt authored some of the key papers on physical activity surveillance issues and challenges

13.2.2 Purpose and Objectives of Physical Activity Surveillance

Surveillance systems are integrated and comprehensive processes of monitoring physical activity and its antecedents in large regions or at the national level. These systems collect information to: inform policymakers about population-level physical activity behavior and its changes over time. As emphasized by Caroline Macera and Michael Pratt (Fig. 13.1), physical activity surveillance should go beyond simply tracking prevalence rates of physical activity and establishing their trends over time [16]. Objectives of 'optimal' physical activity surveillance should include: (i) estimating total, intensity, domain and type-specific physical activity levels; (ii) identifying determinants and correlates of physical activity, including socio-demographic, behavioral, psychological, health, social, and environmental factors; and (iii) monitoring policies, practices and actions that facilitate or impede physical activity across different settings (e.g. local governments, local communities, schools, universities, workplaces).

13.2.3 Principles of Optimal Physical Activity Surveillance

In the following text, we discuss the main principles of population-wide physical activity surveillance, which has different measurement needs to those at the individual or small-scale level. These principles pose specific challenges for implementing objective measures in physical activity surveillance. Further, academic discourse and optimizing of measurement techniques often continues for many years, whilst the surveillance needs of regional and national Governments, and other population health stakeholders are much more acute, and require definitive solutions that can be immediately implemented in surveillance systems for many years.

Fig. 13.2 Paul Loprinzi examined the generalizability of accelerometer-based physical activity estimates among representative samples of US children, adolescents and adults

13.2.3.1 Generalizability

It is essential that physical activity surveillance system provides population-generalizable estimates of physical activity. This is achieved by using appropriate sampling designs and large sample sizes. In order to ensure adequate response rates and adherence to study protocols, data collection methods need to be acceptable to most selected participants. Those who do not adhere to study protocols may systematically differ from participants who remain in the study (as noted by Paul Loprinzi (Fig. 13.2) in an analysis of US population surveys [17].

13.2.3.2 Simplicity

Respondent and researcher burden (e.g. time and complexity of carrying out the measure) should be minimized by selecting simple measurement tools.

13.2.3.3 Data Quality

Health surveillance systems need to provide credible population-level estimates of physical activity, using measures that have shown acceptable reliability and validity.

13.2.3.4 Comprehensiveness

To estimate the population-level adherence to physical activity recommendations [18] selected measures must provide data on the time spent in moderate to vigorous-intensity physical activities. Additionally, weekly frequency of muscle-strengthening exercises should be assessed in surveillance systems, as they are included as a separate recommendation in the recent physical activity guidelines [18]. In an optimal measure, domain-specific data (measures of work, transport, domestic, and leisure-time activity) and type-specific data for common activities, such as walking and cycling, and more recently sedentary behavior (sitting time) should also be assessed. In addition, an optimal surveillance system needs to monitor more than just physical activity behaviors, and ideally should assess personal, societal, environmental, and policy-related determinants of activity [19].

13.2.3.5 Between-Study Comparability

To allow for comparisons among countries, surveillance systems should aim to use comparable measures. For example, some self-report tools have been standardized, such as the International Physical Activity Questionnaire (IPAQ) [20], and the Global Physical Activity Questionnaire (GPAQ) [21] used in the WHO *STEPS* surveillance system [2].

13.2.3.6 Continuity and Sustainability

To allow for monitoring changes in population behavior, physical activity surveillance needs to retain comparable measurement methods over many years in order to observe time trends.

13.2.3.7 Adaptability

Although ideally data collection protocol should always be identical, advances in knowledge and technological and societal development sometimes necessitate slight adjustments [22]. Data collection protocols should be flexible enough to allow for necessary adaptations without compromising data comparability over time.

13.2.3.8 Affordability

Limited funds need to be carefully distributed across physical activity surveillance functions. Increased investment in one component should never lead to reduced funding available for other important components of the surveillance system.

13.3 Use of Objective Measures in Population Surveys: Is There Potential for Surveillance?

13.3.1 Traditional and Current Measures

Initial surveillance systems used self-report questions, asking responders about their physical activity behaviors, participation in different activities, exercise and sports [23]. These questions were asked through telephone or personal interviews, and through self-completed surveys, using paper-based or web-based modes. Strengths of self-report questions included high-level population representativeness, simplicity and brevity of most measures, and the capacity to ask about domains of, and different types of physical activities. Measures such as IPAQ and GPAQ have been used to describe and compare prevalence estimates across countries using similar methods [24, 25]. Self-report measures are amongst the most affordable methods for population measurement, and their use can be sustained over time, with US and Nordic surveys as examples of collecting comparable physical activity data for several decades [22, 26]. However, there are some limitations to self-report measures, including variable reliability and validity among countries and in sub-populations [20]. Further, self-reports of physical activity may be prone to social desirability and reporting biases, and to differential errors in reporting and recall across population groups.

To address these concerns with self-reports, since the 1980s there has been a growing interest in objective measurement of physical activity. Measures of cardiorespiratory fitness were often used as a proxy for the training effects of physical activity [27]. These measures were reliable, but often required special clinic facilities to undertake assessments. Several population studies measured cardiorespiratory fitness, but rarely attempted to repeat measurements over time as part of health surveillance systems [28]. They usually assessed submaximal and maximal oxygen uptake, or the heart rate response to exercise. Whilst these measures proved useful in clinical interventions to enhance fitness, they were less suitable for physical activity surveillance for several reasons. These assessments were expensive to undertake in large samples, did not measure domains of activity or activity-specific participation, and selection effects were seen in the characteristics of individuals that completed the assessments. In addition, fitness is partly genetically determined [29], and does not always reflect recent physical activity *behaviors*, which may partly contribute independently of fitness levels to improved health and disease prevention [30].

More recently, scientists have partnered with engineers to utilize tools that could objectively assess movement. These included a range of models of accelerometers (see earlier chapters), which usually showed very good reliability. In Table 13.1 we present examples of their use in large national samples, all of which aimed to produce representative estimates of population physical activity. Reflecting on the use of accelerometers as surveillance tools involves consideration of the best practice attributes of a surveillance system discussed earlier. These population

Table 13.1 Examples of objective measures of physical activity in population representative surveys

Study/Survey	Sample	Methods of PA assessment	Main findings relevant to objective PA assessment	Surveillance-related attributes
Examples of accelerometer based estimates				
Troiano et al. NHANES [47]	n = 6329 adults (USA) representative sample	Actigraph 7164 accelerometer (7 days wear time)	2–6 % met MVPA recommendations (25–33 % met guidelines as assessed by self-report)	Provided much lower prevalence estimates than self-reports; expensive survey
Evenson et al. [48] NHANES (2003/4, 2005/6)	n = 2630 older adults aged 60+ years	Actigraph 7164 accelerometer different cut points in counts/minutes used	Minutes of MVPA/day varied with cut points used (59′/d for 500 + cpm; down to 9.2′/d for 2000 + cpm used as cut-point), 10′ bouts used. Mean 8.5 hours/day of sedentary behaviour	Different accelerometer cut points led to different prevalence estimates
Hansen et al. [49] Norwegian Survey (2008/9)	n = 3267 adult Norwegians with adequate objective data (30 % of those surveyed)	Actigraph GT1M (7 day wear time) accelerometer	~8000 steps/day estimated by accelerometry (22.7 % met 10,000 steps recommendation) 17.4 % met MVPA guidelines (bouts of 10′ or more) mean 9.1 hours/day of sedentary behavior (62 % of wearing time)	Use of accelerometers for step counts; sedentary time assessed
Colley et al. [50] Health Measurement Survey (2007/9)	Canadian adults (n = 2832) (42 % participation rate)	Actical accelerometer (7 day wear time)	MVPA 27′/d (M), 21′/d (F) Steps/day 9544 (M), 8385 (F) Met MVPA guidelines (30′/d, 10′ bouts): 5.5 % (M), 4 % (F); met 150′/d guidelines 17 % (M), 13.7 % (F) Met 10,000 steps/day 39 % (M), 30 % (F)	Different cut points led to different population prevalence estimates

(continued)

Table 13.1 (continued)

Study/Survey	Sample	Methods of PA assessment	Main findings relevant to objective PA assessment	Surveillance-related attributes
Hagstromer et al. [51, 52] (Survey in 2001)	Swedish adults recruited to participate (34 % with accelerometer data) Comparison with USA data (2010)	Actigraph 7164 (7 day wear time)	Mean 31′/d MVPA (28′ were MPA) Inactivity (SB) ~ 7.7 hours/day Met MVPA guidelines (10′ bouts) 7 % (only 1 % if all bouts ≥10′) SB time (8.3 hours/day Sweden, 7.4–9.0 hours/day USA) MVPA—similar for aged 18–39 and 40–59, but greater in older Swedes than Americans (29′ vs 15′/d)	Assessed MVPA and sedentary time; different cut-points led to different prevalence estimates
Baptista et al. [53] (Survey 2006/8)	Portuguese adults + children (n = 4946) Data available on 3362 (of whom, 24.3 % were children/adolescents)	Actigraph GTIM accelerometer (4 day wear time)	MVPA minutes decline through adolescence and again decline after aged 65+ years ~70 % adults met MVPA guidelines (30′/d), but including 10′ bouts reduced prevalence to 3–9 % across adult age groups	Objective data confirm decline in MVPA through adolescence
Ekelund et al. [54] Multi-country data base (ICAD)	14 accelerometer-based data sets on children (not all representative) (pooled n = 20,871)	Pooled accelerometer data using Kinesoft software to obtain comparable estimates	Mean MVPA time 30′/d, mean SB time 5.9 hours/day	Multi-country comparisons possible
Pedometer based estimates				
Craig et al. [36] Craig et al. [55] Continuous monitoring surveys of	Baseline 2005/7 (n = 11,669 children) (~2/3 of those approached) Long-term trend	Yamax SW-200 pedometer	Provided age-sex norms for pedometer steps distribution; Boys 5–19 years, mean 12,259 steps/day; girls 10,906	Pedometers are simple, feasible in schools; methods comparable; step count is only outcome;

(continued)

Table 13.1 (continued)

Study/Survey	Sample	Methods of PA assessment	Main findings relevant to objective PA assessment	Surveillance-related attributes
children and adolescents (2005–2014)	data to be reported		steps/day ~40 % boys, half of girls did not meet 10,000 steps/day	sustained PA surveillance
Inoue et al. [37] Annual Japan Health and Nutrition surveys (1995-2007)	~8000 adults surveyed per year	Yamax AS200 pedometer	Mean 7321 steps/day (M), 6267 steps/day (F) [2007]; ~19 % met 10,000 steps/day Decline since year 2000 of around 500 steps/day (M) and 850 steps/day (F)	Part of national surveys, provide useful trends demonstrating slight decreases in adult PA in Japan
Matthiessen et al. [56] Danish population surveys (2007/8 and 2011/12)	Adults 18–75 years, ~55 % response rate 2007/8 (n = 205) and 2011/12 (n = 1419)	Yamax SW-200 pedometer	Steps/day decreased from 8788 to 8341 over 5 year period	Small initial sample, but representative; showed feasible for Danish adults

Legend: *PA* physical activity, *MVPA* moderate-to-vigorous physical activity, *′/d* minutes per day

surveys (Table 13.1) were sometimes intended as the start of physical activity surveillance systems, but few assessed populations on more than one occasion using identical methods and accelerometer models. For example, this occurred in the NHANES in the USA, where different models and placement were used on two measurement occasions 5 years apart, leading to comparability problems [31].

Accelerometers are expensive, and continuous technological upgrades maintain their high costs, while providing less comparable between-device data. Accelerometer data may be collected from population samples, but selection issues occur in the sub-set of the responders who complete sufficient time wearing the accelerometers. This is due to the complexity of wearing the devices for several days, and the associated respondent burden. Accelerometer data provided useful information on moderate and vigorous activity time and intensity, but less good differentiation between sitting and standing time, and no information on domains of activity or activity types. More recent algorithms are being developed to use comparable raw accelerometer data, which may overcome academic debate about where the count-specific cut points lie between different intensities of activity [31]. However, some accelerometer raw data are not available, with proprietary methods introduced by the manufacturing companies that preclude their access.

One key finding from accelerometer-based population surveys is the difference in observed prevalence of physical activity, compared to self-report measures. Accelerometer data indicate much smaller prevalence rates of meeting physical activity guidelines among adults, ranging from one half to one –tenth of the self-report prevalence. Different accelerometer cut-points produce very different estimates [32] and their most stringent estimates typically demonstrate that only 5 % of adults or fewer meet the 'five days a week with 30 minutes of MVPA, in bouts of at least 10 minutes' physical activity guideline (see Table 13.1). Accelerometer measurement among children is also variable. Guinhouya [33] in a systematic review of 40 studies that used accelerometry in children identified that only one third of studies provided the percentage meeting physical activity guidelines, and noted that different cut points for accelerometers resulted in different estimates; the most often used cut point (>2000 counts per minute) classified almost all children as sufficiently active (87 %). Further, markedly lower prevalence rates of meeting physical activity guidelines were seen with different cut points, ranging from almost all children meeting a threshold, down to only 3–5 % classified as "active".

Further, accelerometers provide good currently available data on movement, but miss some activities and patterns of activity. Self-reports may also over-estimate or under-estimate activity, and this may be differential between populations [34]. Therefore, the best approach may be to combine objective and self-report estimates to account for different sources of measurement error [35].

One alternative approach is to use pedometers to assess and monitor population step counts (see Table 13.1; [36, 37]). Step-counts have been proposed as a health indicator through an examination of cadence (steps/minute), with those achieving 'high' or 'very high' daily step goals (such as 10,000 or 12,500 steps/day) being much more likely to meet physical activity recommended levels than those with lower step counts [38]. The pedometer-based surveys have broad reach, are less expensive than accelerometry, and have demonstrated successful use in serial surveillance systems among children in Canada, and among adults in Japan and Denmark. The main limitations of the pedometer are that they do not provide data on MVPA and time spent in different intensities, and that some prevalent activities are not well captured (for example, cycling). Nonetheless, pedometers are a feasible and affordable method for surveillance of walking-related activities, consistently the most prevalent activity type across the past decades and throughout adulthood [39].

13.3.2 Emerging Wearable Technologies

New technologies provide high population-reach devices. Many have the capacity to measure and assess physical activity and exercise. The most recent group, wearable technology ("wearables"), defines a range of usually waist or wrist-

Table 13.2 Examples of the popularly used wearable technology for assessing physical activity [2011–2014][a]

Company	Model	Hardware	Data recorded	Data output/second (interface)
Fitbit	Charge (Wrist worn)	Triaxial accelerometer Altimeter	Activity (1 minute epochs) (Sampling rate data unavailable)	– Steps, Distance, Energy expenditure, Stairs climbed (Device, Smartphone app, Online) – Activity minutes, Sleep time & efficiency (Smartphone app, Online) – Sedentary minutes (Online)
Fitbit	Charge + HR (Wrist worn)	Triaxial accelerometer Altimeter Optical heart rate monitor	Activity (1 minute epochs) (Sampling rate data unavailable) HR (1–5 second epochs)	– Steps, Distance, Energy expenditure, Stairs climbed, Heart rate (Device, Smartphone app, Online) – Activity minutes, Sleep time & efficiency (Smartphone app, Online) – Sedentary minutes (Online)
Fitbit	Flex (Wrist worn)	MEMS Triaxial accelerometer	Epoch length and sampling rate data unavailable	– Steps, Distance, Energy expenditure (Smartphone app, Online) [One output, as chosen by user, is viewable on the device, expressed in 20 % increments of user set goal] – Activity minutes, Sleep time & efficiency (Smartphone app, Online) – Sedentary minutes (Online)
Fitbit	One (Waist worn)	MEMS Triaxial accelerometer	Epoch length and sampling rate data unavailable	– Steps, Distance, Energy expenditure, Stairs climbed (Device, Smartphone app, Online) – Activity minutes, Sleep time & efficiency (Smartphone app, Online) – Sedentary minutes (Online)
Fitbit	Surge (Wrist worn)	GPS Triaxial accelerometer Triaxial gyroscope Digital compass Optical HR monitor Altimeter Ambient light sensor	Activity (1 minute epochs) (Sampling rate data unavailable) HR (1–5 second epochs) GPS sampling rate (1 Hz)	– Steps, Distance, Energy expenditure, Stairs climbed, Heart rate (Device, Smartphone app, Online) – Activity minutes, Sleep time & efficiency (Smartphone app, Online) – Sedentary minutes (Online)

(continued)

Table 13.2 (continued)

Company	Model	Hardware	Data recorded	Data output/second (interface)
Fitbit	Zip (Waist worn)	MEMS Triaxial accelerometer	Epoch length and sampling rate data unavailable	– Steps, Distance, Energy expenditure, Stairs climbed (Device, Smartphone app, Online) – Activity minutes, Sleep time & efficiency (Smartphone app, Online) – Sedentary minutes (Online)
Garmin	Vivoactiv (Wrist worn)	Accelerometer GPS	Epoch length and sampling rate data unavailable	– Steps, Distance, Energy expenditure, Sedentary period alerts (Device, Smartphone app, Online) – Sleep time & efficiency (Smartphone app, Online) – Heart rate (Requires additional accessory)
Garmin	Vivofit (Wrist worn)	Accelerometer	Epoch length and sampling rate data unavailable	– Steps, Distance, Energy expenditure, Sedentary period alerts (Device, Smartphone app, Online) – Sleep time & efficiency (Smartphone app, Online)
Garmin	Vivofit 2 (Wrist worn)	Accelerometer	Epoch length and sampling rate data unavailable	– Steps, Distance, Energy expenditure, Sedentary period alerts (Device, Smartphone app, Online) – Sleep time & efficiency (Smartphone app, Online)
Garmin	Vivosmart (Wrist worn)	Accelerometer	Epoch length and sampling rate data unavailable	– Steps, Distance, Energy expenditure, Sedentary period alerts (Device, Smartphone app, Online) – Sleep time & efficiency (Smartphone app, Online) – Heart rate (Requires additional accessory)
Jawbone	UP3 (Wrist worn)	Triaxial accelerometer Bioelectrical impedance monitor/second Thermometer	Epoch length and sampling rate data unavailable	– Steps, Distance, Energy expenditure, Activity minutes, Sleep time & efficiency, Heart rate, Galvanic skin response, Respiration, Skin temperature, Ambient temperature (Smartphone app) – Sedentary alerts optional (Device vibrates)

(continued)

Table 13.2 (continued)

Company	Model	Hardware	Data recorded	Data output/second (interface)
Jawbone	UP24 (Wrist worn)	Triaxial accelerometer	Epoch length and sampling rate data unavailable	– Steps, Distance, Energy expenditure, Activity minutes, Sleep time & efficiency (Smartphone app) – Sedentary alerts optional (Device vibrates)
Jawbone	UP Move (Waist worn)	Triaxial accelerometer	Epoch length and sampling rate data unavailable	– Steps, Distance, Energy expenditure, Activity minutes, Sleep time & efficiency (Smartphone app) – Device outputs progress towards some measure using 12 LEDs
Misfit	Flash	Triaxial accelerometer	Epoch length and sampling rate data unavailable	– Proprietary accumulated activity measure (Device, Smartphone app) – Steps, Distance, Energy Expenditure, Sleep time & efficiency (Smartphone app)
Misfit	Shine	Triaxial accelerometer	Epoch length and sampling rate data unavailable	– Proprietary accumulated activity measure (Device, Smartphone app) – Steps, Distance, Energy Expenditure, Sleep time & efficiency (Smartphone app)
Nike	Fuel Band	Triaxial accelerometer	Epoch length and sampling rate data unavailable	– Proprietary accumulated activity measure, Steps, Energy Expenditure (Device, Smartphone app) – Distance (Smartphone app)

[a]References to this table are shown in the Appendix, at the end of the reference list, as these are not published sources

worn, consumer-oriented devices that aim to objectively assess physical activity (PA) of the wearer. The majority utilize types of accelerometry to determine various physical activity measures, typically offering consumers estimates of their total and hourly number of steps, distance travelled, energy expenditure (EE) and active minutes. In addition, some devices are advertised as measuring sleep efficiency, stairs climbed, heart rate and even galvanic skin response. The range of these devices available between 2011 and 2014 is shown in Table 13.2. This table shows the hardware, the data capture method, and the range of health-related measures that each device offers to individual consumers.

Since around 2007, the range and functionality of wearable devices has markedly increased. The major companies currently marketing wearables include Fitbit, Garmin, Jawbone, Misfit and Nike. Each company has an accompanying smartphone application or operating platform to facilitate downloading of physical activity data from the device. The applications provide an interface that allows the user to observe real-time presentation of physical activity data as it is collected, usually retrieved from a proprietary storage location. It seems that the marketing and sales of these wearables increased exponentially between 2012 and 2014. Determining population reach is difficult as sales data are either not disclosed or substantially limited. Nonetheless, marketing data suggest that during 2014, more than 10 million wearable devices were sold, with Fitbits comprising half of them [40]. Increases each year were marked, especially for Fitbit and Jawbone devices.

Smartphones have inbuilt accelerometers and sometimes GPS devices, and also pose a vast market of phones capable of measuring physical activity. The number of smartphones vastly exceeds wearable devices, with up to a billion devices sold between 2007 and 2014 (based on commercial sales estimates) [41]. Additionally, Apple has started including proprietary accelerometers and health applications directly into their smartphones, and has moved into the wearable market by producing a smart Apple watch in 2015. Google has also developed an Android watch platform. Further, a broad range of Apps are available for assessing physical activity under both Apple and Android platforms; more sophisticated examples, such as *Moves, Nike + Running* and *RunKeeper*, utilize both the GPS and location services available on smartphones to track physical activity and map the location where the physical activity was conducted. Apple data sales numbers are not released through iTunes, but Google Play provides broad ranges listing *Moves* as having between 1 and 5 million installs and *Nike + Running* and *RunKeeper* as having had between 10 and 50 million installs. The duration of use after download is not known, and the point prevalence of current use is also not known for specific applications.

Wearable devices are heavily marketed for their potential to measure physical activity and provide feedback to consumers. Given the market share of leading brands (Apple, Android phone platforms; Fitbit and Jawbone wearables) these devices have the potential to generate 'big data' and are already measuring physical activity among millions of people. Their potential as surveillance tools depends on their measurement properties, the representativeness of their users, and the capacity for data sharing (given that data collection and storage are generally limited by the commercial-in-confidence practices of companies marketing these devices).

The measures generated by wearables and phone apps may inform individuals, but information on their reliability, validity and utility is less well known. Given their sales-oriented focus, companies release very little research examining measurement aspects of wearable technologies. The rapid turnover and proactive marketing of new devices lead to discontinuation of older devices, and make any validation research rapidly obsolete.

Although limited, the publicly available evidence suggests that wearable devices show variable measurement properties. An analysis of the Fitbit One, during treadmill walking, by Takacs et al. [42] found that it produced step counts that did not differ significantly from observer counts. Furthermore, each participant wore three devices, all of which demonstrated inter-device consistency. However, simultaneously, each device significantly over-estimated the distance travelled. Ceaser [43] and Stackpool [44] also found varied validity when examining multiple devices during various activity assessments. The former study [43] found that energy expenditure estimates by the Nike Fuel Band differed significantly from the indirect calorimetry for the majority of activities examined. Furthermore, the differences included both under- and over-estimations. A similar finding was shown for the Fitbit. The latter study [44] found the energy expenditure estimates produced by the Fuel Band were significantly different from criterion measures. The Jawbone UP showed significant differences in half the activities, with varied direction and the Fitbit Ultra only differed significantly for one activity type. In the same study, when examining the validity of step counts, the Fuel Band underestimated step counts for half of the activity types, as did the Fitbit Ultra. The Jawbone UP did not mis-estimate step counts for any activity. However, in a similar issue to Ceaser [43], both the Fitbit Ultra and Jawbone UP have been superseded and are no longer sold directly by their respective companies.

Case et al. [45] also conducted a multiple device study, including the Fitbits Flex, One and Zip, Jawbone UP24, Nike Fuel Band and several apps, including the Moves app on both an iPhone and Samsung, and the Fitbit and HealthMate app on the iPhone alone. Whilst their results are limited by a small sample, they found relatively low errors in step count calculations for most devices and apps, excepting the Fuel Band. However, all of these studies were conducted under controlled circumstances where, arguably, the greater interest is in the physical activity in free-living conditions.

One study, conducted by Tully et al. [46] examined the step counts produced over seven free-living days by a Fitbit Zip and a Yamax CW700 pedometer, using the ActiGraph GT3X as a well-validated criterion measure. Whilst they found the Fitbit Zip significantly overestimated steps relative to the ActiGraph, there was no bias present in the Bland-Altman analysis and no difference relative to the Yamax pedometer. Hence, the evidence for the validity of wearable physical activity monitors is mixed, but simultaneously quite sparse. Whilst further research is warranted, as in the cases above, in a matter of years these specific devices are likely to become outdated.

Regardless of their current shortcomings, wearables may offer the community many of the measurement benefits of accelerometry, at low initial cost, and personalized motivational feedback using an individual's own data. More work and scrutiny are required to evaluate the validity, reliability and population reach of wearables. We also need to understand the demographic distribution of users, the medium-term adherence of users to specific models, and the cross-model validity as many devices are annually updated. Given their reach, their potential is substantial to assess physical activity at the population level, if the issues of the representativeness of users can be further confirmed, and the issues of data confidentiality, data privacy and model upgrades can be overcome.

13.4 Conclusions

Ideally, surveillance systems use measures that are reliable and valid and provide comprehensive data on all dimensions of physical activity, whilst being inexpensive and causing a minimal burden for researchers and participants. Physical activity is a complex behaviour and, for the time being, such 'ideal' tools are not available. Each available measure has comparative advantages and disadvantages with regards to its use for physical activity surveillance. Moreover, these attributes may substantially vary across different age groups. Therefore, measures need to be selected to suit the purpose of the measurement and the target population, while trying to find an optimal balance between their qualities.

Table 13.3 summarizes this chapter, appraising the suitability of objective measures for physical activity surveillance by comparing them with self-reports as the reference group of measures. This table shows that the relative advantages of objective measurements for surveillance are still not substantial or well enough understood to replace self-report surveillance measures. In particular, the simplicity and affordability of self-report measures make them suitable and sustainable for surveillance systems, whilst the challenge for objective measurement is to establish standardized measurement protocols and sustain them over time. Among the available options, pedometers seem to have more scalable properties that make them suitable for population assessment. Finally, wearable technology has the greatest potential for population reach, but the least is known about their reliability, validity and sustainability.

To conclude, further development of objective measures is needed to improve their suitability for physical activity surveillance. This refers particularly to standardization of measurement protocols and their sustainable use over time. Due to recent technological advances and increased availability of physical activity monitors, researchers may face a dilemma between the consistency in using

Table 13.3 Advantages and disadvantages of objective measures for physical activity surveillance when compared to self-reports

Attribute	CR fitness measures	Accelerometers	Pedometers	Wearable technologies
Generalizability	?	–	?	?
Simplicity	– –	– –	–	+/–
Data quality	+	+	++	?
Comprehensiveness (activity types, domains)	– –	–	– –	–
Between-study comparability	++	+/–	+	?
Continuity and sustainability	–	–	+/–	– –
Adaptability	– –	–	–	–
Affordability	– –	– –	–	+/–

Legend: CR cardio-respiratory; "?" unknown; "–" worse than self-reports; "+/–" equal to self-reports; "+" better than self-reports; more pluses or minuses denote greater advantage/disadvantage when compared to self-reports questionnaires and switching to new methods. Based on current evidence, it seems that objective measures should not be used to substitute, but only to complement self-reports in physical activity surveillance systems. Finally, the potential for wearable technology generates much enthusiasm, but for population surveillance, it should be tempered by more informed consideration of the measurement attributes of these devices, and by the problems posed by ever-changing technology on comparable monitoring of physical active prevalence estimates.

Appendix: Specific References to Wearable Technology

1. Fitbit Website, 2015, Fitbit, San Francisco, viewed 18 April 2015, http://www.fitbit.com/
2. Fitbit About Page, 2015, Fitbit, San Francisco, viewed 18 April 2015, http://www.fitbit.com/au/about
3. The United States Securities and Exchange Commission [SEC], 2000, *Form S-1 Registration Statement, Garmin,* viewed 18 April 2015, https://www.sec.gov/Archives/edgar/data/1121788/000101381600000038/0001013816-00-000038-0001.txt
4. Garmin, 2003, *Garmin® Unveils Low-Cost, GPS Fitness Product—the Forerunner™ 101,* viewed 18 April 2015, https://www8.garmin.com/pressroom/outdoor/030204.html
5. Jawbone Press Release Page, 2015, Jawbone, San Francisco, viewed 18 April 2015, https://jawbone.com/press
6. The United States Securities and Exchange Commission [SEC], 2000, *Form D Notice of Exempt Offering of Securities, Misfit,* viewed 18 April 2015, https://www.sec.gov/Archives/edgar/data/1535557/000153555712000001/xslFormDX01/primary_doc.xml
7. PRNewswire, 2012, *Misfit Unveils Shine: World's First All-Metal Wireless Fitness Device*, viewed 18 April 2015, http://www.prnewswire.com/news-releases/misfit-unveils-shine-worlds-first-all-metal-wireless-fitness-device-179367061.html
8. The United States Patent and Trademark Office [USPTO], 2015, *Trademark Status & Document Retrieval (TSDR),* viewed 18 April 2015, https://tsdr.uspto.gov/#caseNumber=72414177&caseType=SERIAL_NO&searchType=statusSearch
9. CNet, 2014, *Nike FuelBand—out of gas already?*, viewed 18 April 2015, http://www.cnet.com/news/is-nike-fuelband-out-of-gas-already/
10. Business Insider Australia, 2014, *Just 3.3 Million Fitness Trackers Were Sold In The US In The Past Year*, viewed 18 April 2015, http://www.businessinsider.com.au/33-million-fitness-trackers-were-sold-in-the-us-in-the-past-year-2014-5
11. Re/code.net, 2014, *As Nike Downsizes Digital Division, Data Shows It Trailing Fitbit, Jawbone,* viewed 18 April 2015, https://recode.net/2014/04/24/as-nike-downsizes-digital-division-data-shows-it-trailing-fitbit-jawbone/

12. Google Play Online Store, 2015, *Moves App Page*, viewed 18 April 2015, https://play.google.com/store/apps/details?id=com.protogeo.moves&hl=en
13. Google Play Online Store, 2015, *Nike + Running App Page,* viewed 18 April 2015, https://play.google.com/store/apps/details?id=com.nike.plusgps&hl=en
14. Google Play Online Store, 2015, RunKeeper App Page, viewed 18 April 2015, https://play.google.com/store/apps/details?id=com.fitnesskeeper.runkeeper.pro&hl=en
15. Fitbit Inc. (Not Dated) *Fitbit charge product manual version 1.1.* http://help.fitbit.com/resource/manual_charge_en_US
16. Fitbit Inc. (Not Dated) *Fitbit zip product manual.* http://help.fitbit.com/resource/manual_zip_en_US
17. Garmin Ltd. (2015) *Garmin Vivofit Product Information Webpage "Specs" Sub tab.* Retrieved 15 January 2015 from https://buy.garmin.com/en-US/US/into- sports/health-fitness/vivofit-/prod143405.html
18. Jawbone (2015) *Jawbone UP24 User Manual: Data Accuracy.* Retrieved 15 January 2015 from https://help.jawbone.com/articles/en_US/PKB_Article/data-accuracy- up24
19. Misfit Inc. (2015) *Misfit Shine Product Information Webpage "Specs" Subsection.* Retrieved 15 January 2015 from http://misfit.com/products/shine?locale=en#specs

References

1. World Health Organization. Global strategy on diet, physical activity and health. Geneva, CH: World Health Organization; 2004.
2. World Health Organization: NCD Global monitoring framework. http://www.who.int/nmh/global_monitoring_framework/en/ (2012)
3. Lozano R, Naghavi M, Foreman K, et al. Global and regional mortality from 235 causes of death for 20 age groups in 1990 and 2010: A systematic analysis for the Global Burden of Disease Study 2010. Lancet. 2012;380(9859):2095–128.
4. World Health Organization. WHO STEPS surveillance manual: the WHO STEPwise approach to chronic disease risk factor surveillance/Noncommunicable Diseases and Mental Health. Geneva, CH: WHO Press; 2005.
5. World Health Organization. Global School-based Student Health Survey (GSHS): World Health Organization. http://www.who.int/chp/gshs/en/ (2013). Accessed 9 Sept 2013.
6. World Health Organization. Social determinants of health and well-being among young people. Health Behaviour in School-aged Children (HBSC) study: International report from the 2009/2010 survey. Copenhagen, DK: WHO Regional Office for Europe; 2012.
7. World Health Organization. Review of physical activity surveillance data sources in European Union member states. Copenhagen, DM: WHO Regional Office for Europe; 2010.
8. National Obesity Observatory (NOO). Physical activity surveillance in England: what is measured and where are the gaps? Oxford: NOO; 2009.
9. Scottish Health Survey Team. The Scottish Health Survey 2011. Edinburgh: The Scottish Government; 2012.
10. Sadler K, Doyle M, Hussey D, Pickering K, Stafford R. Welsh Health Survey 2011. London: NatCen Social Research; 2012.

11. Central Survey Unit. Northern Ireland Health and Social Wellbeing Survey 2005/06. Belfast: Department of Health, Social Services and Public Safety; 2007.
12. Department of Health SSPS. Health Survey Northern Ireland: First results from the 2010/11 survey. Belfast: Public Health Information & Research Branch Information & Analysis Directorate; 2011.
13. NatCen Social Research. National Diet and Nutrition Survey headline results from years 1, 2 and 3 (combined) of the rolling programme (2008/2009–2010/11). London: Department of Health and the Food Standards Agency; 2011.
14. Cordell WR. Ageing trends. Hamilton, Bermuda: Government of Bermuda; 2008.
15. Bermuda Health Council and Department of Health. Health in review: an international comparative analysis of Bermuda health system indicators. Hamilton, Bermuda: Government of Bermuda; 2011.
16. Macera CA, Pratt M. Public health surveillance of physical activity. Res Q Exerc Sport. 2000;71(S2):97–103.
17. Loprinzi PD, Cardinal BJ, Crespo CJ, et al. Differences in demographic, behavioral, and biological variables between those with valid and invalid accelerometry data: Implications for generalizability. J Phys Act Health. 2013;10(1):79–84.
18. World Health Organization. Global recommendations on physical activity for health. Geneva, Switzerland: World Health Organization; 2010.
19. Stokols D. Translating social ecological theory into guidelines for community health promotion. Am J Health Promot. 1996;10(4):282–98.
20. Craig CL, Marshall AL, Sjöström M, et al. International physical activity questionnaire: 12-Country reliability and validity. Med Sci Sports Exerc. 2003;35(8):1381–95.
21. Bull FC, Maslin TS, Armstrong T. Global physical activity questionnaire (GPAQ): nine country reliability and validity study. J Phys Act Health. 2009;6(6):790–804.
22. Mokdad AH. The behavioral risk factors surveillance system: past, present, and future. Annu Rev Public Health. 2009;30:43–54.
23. Craig CL, Russell SJ, Cameron C, Bauman A. Twenty–year trends in physical activity among Canadian adults. Can J Publ Health. 2004;95(1):59–63.
24. Bauman A, Bull F, Chey T, Craig CL, et al. The International Prevalence Study on Physical Activity: results from 20 countries. Int J Behav Nutrit Phys Act. 2009;6:21.
25. Guthold R, Ono T, Strong KL, et al. Worldwide variability in physical inactivity: a 51-country survey. Am J Prev Med. 2008;34(6):486–94.
26. Harro M, Oja L, Tekkel M, et al. Monitoring physical activity in Baltic countries: The FINBALT study, HBSC and other surveys in young people. J Public Health. 2006;14:103–9.
27. Shephard RJ, Aoyagi Y. Measurement of human energy expenditure, with particular reference to field studies: an historical perspective. Eur J Appl Physiol. 2012;112(8):2785–815.
28. Kaminsky LA, Arena R, Beckie TM, et al. The importance of cardiorespiratory fitness in the United States: the need for a national registry: a policy statement from the American Heart Association. Circulation. 2013;127(5):652–62.
29. Tucker R, Collins M. What makes champions? A review of the relative contribution of genes and training to sporting success. Br J Sports Med. 2012;46(8):555–61.
30. Sherman SE, D'Agostino RB, Silbershatz H, et al. Comparison of past versus recent physical activity in the prevention of premature death and coronary artery disease. Am Heart J. 1999;138(5):900–7.
31. Pedišić Ž, Bauman A. Accelerometer-based measures in physical activity surveillance: current practices and issues. Br J Sports Med. 2015;49(4):219–23.
32. Loprinzi PD, Lee H, Cardinal BJ, et al. The relationship of actigraph accelerometer cut-points for estimating physical activity with selected health outcomes: results from NHANES 2003-06. Res Q Exerc Sport. 2012;83(3):422–30.
33. Guinhouya B, Samouda H, de Beaufort C. Level of physical activity among children and adolescents in Europe: a review of physical activity assessed objectively by accelerometry. Public Health. 2013;127(4):301–11.

34. Prince SA, Adamo KB, Hamel ME, et al. A comparison of direct versus self-report measures for assessing physical activity in adults: a systematic review. Int J Behav Nutr Phys Act. 2008;5:56.
35. Troiano RP, Pettee Gabriel KK, et al. Reported physical activity and sedentary behavior: why do you ask? J Phys Act Health. 2012;9 Suppl 1:S68–75.
36. Craig CL, Cameron C, Griffiths JM, et al. Descriptive epidemiology of youth pedometer-determined physical activity: CANPLAY. Med Sci Sports Exerc. 2010;42(9):1639–43.
37. Inoue S, Ohya Y, Tudor-Locke C, et al. Time trends for step-determined physical activity among Japanese adults. Med Sci Sports Exerc. 2011;43(10):1913–9.
38. Tudor-Locke C, Johnson WD, Katzmarzyk P. Accelerometer-determined steps per day in US adults. Med Sci Sports Exerc. 2009;41(7):1384–91.
39. DiPietro L, Williamson DF, Caspersen CJ, et al. The descriptive epidemiology of selected physical activities and body weight among adults trying to lose weight: the Behavioral Risk Factor Surveillance System survey, 1989. Int J Obes Relat Metab Disord. 1993;17(2):69–76.
40. Canalys.com website. http://www.cnet.com/au/news/fitbit-rules-50-percent-of-the-worlds-wearable-market/ and http://www.canalys.com/newsroom/fitbit-accounted-nearly-half-global-wearable-band-shipments-q1-2014. Accessed May 2015.
41. http://www.statista.com/topics/840/smartphones/. Accessed May 2015.
42. Takacs J, Pollock CL, Guenther JR, et al. Validation of the Fitbit One activity monitor device during treadmill walking. J Sci Med Sport. 2014;17(5):496–500.
43. Ceaser TG. The estimation of caloric expenditure using three triaxial accelerometers. PhD dissertation. University of Tennessee, Knoxville, TN; 2012. http://trace.tennessee.edu/utk,_graddiss/1514
44. Stackpool, CM. The accuracy of various activity trackers in estimating steps taken and energy expenditure. Master's thesis. Çollege of Science and Health, University of Wisconsin-La Crosse; 2013. 42pp.
45. Case MA, Burwick HA, Volpp KG, et al. Accuracy of smartphone applications and wearable devices for tracking physical activity data. JAMA. 2015;313(6):625–6.
46. Tully MA, McBride C, Heron L, et al. The validation of Fibit ZipTM physical activity monitor as a measure of free-living physical activity. BMC Res Notes. 2014;7(1):952.
47. Troiano RP, Berrigan D, Dodd KW, et al. Physical activity in the United States measured by accelerometer. Med Sci Sports Exerc. 2008;40(1):181–8.
48. Evenson KR, Buchner DM, Morland KB. Objective measurement of physical activity and sedentary behavior among US adults aged 60 years or older. Prev Chronic Dis. 2012;9, 110109.
49. Hansen BH, Kolle E, Dyrstad SM, et al. Accelerometer-determined physical activity in adults and older people. Med Sci Sports Exerc. 2012;44(2):266–72.
50. Colley RC, Garriguet D, Janssen I, et al. Physical activity of Canadian adults: accelerometer results from the 2007 to 2009 Canadian Health Measures Survey. Health Rep. 2011;22 (1):7–14.
51. Hagströmer M, Oja P, Sjöström M. Physical activity and inactivity in an adult population assessed by accelerometry. Med Sci Sports Exerc. 2007;39(9):1502–8.
52. Hagströmer M, Troiano RP, Sjöström M, et al. Levels and patterns of objectively assessed physical activity—a comparison between Sweden and the United States. Am J Epidemiol. 2010;171(10):1055–64.
53. Baptista F, Santos DA, Silva AM, et al. Prevalence of the Portuguese population attaining sufficient physical activity. Med Sci Sports Exerc. 2012;44(3):466–73.
54. Ekelund U, Luan JA, Sherar LB, et al. Moderate to vigorous physical activity and sedentary time and cardiometabolic risk factors in children and adolescents. JAMA. 2012;307 (7):704–12.
55. Craig CL, Cameron C, Tudor-Locke C. CANPLAY pedometer normative reference data for 21,271 children and 12,956 adolescents. Med Sci Sports Exerc. 2013;45(1):123–9.
56. Matthiessen J, Andersen E, Raustorp A, et al. Reduction in pedometer-determined physical activity in the adult Danish population from 2007 to 2012. Scand J Public Health. 2015. [Epub ahead of print]. doi:10.1177/1403494815578321.

Chapter 14
Self-Report and Direct Measures of Health: Bias and Implications

Sarah Connor Gorber and Mark S. Tremblay

Abstract Much of the world's population health, public health and clinical information is based on self-reported data. However, significant and meaningful bias exists across a broad range of health indicators when self-reported data are compared to direct measures. This bias can lead to over- and underestimation of risk factor and disease prevalence and burden. Understanding the implications of such bias for health surveillance, research, clinical practice and policy development may provoke adjustments to current epidemiological practice and may assist in understanding and improving the health of populations.

14.1 Introduction

Measuring the state of health within a population is crucial for health surveillance, research, clinical practice and policy development. It provides a current picture of a population's status, allows for monitoring changes over time and indicates inequities between population sub-groups and among countries. Adequate measurement strategies are essential to ensure that evidence upon which resources will be allocated and interventions designed is reliable and valid.

Occasionally, epidemiologists who seek to relate physical activity and health may have access to relatively accurate data, such as clinical measurements of height, weight, and systemic blood pressure. But much of our health information is based on subjective or self-reported measures of health, because most population data come from surveys that rely on self-reports of participants' health status and disease experience. Self-reports are often used because of their practicality, low

S. Connor Gorber, PhD (✉)
Research, Knowledge Translation and Ethics Portfolio, Canadian Institutes of Health Research, Ottawa, ON, Canada
e-mail: Sarah.ConnorGorber@cihr-irsc.gc.ca

M.S. Tremblay
Department of Pediatrics, University of Ottawa, Ottawa, ON, Canada

R.J. Shephard, C. Tudor-Locke (eds.), *The Objective Monitoring of Physical Activity: Contributions of Accelerometry to Epidemiology, Exercise Science and Rehabilitation*, Springer Series on Epidemiology and Public Health,
DOI 10.1007/978-3-319-29577-0_14

cost, low participant burden, and general acceptance in the population [1]. Increasingly, however, the accuracy of self-reported data has been called into question and there has been a push to include more objective measures in our health information system [2]; a trend that is facilitated by advances in technology allowing for more feasible direct measurement. This brief analysis examines the bias in self-reported information across a range of population, public health and clinical conditions and, using obesity as an example, discusses the implications of this bias for Canadian policy and practice.

14.2 Self-Report vs. Direct Measures Bias

A recent series of systematic reviews has highlighted the bias in self-reported measurements for a variety of health conditions and determinants in both children and adults. Reviews have examined the relationship between reported and measured height, body mass, and body mass index (BMI) (64 studies) [3], smoking (67 studies) [4], hypertension (144 studies) [5] and physical activity in adults (173 studies) [6] and in children (83 studies) [7]. These reviews have consistently

Table 14.1 Differences between reported and measured estimates of health variables from 5 published systematic reviews [3–7]

	Studies with males	Studies with females	Studies with male and female data combined[a]
Height (mean difference, mm)	12	10	1.7 cm
Body mass(mean difference, kg)	−0.7	−1.3	−1.1
Body Mass Index (mean difference, kg/m^2)	−0.6	−0.6	−0.9
Physical activity—vs, accelerometer measurements			
Adults (mean percent difference)	44 %	138 %	44 %
Children/Youth (mean percent difference)	114 %	584 %	147 %
Hypertension			
Awareness of hypertensive status at 140/90 mmHg			58 %
Awareness of hypertensive status at 160/95 mmHg			62 %
Smoking			
Sensitivity vs. cotinine concentrations measured in saliva			86 %
Sensitivity vs. cotinine concentrations measured in blood			76 %
Sensitivity vs. cotinine concentrations measured in urine			75 %

[a]Mean estimates include data from different studies, depending on whether studies report data for males and females separately or together—many studies only reported data for males and females combined

demonstrated that reported data under- or over-estimated measured values (Table 14.1). For example, self-reported height was consistently overestimated, while body mass and BMI were consistently underestimated in adults, which led to an underestimation in obesity prevalence [3]. Smoking [4] and hypertension [5] prevalence were also underestimated when data were based on individuals' self-reports. Furthermore, if a standard clinically-determined systemic blood pressure of 140/90 mmHg was used to diagnose hypertension, just over half of respondents in the studies, which included data on more than 1 million people, were aware of their hypertensive status [5].

Low to moderate correlations were found between direct measures of physical activity (e.g. accelerometers, doubly labelled water) and self-reports (e.g. surveys, questionnaires, diaries) [6, 7]. In pediatric populations (less than 19 years of age) the self-reported measures of physical activity overestimated children's activity levels, implying that children and youth were much less active than they believed they were (overall mean percent difference of 147 %) [7]. In adults both under- and over-reporting were present and varied according to the sex of the participants and the level of physical activity measured, with greater discrepancies seen at higher levels of exertion or with more vigorous exercise [6].

Katzmarzyk and Tremblay [8] discussed the apparent contradiction in Canadian health surveillance data that indicated a temporal decrease in physical inactivity and a decrease in food intake, yet an increase in obesity and obesity/inactivity-related chronic disease. They concluded that inherent short-comings of self-report data and inconsistencies in data analyses likely contribute to these contradictory findings and they suggested the use of direct measures. The recent reports on the fitness of the nation from the Canadian Health Measures Survey (CHMS) [9, 10] strongly suggest the physical inactivity trend data are misleading and likely incorrect.

Other recent Canadian data have confirmed the bias between reported and measured health conditions such as obesity. For instance, Shields and colleagues [11] found that the prevalence of obesity based on measured data was 7 percentage points higher than the estimate based on self-reported data (22.6 % versus 15.2 %). They also found that the extent of under-reporting rises as BMI increases, so the greatest bias was seen in individuals who were overweight or obese [11].

14.3 Implications for Health Surveillance

Underestimating disease prevalence is one consequence of the reporting bias discussed above, but the misclassification that results from using reported data can have further implications for understanding the burden associated with specific health conditions. Using obesity as an example, a study using data from the 2005 Canadian Community Health Survey found that for adults aged 40 years and older who were classified as obese based on self-reported data 360,000 were also classified as having diabetes. If, however, measured data were used to classify respondents as obese, then 530,000 adults (nearly 50 % more) had diabetes [12].

With self-reported data, therefore, the burden of disease due to obesity is significantly underestimated.

Research has also shown that when estimates of obesity are based on self-reported data, the relationship between obesity and obesity-related health conditions such as diabetes, hypercholesterolemia, hypertension, arthritis and heart disease is substantially exaggerated [12–14]. One study [12] found that the odds ratios for associations between measured overweight, obesity class I and obesity class II or III and diabetes were 1.4, 2.2, and 7.0 respectively, but when the reported BMI was used to classify respondents into obesity categories the odds ratios increased to 2.6, 3.2, and 11.8. This distortion occurs because fewer respondents are classified as overweight or obese when the classification is based on reported data, since many of the population are classified into a lower weight category. Yet, the average weights of those who do self-report as being overweight or obese are higher than the average weights of those whose measured data place them in the overweight or obese categories. As a result, a stronger association with morbidity is observed when overweight and obese categories are based on self-reported data, because the respondents in these categories are actually heavier (Fig. 14.1).

Researchers have attempted to correct self-reported data statistically to determine if the reported numbers could be adjusted to approximate the measured values more closely [15–18] (Fig. 14.2). This was successfully accomplished in a Canadian study in which the reported prevalence of excessive body mass was corrected sufficiently so that the prevalence of overweight and obesity was no longer statistically different from the corresponding measured estimates [18]. In addition, sensitivity (the proportion of the population correctly classified as obese) for males increased from 59 % using reported data to 74 % using corrected data, and from 69 % to 86 % in females.

The generalizability of these correction equations, however, is questionable; the reporting bias in Canada, for example, has varied over time, doubling in the last decade [19]. If the bias was constant, or at least changed systematically over time as it has in the United States, it is more likely that a standardized statistical adjustment could be successful. Therefore, the most effective way to deal with reporting bias

Fig. 14.1 Arnaud Chiolero of Lausanne is a Swiss epidemiologist who has written on the discrepancy between reported and actual body mass

Fig. 14.2 Michael Plankey is among epidemiologists who questioned whether prediction equations can correct errors in the self-reporting of body mass

may not be by making post-collection data corrections, but rather by increasing the epidemiologist's capacity to collect directly measured data.

14.4 Implications for Research

Though self-report methods of assessing health indicators are convenient for research purposes, they must be employed with caution, especially when they are related to socially desirable behaviours, and due consideration should be given to the fact that results may lack reliability and validity. Accordingly, future research should:

- further compare self-report and direct measures across different variables and in different populations;
- where possible use directly measured data to reassess behaviour—health relationships that have been examined previously using self-report data;
- work to advance direct measurement methods to reduce their cost, respondent burden and reactivity; and
- if subjective measures are used, ensure that a subset of research respondents are assessed by both self-report and direct measurements, allowing for study-specific correction factors to be developed and used.

14.5 Implications for Clinical Practice

Standard clinical practice regularly uses a variety of biomarkers to inform diagnoses and monitor treatment progress. These measurements are generally collected using carefully validated procedures and analytical techniques that are both accurate and precise. Such data quality assurance is not mirrored when behavioural information is collected. The systematic reviews summarized above [3–7] clearly indicate cause for concern when relying on self-reported data to assess health-related behaviours and their outcomes. Consequently, indicators related to common chronic diseases should be tracked with direct measurements (e.g. height, body mass, systemic blood pressure). Pedometers or accelerometers can be used to measure daily activity objectively, and because there is as yet no equivalent direct measurement procedure for diet, the development of an appropriate technique should be a high priority for future research. Assessment of behaviour modification treatments that are based on self-report or subjective data may result in misleading findings and suboptimal clinical care.

14.6 Implications for Policy

Research based on data with measurement biases as described above may not contribute meaningfully to the health research literature. Indeed, it may utilize finite research resources ineffectively. It may also misinform policy directions, and could even cause harm. These side-effects of poor measurement can occur at the individual patient/respondent level (misinforming, misdiagnosing, misadvising) as well as at the population level (misinforming policy, expenditure allocations, burden of disease planning). Using self-reported data to determine obesity status, the estimated number of Canadian adults with diabetes was underestimated by nearly 50 % [12].

14.7 Limitations of Direct Measurements

Direct measures are not without their limitations. For example, the estimation of the intensity and total volume of weekly energy expenditures has become practical for epidemiologists with the replacement of questionnaires by relatively low cost objective monitoring devices. Step counting has become progressively more sophisticated, with an ability to classify the intensity of impulses and accumulate activity data over long periods. Pedometers/accelerometers yield precise data for standard laboratory exercise, and in groups where steady, moderately paced walking is the main form of energy expenditure they can provide very useful epidemiological data. Nevertheless, such instruments remain vulnerable to external vibration and they fail to reflect adequately the energy expenditures incurred in hill climbing and isometric activity, as well as many of the everyday activities of children and younger adults. Multi-phasic devices hold promise as a means of assessing atypical activities, but appropriate and universally applicable algorithms based on such equipment have as yet to be developed. Moreover, the multiphasic equipment is at present too costly and complex for epidemiological use.

14.8 Conclusions

Much of the world's epidemiological research and evidence is based on self-reported data. Such data have systematic biases and limitations, and the reported values often deviate significantly and meaningfully from more robust direct (objective) measurements. This bias can lead to both over- and under-estimation of risk factor and disease prevalence and burden. Understanding the implications of such bias on health surveillance, research, clinical practice and policy development may provoke adjustments to current epidemiological practice that can assist in understanding and improving the health of populations.

References

1. Singleton RA, Straits BC, Straits MM. Approaches to social research. 2nd ed. New York: Oxford University Press; 1993.
2. Tremblay MS. The need for directly measured health data in Canada. Can J Public Health. 2004;95:165–6.
3. Connor Gorber S, Tremblay M, Moher D, et al. A comparison of direct vs. self-report measures for assessing height, weight and body mass index: a systematic review. Obes Rev. 2007;8:307–26.
4. Connor Gorber S, Schofield-Hurwitz S, Hardt J, et al. The accuracy of self-reported smoking: a systematic review of the relationship between self-reported and objectively assessed smoking status. Nicotine Tob Res. 2009;11:12–24.
5. Connor Gorber S, Tremblay M, Campbell N, et al. The accuracy of self-reported hypertension; a systematic review and meta-analysis. Curr Hypertens Rev. 2008;4:36–62.
6. Prince SA, Adamo K, Hamel ME, et al. A comparison of direct versus self-report measures for assessing physical activity in adults: a systematic review. Int J Behav Nutr Phys Act. 2008;5:56.
7. Adamo KB, Prince SA, Tricco AC, et al. A comparison of indirect vs. direct measures for assessing physical activity in the pediatric population: a systematic review. Int J Pediatr Obes. 2009;4:2–27.
8. Katzmarzyk PT, Tremblay MS. Limitations of Canada's physical activity data: implications for monitoring trends. Appl Physiol Nutr Metab. 2007;32 Suppl 2:S185–94.
9. Shields M, Tremblay MS, Laviolette M, et al. Fitness of Canadian adults: results from the 2007-2009 Canadian Health Measures Survey. Health Rep. 2010;21(1):21–36.
10. Tremblay MS, Shields M, Laviolette M, et al. Fitness of Canadian children: results from the Canadian Health Measures Survey. Health Rep. 2010;21(1):7–20.
11. Shields M, Connor Gorber S, Tremblay MS. Estimates of obesity based on self-report versus direct measures. Health Rep. 2008;19(2):61–76.
12. Shields M, Connor Gorber S, Tremblay MS. Effects of measurement on obesity and morbidity. Health Rep. 2008;19(2):77–84.
13. Chiolero A, Peytremann-Bridevaux I, Paccaud F. Associations between obesity and health conditions may be overestimated if self-reported body mass index is used. Obes Rev. 2007;8:373–4.
14. Shields M, Connor Gorber S, Tremblay MS. Associations between obesity and morbidity: effects of measurement methods. Obes Rev. 2008;9:501–2.
15. Rowland ML. Self-reported weight and height. Am J Clin Nutr. 1990;52:1125–33.

16. Kuskowska-Wolk A, Bergstrom R, Bostrom G. Relationship between questionnaire data and medical records of height, weight and body mass index. Int J Obes. 1992;16:1–9.
17. Plankey MW, Stevens J, Flegal KM, et al. Prediction equations do not eliminate systematic error in self-reported body mass index. Obes Res. 1997;5:308–14.
18. Connor Gorber S, Shields M, Tremblay MS, et al. The feasibility of establishing correction factors to adjust self-reported estimates of obesity in the Canadian community health survey. Health Rep. 2008;19(3):71–82.
19. Connor Gorber S, Tremblay MS. The bias in self-reported obesity from 1976 to 2005, Canada—U.S. comparison. Obesity. 2010;18:354–61.

Chapter 15
Conclusions and Future Directions

Roy J. Shephard

15.1 The Physical Activity/Health Association

Beginning with the stimulus of the cardiac epidemic in the mid-twentieth century, epidemiologists have shown an ever-increasing interest in the interactions between habitual physical activity and health. Initial enquiries, based on occupational classifications and studies of athletes, were stimulated by the apparent epidemic of ischaemic heart disease. The issues of self-selection and atypical body build were quickly recognized as problems in studies of athletes, and occupational investigations were soon compromised by the declining energy expenditures demanded in most industries. Nevertheless, the search for associations between habitual physical activity and health continued, using activity diaries and physical activity questionnaires of greatly varying complexity and sophistication. Many such instruments proved able to classify target populations into 3 or 4 groups with differing levels of habitual activity. Thus, it became possible to demonstrate statistically and clinically significant associations between the volume and/or the intensity of regular physical activity and protection against not only ischaemic heart disease, but also a wide range of other chronic conditions which to that point had lacked effective prophylaxis.

Roy J. Shephard (✉)
Faculty of Kinesiology & Physical Education, University of Toronto, Toronto, ON, Canada
e-mail: royjshep@shaw.ca

R.J. Shephard, C. Tudor-Locke (eds.), *The Objective Monitoring of Physical Activity: Contributions of Accelerometry to Epidemiology, Exercise Science and Rehabilitation*, Springer Series on Epidemiology and Public Health,
DOI 10.1007/978-3-319-29577-0_15

15.2 The Danger of Reliance upon Self Reports of Health

Much of the world's population health, public health and clinical information, including the reported association between physical activity and health, is based on self-reported data on health status. However, recent research has demonstrated a significant and meaningful bias across a broad range of health indicators when self-reported data are compared to direct measures. This bias can lead to either an over- or an underestimation of risk factor and disease prevalence, and it is important that those concerned with health surveillance, research, clinical practice and policy development understand and allow for such biases.

15.3 Public Health Norms of Physical Activity and Population Compliance

Faced with the demonstration of the importance of regular physical activity to the continued health and well-being of aging populations, health agencies soon were anxious to publicize evidence-based recommendations highlighting the lifestyle compatible with optimal sustained health. Many physiologists and epidemiologists, very conscious of the limitations in their data-base, were reluctant to make firm recommendations on the minimum required volumes of physical activity. Nevertheless, public health agencies pursued the search for this information, and reached conclusions that were based more upon a consensus of expert opinion than on defensible experimentation. For adults, one common conclusion was to advise engaging in moderate to vigorous physical exercise for at least 30 minutes on 5 or more days per week.

Having established such norms, a further interest of the public health agencies was to establish how effective their publicity was in achieving such levels of physical activity on a population basis, and on-going surveillance programmes were established. Although objective monitors were becoming more widely available, such surveys continued to be based largely upon questionnaire responses. The information thus derived provided moderate comfort to programme organizers, often showing a half or more of the population meeting the recommended levels of physical activity.

However, critics began to point out that such claims did not jibe with everyday observations of population behaviour, and suspicion was aroused that for various reasons, many of those questioned were exaggerating the extent of their habitual physical activity, possibly by a factor as large as 2 or 3. Comparisons with the findings from objective monitors were thus initiated, and the fears of inaccurate and exaggerated reports were confirmed. While questionnaires could create 2 or 3-level gradations of personal activity with reasonable reliability, confirming the health associations that had been postulated, they had little ability to measure the absolute volumes and intensities of physical activity that were required. Moreover, very elaborate and time-consuming questionnaires often yielded less valid information than very simple instruments, with just a handful of well-chosen questions, anchored to easily recognized physical manifestations of exercise.

15.4 The New Contributions of Objective Physical Activity Monitors

15.4.1 The Promise of the Objective Monitor

Epidemiologists were hopeful that the use of pedometers and more sophisticated accelerometers would overcome the problems associated with reliance upon physical activity questionnaires. Modern objective monitors would provide more convincing evidence of causality by providing an accurate measure of the intensity, frequency, volume and sequence of an individual's daily physical activity, while at the same time identifying those periods during a day when the activity was occurring. Information would also now be garnered concerning sedentary behaviours and sleep, together with their interactions with physical activity and their combined relationships to health indicators.

15.4.2 Reliability and Validity

The expectation was that objective monitors would provide accurate and bias-free information, and indeed Japan has enacted legislation setting such standards for objective monitors. When tested under standardized conditions, such as running on a laboratory treadmill, the better forms of pedometer have a high level of reliability, and the step counts indicated by the instrument have shown their validity by the close correlation of the output with direct observations of the individual's behaviour. Correlations with external criteria have been much less close under free-living conditions, when the individual's stride length, force and incidental movements are less well controlled. Further, unless ancillary information has been collected by such techniques as an activity diary, a GPS recorder or a posture indicator, a number of potential daily activities such as cycling, swimming, hill climbing and resistance exercise have not been detected adequately. Debate continues as to how important a fraction of total activities is unrecorded in this way.

15.4.3 Achievements of Objective Monitoring

The use of objective monitors has allowed a more accurate recalibration of the previously identified physical activity/health relationships for a wide range of conditions, including not only all-cause mortality, cardiac death, cardiovascular disease, stroke, peripheral and vascular disease, but also hypertension, cardiac and metabolic risk factors, diabetes mellitus, obesity, low back pain. osteoarthritis, osteoporosis, chronic chest disease, cancer, depression, quality of life and the capacity for independent living. Attempts have also been made to rephrase public health recommendations in terms of objective monitor outputs, for instance by defining recommended ranges of daily step count, or minutes of physical activity at specified MET rates. However, there is still a need to develop unanimity in specifying recommended step-rate ranges; because of frequent changes in monitor design and internal algorithms, rankings of activity and recommended behaviours have unfortunately remained instrument-specific.

The prevalence of physical activity in the general population has also been re-evaluated. The new objective surveys have revealed disappointing and sometimes decreasing levels of physical activity in many communities. The more precise measurement of individual activity levels has also offered the potential to settle such issues as the shape of the dose/response relationship linking physical activity to the prevention of various chronic conditions. In general, there does not seem to be a minimum threshold of physical activity for health benefit, although for some conditions there is evidence of a ceiling in response, beyond which the exerciser gains no additional benefits. Moreover, because relatively small samples have as yet been tested using objective monitors, in many areas of enquiry the presently available pedometer and accelerometer data have not added greatly to the understanding of dose/response relationships previously derived from multi-level questionnaire classifications of physical activity in large populations.

15.4.4 Advances in Health Economics

Objective monitoring is now offering the potential to make a more precise gradation of the impact of various levels of habitual physical activity upon the incremental costs attributable to various chronic diseases. Rather than assuming a generic economic benefit from "activity" vs. inactivity, it is now possible for the health economist to quantitate with respect to each of a range of chronic diseases the magnitude of the fiscal benefits likely to accrue from the small increases of physical activity likely to be achieved in sedentary populations.

15.4.5 *Monitoring of Rehabilitation*

Objective monitoring is a growing trend in many areas of rehabilitation. Simple designs of pedometer/accelerometer have proven useful motivational tools, and they have provided investigators with well-documented data on increments in weekly activity that can be tallied against responses to rehabilitation.

15.4.6 *Inactivity in the Etiology of Obesity*

Pedometer data have helped to resolve the over-eating/sedentarity controversy in relation to the etiology of obesity. Such data provide convincing evidence that those who are obese take 2000–3000 fewer steps per day than those who have a normal body mass.

15.5 Choice of Instrumentation; Pedometers vs. Accelerometers

Accelerometers can provide detailed three-dimensional information on the accelerational forces developed by the individual. They might thus seem much superior to even the most modern type of pedometer, where the development of a pre-determined force is registered as a step, and in some instances the instantaneous energy consumption is inferred from the rate of stepping. But in practice, the greater sophistication of the accelerometer has added to both the cost of the equipment and its maintenance. Data handling is also much more complicated, and it has seemed difficult to realize the greater potential of the accelerometer relative to the pedometer. In many cases, the accelerometer output is based upon proprietary algorithms that manufacturers are reluctant to disclose to investigators, and with some accelerometers it has proven difficult to obtain consistent values even for basic daily step counts.

The addition of information derived from other physiological sensors has yet to prove its value. Other potential sources of additional information include more complex statistical treatments of raw accelerometer data, and combinations of accelerometry, postural and GPS data. These new approaches will probably prove their worth in the future, but for the moment the latest type of piezo-electric pedometer seems the optimum instrument choice in many applications (particularly in populations where the main activity is walking).

15.6 The Importance of Sedentary Time and Sleep

Recent research has underlined the importance of physical inactivity from sedentary time and sleep in relation to health. The latter two variables seem not simply the converse of physical activity, but also to have an independent action upon health and well-being. This is a rapidly growing area of epidemiology, with opportunities to parse the nature and context of sedentary time much as has been done for physical activity, and to assess the impact upon future health. Investigators will need to clarify how far the effects of sedentary time have an independent influence upon health, how far they are attributable to associated behaviours such as snacking, and how far responses are influenced by the activity in which the individual also engages.

15.7 Data Organization

An important recent development has been the systematic improvement in protocols for data collection, management and treatment. Attention has focussed upon the types of instrumentation that are appropriate to answer various research questions and the budgetary implications of such choices. Well-organized investigators now specify systematic processes for quality control, data cleaning, data organization and storage. They also prescribe decision rules that shape the information that is accumulated, including algorithms for the computation of derived variables.

Data interpretation is facilitated by a growing array of normative data for various objectively measured physical activity metrics, coupled with standards, checklists, and flowcharts to support the clear, complete, and transparent reporting of information.

15.8 Continuing Controversies

One continuing topic of controversy is the minimum number of days recording of physical activity that are needed for data validity. The instrumentation of subjects has now generally increased from 14–16 h/day to 24 h/day, but some reports are still based upon 4–7 days of reasonably complete recording. If both weekday and weekend days are included in such sampling, this may provide reasonably representative data for a particular season, but a larger number of randomly chosen days or seasonally selected recordings are necessary to gauge physical activity throughout an entire year, and neglect of this precaution can lead to a substantial biassing of data in parts of the world that face seasonal extremes of climate.

Many objective monitors also provide measures of the instantaneous intensity of activity. The accuracy of such estimates requires further validation, and there is a need to decide upon age-specific levels of intensity that are of interest from an epidemiological point of view. For instance, activities in the range 3–6 METs may be the main focus of interest in an elderly population, but a substantially higher range of intensities will need to be recorded in young adults.

15.9 Future Opportunities

Much of the objective monitor information that is currently available remains cross-sectional in type, and there remains a need for well-designed longitudinal trials, using objective monitors to follow the impact of changes in habitual activity on the health of populations, thus strengthening evidence of the causal nature of the associations that have been described.

Automated pattern recognition programmes that carry out the very necessary checking, verifying and classification of objective monitor outputs is now allowing much larger populations to be studied. In the future, this process should be further complemented by a greater cooperation between laboratories, and a standardization of techniques that will allow comparisons and even a pooling of data.

Future research will undoubtedly focus on a closer integration of information on physical activity, sedentary time and sleep, in order to determine their individual and combined influence upon future health. There also seems a need to determine dose/response relationships, developing evidence-based public health guidelines, specifying appropriate periods of sleep and sedentary behaviour, and devising effective tactics to encourage the adoption of such guidelines by the general public.

The growing memory capacity of personal monitors, the ability to transmit information to I-Cloud storage, and the linking of pedometer/accelerometer records to GPS and position sensors is opening up a new vista for the collection of massive personal behaviour data banks on very large populations. Even a hundred-fold increase over the subject numbers used in some recent surveys seems likely to overcome "small-sample" problems in defining the shape of dose/response relationships. The big challenge to future generations of epidemiologists will be to conceptualize and handle such a vast array of data appropriately, and in a consistent manner from one laboratory to another.

Printed in the United States
By Bookmasters